GENERALIZED THERMODYNAMICS

GENERALIZED THERMODYNAMICS

Laszlo Tisza

THE M.I.T. PRESS

Massachusetts Institute of Technology

Cambridge, Massachusetts, and London, England

Library of Congress Catalog Card Number: 66-16555
Printed in the United States of America

PREFACE

One of the principal objects of theoretical research in any department of knowledge is to find the point of view from which the subject appears in its greatest simplicity.

J. W. Gibbs

This volume is a collection of papers of which some have already appeared in print, whereas others have been prepared from unpublished lecture notes. The material is not entirely homogeneous, and the lecture notes are written in a lighter vein than the published papers. I believe nevertheless that the volume has a certain coherence and conveys a reasonably rigorous, though open-ended, theoretical structure.

It is a noncontroversial statement that thermodynamics is a theory in close contact with measurement. Yet the specific implications of this statement have changed beyond recognition since the formative years of the classical theory. The dominating fact of twentieth century physics is the opening up of the atomic and subatomic domains to experimental study. It is appropriate that thermodynamics should catch up with these advances. This is, of course, a major program, and the present volume can be considered as a report on its progress.

The widening of the scope of thermodynamics over its traditional limits is attempted in a series of steps in the course of which the macroscopic theory is deepened in the microscopic direction. There results a multiplicity of theories, the organization of which is itself a nontrivial problem. Such questions of organization are studied in some generality in Papers 11 and 12. The results of this study can be summed up in the following recommendations: For conceptual clarity it is advisable to consolidate theories in the form of deductive systems. I hasten to point out that the suggested use of deductive systems has little if any of that narrowness and rigidity commonly associated with this method. (See page 344.)

Due account is taken of the fact that such systems have so far always proved to be of limited scope, and we use several of them to widen and deepen our view. The use of a multiplicity of systems raises in turn questions of mutual consistency. In situations that are already well understood, the harmonization of systems can be ensured with surprising ease by staking out the scope and limit of validity for each system, and by the establishment of priorities between them. In such a hierarchy of systems, flexibility is reconciled with taut organization. A number of major theories of physics are already sufficiently organized for this hierarchic structure to be readily recognized. Newtonian and relativistic mechanics and Maxwell's theory of the electromagnetic field can be mentioned as such reasonably well constructed systems. There are many reasons, however, why the situation should be less favorable in the case of the theories dealing with the structural properties of matter where the superabundance of results renders the scope of effective organization particularly wide and the task more difficult. The above-mentioned recommendations constitute here a program that is at present being executed. The most solid progress to date consists in the axiomatic development of two theories referred to as MTE and STE.* These theories overlap considerably with Gibbsian thermodynamics and statistical mechanics, respectively. However, MTE and STE are in a closer and more transparent relation to each other than the original theories, a fact that demonstrates the advantages of deliberate coordination.

The core of MTE is developed primarily in Paper 3 of Part A; however, insofar as applications are concerned, the main emphasis in this paper is on the theory of stability and of phase equilibrium. The presentation of MTE is rounded out in Papers 2, 5, and 6 by supplementary material.† An earlier version of the theory (Paper 4) is reproduced because its qualitative approach to the microscopic interpretation of phase transitions might still be of interest.

Although an effort was made to line up the papers in a logical order, this sequence may be varied according to taste and interest. The reader is urged, however, to read as an introduction the quasi-historical Paper 1, which is meant to soften the harshness of deduc-

* This is short for *"macroscopic"* and *"statistical thermodynamics of equilibrium."* Generally, I propose to use capitalized abbreviations of theories to designate deductive systems developed from an explicitly stated postulational basis. At the same time such expressions as *"thermodynamics"* and *"thermostatics"* will be used in the traditional sense admitting a broad spectrum of meaning.

† Some of the ideas underlying MTE were developed in a more pedagogical vein in H. B. Callen, *Thermodynamics* (John Wiley & Sons, Inc., New York, 1960).

tive organization. The natural requirements of the learning process would make it desirable to introduce only one major concept at a time. Unfortunately, this physchological preference cannot be reconciled entirely with the hard fact of logic that a precise argument can be initiated only if a sufficiently rich conceptual context is already available. Of course, the one-by-one introduction of concepts is constantly happening in the course of historical evolutions, and a selective and somewhat stylized historical approach turns out to provide a reasonably good introduction to the subject. Far from being opposites, the inductive-historical and the logical-deductive methods can be arranged to supplement each other in a very natural fashion. The intuitive understanding of the basic concepts and a preliminary experimental motivation are provided by the first method, whereas a rigorous exposition and a closer scrutiny of the extent of experimental verification are within the purview of the second. I note also that the historical form of Paper 1 is only a means to an end, which is the intuitive explanation of the conceptual basis of the modern theory. The discussion is confined to the highlights of evolution that mark the emergence of new concepts, with a brief reference to the widening empirical background. Simplicity of exposition takes precedence over a meticulous distribution of credit, particularly in cases of multiple independent discovery. The presentation is different from conventional history writing in that modern mathematical expression is preferred to archaic language. However, I do make an attempt to avoid or minimize anachronisms in matters of substance rather than letter, and discuss the solution of conceptual snags in terms of information available at any particular juncture.

Another objective of Paper 1 is to clarify the relation of MTE to classical thermodynamics, which appears here as a stage in the historical evolution rather than as its culmination.

Turning to Part B, where statistical theories are discussed, we have to recall that statistical phenomena exhibit, in general, an interplay of randomness with correlations. However, in order to develop the equilibrium theory STE, we have to deal explicitly with pure randomness only. In its present form, STE provides a sufficient foundation for most of the standard quantum-statistical calculations pertaining to equilibrium, with the exclusion of cases in which the density matrix is relevant due to the importance of quantum correlations.

Irreversible processes are included in STE only to the extent that they can be considered as ordered sequences of equilibrium states. A more penetrating theory of irreversibility requires the explicit handling of space-time correlations stemming from microscopic dynamics. This is a most widely discussed topic in current literature, and there are many results that could be considered as extensions of

STE. Yet, in view of the considerably greater difficulty of the problem, it is not surprising that the correlation theories have neither the formal generality nor the compactness of STE. They are presented here only to a very limited extent in Papers 8 through 10 (Paper 9 is a simplified version of Paper 10). These papers are reproduced because I believe they contain elements that are not in common use but that may become part of a more definite theory.

My papers on the theory of liquid helium, in particular on the two-fluid model, could be classified as belonging to generalized thermodynamics, but they are not included here because their contents are available in improved form.

The papers are reproduced in their original form, apart from slips and misprints corrected, particularly in Papers 3 and 11. A few shortcomings of Papers 3 and 4 are corrected and brought up to date in Sections 4 and 6 of Paper 2.

A consolidation of all this material would evidently be desirable. Yet it seems to me that the point of departure best suited for this purpose has not yet been reached.

Cambridge, Massachusetts
December, 1965

LASZLO TISZA

ACKNOWLEDGMENTS

The writing of the previously unpublished part of this volume was begun during my leave of absence in 1962–1963 as a John Simon Guggenheim Memorial Fellow and visiting professor at the University of Paris. The afforded opportunity is gratefully appreciated.

Most of the original research reported in this volume was carried out in affiliation with the Research Laboratory of Electronics of the Massachusetts Institute of Technology, and was supported in part by their sponsoring agencies. More specific support was offered by the U.S. Air Force, Office of Scientific Research, as indicated in the acknowledgement of the individual papers.

Most of all I wish to thank my former students whose interest provided a powerful stimulus for the development of this logical structure.

My thanks are due to the publishers and to my co-authors who kindly consented to the reproduction of previously published papers.

CONTENTS

PART A

Thermostatics

PAPER 1

THE EVOLUTION OF THE CONCEPTS OF THERMODYNAMICS

CHAPTER 1. PRECLASSICAL THERMODYNAMICS

1.1 Thermometry and Calorimetry

Theoretical thermodynamics is anchored in observations and quantitative experiments pertaining to phenomena commonly denoted as *thermal.* These experiments always involve the use of *thermometers;* hence the first, or rather the preparatory, stage of the evolution of thermodynamics is thermometry.

The magnitude measured by thermometers is the *temperature,* a quite subtle concept, the proper definition of which can be given only within the framework of thermodynamic theory. The obvious circularity of this situation did not block the evolution of thermometry, but it amply explains the stumbling character of the early developments.

The pioneers of thermometry were not guided by a clear understanding of the property they intended to measure; they proceeded by trial and error to find an objective, quantitative counterpart to such subjective, qualitative experiences which are described as "cold," "warm," "hot." Was such an effort worth while? Was there a reasonable hope for success? The answers to these questions are far from obvious. But in the Europe of the seventeenth century, the interest in quantitative measurements was "in the air." At any rate, there were enough people making an effort to prove their point with regard to the measurement of temperature in order to bring the project to a successful conclusion in little more than a century.

Generally speaking, thermometry is based on the observation that the physiological variations in the "degree of hotness" are paralleled by various physical changes in appropriate systems. The simplest "thermometric property" is the volume of a fluid which may be considered as a measure of temperature. This term is used here in a vague intuitive meaning equivalent to "degree of hotness." The problem is to follow the gradual refinement of this term into the precise thermodynamic concept of temperature.

The expansion of air on heating was known already in Hellenistic times, but the idea of utilizing this phenomenon for thermometric purposes probably occurred first to Galileo (about 1600). Galileo's instrument consisted of a glass bulb containing air and having a long stem extended downward into a vessel of water. The change of volume was taken as an indication of a change of temperature. The device had an arbitrary scale and measured only some combined effect of pressure and temperature. It was more a "barothermoscope" than a thermometer. However, this fact could hardly have been appreciated before the invention of the barometer by Galileo's student Torricelli in 1643.

The preparation of narrower capillaries made possible the use of liquids as thermometric substances. The thermal expansion of these is smaller than that of gases, but it is hardly affected by changes of pressure. Later, about 1700, when the properties of gases were better understood and the pressure could be controlled, the gas thermometer came back into its own.

The first thermometric liquid was water (1631), an obvious, but unfortunate choice because of the anomalous expansion properties of this substance. Alcohol was used ten years later, and the open stem was sealed to prevent evaporation.

The transition from qualitative thermoscopes to quantitative thermometers called for the introduction of reproducible fiducial points, usually called fixed points, in order to render the definition of an objective scale possible. It is not obvious that nature provides us with such fixed points, and the first temperature scales were based on quite unsatisfactory points of reference. A few typical examples were: "the most severe winter cold," "the greatest summer heat," "the warmest water the hand can stand," "the temperature of the human body," reflecting the continued closeness of thermometry to its physiological beginnings.

The choice of the melting point of butter and the freezing of aniseed oil marked a step toward objectivity, but these temperatures were neither sharp nor exactly reproducible. It was a great discovery that the freezing and boiling points of water at a specified pressure (the ice and steam point) are both sharp and reproducible. Their selection as fixed points was suggested in 1694. This convention continued until recently;* it has a firm theoretical basis in the theory of phase equilibrium developed much later. The role of water is, of course, conventional. The important matter is to choose a chemically pure substance.

Given the fixed points, a scale can be extablished by dividing the volume expansion between the fixed points into a certain number of equal divisions. The Reaumur, Celsius (centigrade), and Fahrenheit scales are obtained for 80, 100, or 180 equal parts.

Such temperature scales are, of course, relative to the choice of the thermometric substance. If the temperature scale is based on the ex-

* The one fixed-point scale was adopted by international convention in 1954. The arguments supporting and explaining this choice cannot yet be stated here, but will be considered in Paper 2 (page 77).

pansion of alcohol, the expansion of mercury will not appear quite uniform and vice versa. From a practical angle, there is no serious objection against this situation since the different scales can be compared with each other. The simple device of bringing two bodies in contact until they reach a common final temperature allows us to determine whether their temperatures were originally different and which of them was higher.* In such a way, we can establish markings corresponding to the same temperature on different scales. The temperatures represent an ordered sequence, but this order of temperature markings cannot be made quantitative since we have nothing to choose between the various temperature scales and neither of them can be expected to provide a "true" temperature. Of course, one might think that there is no "true" temperature, and it takes the full power of thermodynamic theory to show that there *is* one. This is a subtle question that we can and must ignore at this time.

We may conclude this survey by stating that by the first decades of the eighteenth century the available thermometers were about as good as could be devised without a deeper theoretical understanding. The thermometers were certainly adequate for the experiments which were to make this theoretical understanding possible.

In order to study a thermodynamic system under definite reproducible conditions, we have to isolate the system from the surroundings or, more generally, we have to create appropriate boundary conditions for it. From the experimental point of view, this is frequently achieved by a *calorimeter*, which is among the most important devices of experimental thermodynamics. We proceed now to discuss in a schematic form the principles underlying calorimetry.

The various devices, such as containers, pistons, membranes, partitions, etc., which are used to enforce certain boundary conditions on the system, are traditionally referred to as *walls*. It is a common experience that the available walls vary greatly in their heat conductivity. Consequently, by manipulating walls of different kinds we can trap and release the heat content of the system at will. The concept of heat conductivity, i.e., the flux of heat quality, would enable us to classify the types of walls in the simplest manner. Such a procedure would be not be without circularity, since we want to use calorimetry to establish the concept of heat quantity and its proper relation to temperature. Therefore, even at the risk of appearing somewhat long-winded, we shall show how thermometer readings alone are sufficient to describe the different types of walls in thermodynamics.

* This fact is deeply rooted in common experience, and until recently it has not been recognized as a distinct principle. The importance of the fact of *temperature equilibrium* has been emphasized by Fowler who called it the *zeroth law of thermodynamics.*

Consider a large system and insert a small vessel containing a system at a different temperature. This temperature will change to reach asymptotically the temperature of the large system. This is again an instance of the zeroth law of thermodynamics. The general character of the curves describing the variation of the temperature is represented in Figure 1. The various curves correspond to containers made of different materials. While the qualitative nature of the curves is unchanged, the time rate of the process is decreased as one passes from, say, a copper sheet to

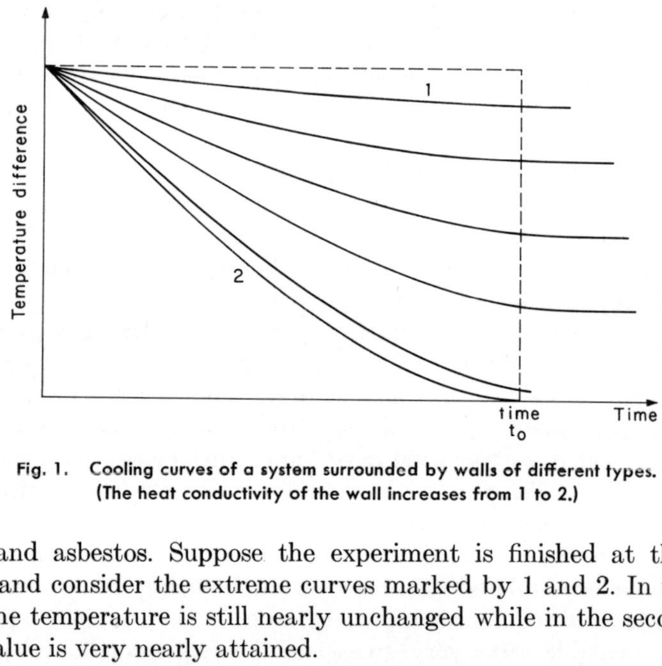

Fig. 1. Cooling curves of a system surrounded by walls of different types. (The heat conductivity of the wall increases from 1 to 2.)

wood and asbestos. Suppose the experiment is finished at the time $t = t_0$ and consider the extreme curves marked by 1 and 2. In the first case, the temperature is still nearly unchanged while in the second, the final value is very nearly attained.

This situation suggests the introduction of two idealized concepts: *adiabatic* and *diathermic* walls. A system completely surrounded by adiabatic walls can be maintained indefinitely at a given temperature regardless of the temperature of the surroundings. On the other hand, diathermic walls offer no obstacle to the equalization of temperatures.

One of the basic calorimetric experiments can be described as follows: The calorimeter is adiabatically isolated. In the initial state an adiabatic partition subdivides the system into two subsystems. The adiabatic partition is replaced by a diathermic one or else it is removed altogether (mixing). The thermometer reading is taken both in the initial and the final state.

It is to be remembered that adiabatic and diathermic walls are idealized concepts. In actual cases the experimentalist has to watch out for heat leaks through supposedly adiabatic walls and for an inadequate heat exchange through diathermic ones.

1.2 Caloric Thermodynamics

The joint use of the thermometer and the calorimeter opened up a wide range of questions to experimental investigation. Among the first problems studied was the mixing of different amounts of water initially at different temperatures. The result of these experiments could be summed up in the following formula for the final temperature:

$$t = \frac{m_1 t_1 + m_2 t_2}{m_1 + m_2} \tag{1}$$

where m_1, m_2 are the masses (or weights) of water and t_1, t_2 are the initial temperatures.

Equation 1 can be considered as the consequence of a conservation law and has been derived theoretically on such a basis. In fact, let us define a heat quantity

$$\delta Q = m\,\delta t \tag{2}$$

Then Equation 1 is equivalent to the conservation law

$$\delta Q_1 + \delta Q_2 = 0 \tag{3}$$

expressing the well-known "heat gained equals heat lost" relation, with

$$\begin{aligned} \delta Q_1 &= m_1\,\delta t_1 = m_1(t - t_1) \\ \delta Q_2 &= m_2\,\delta t_2 = m_2(t - t_2) \end{aligned} \tag{4}$$

Equations 3 and 4 imply Equation 1. Of course, the converse is not true; thus the definition of Q could be modified by any constant factor and still be in agreement with the experimental relation 1.

One particular alternative definition of the heat quantity appeared plausible to the early investigators. The mixing law could have been equally well expressed in terms of volume rather than in terms of mass:

$$t = \frac{v_1 t_1 + v_2 t_2}{v_1 + v_2} \tag{1a}$$

suggesting the alternative definition

$$\delta Q = v\,\delta t \tag{2a}$$

There was nothing to choose between definitions 2 and 2*a* as long as a single substance was involved in the experiment, but it was believed that one of these relations would hold also for the mixing of different substances. [Boerhaave, *Elementa Chemiae,* 1732, reference 1, page 155; reference 2, page 23].

Boerhaave induced Fahrenheit to study the matter experimentally. The calorimetric mixing of water and mercury proved incompatible with Equations 1 and 2 but Equations 1*a* and 2*a* predicted the correct order of magnitude although no quantitative agreement was obtained.*

In view of this situation, the volume hypothesis expressed in Equations 1*a* and 2*a* was accepted as the best available theoretical account of the experimental facts. This interpretation remained unchallenged for about 30 years.

A more satisfactory agreement with experiment could not be achieved without the introduction of basically new concepts. It took the boldness and insight of the physician-chemist Joseph Black to realize that this step was warranted even on the basis of the available scanty experimental material.

Black offered his new analysis of Fahrenheit's experiment around 1760 in his Glasgow lectures on chemistry. He interpreted the experimental discrepancies as a sign that the relation between heat quantity and temperature was influenced by the specific properties of the individual substances. He replaced hypothesis 2*a* by

$$\delta Q = mc\,\delta t = C\,\delta t \qquad (2b)$$

The constant c was at first referred to as "affinity for heat" of "faculty for receiving heat" and was finally named "specific heat." The constant $C = mc$ is the "heat capacity."

Black's innovation marked the beginning of theoretical thermodynamics, and it is worthwhile to compare the meaning of Equations 2, 2*a*, and 2*b* in some detail.

All three equations are based on the idea of the conservation of heat quantity expressed in Equation 3. The differences appear in the interpretation of the thermometer reading.

According to Equation 2, the thermometer reading is heat per unit mass; according to Equation 2*a*, it is heat per unit volume. Hence a single quantity, the thermometer reading, called *calor* (heat) at that time, seemed sufficient to describe both the thermal *and* the caloric state of a body. According to Black's formula 2*b*, this undifferentiated concept of heat splits into an *intensity of heat*, to be called *temperature*, and a *quantity* of heat, to be called *caloric*.

* Boerhaave and Fahrenheit did not consider Equation 1, established only around 1750, but used rather the special case in which $m_1 = m_2$ and the final temperature is $t = (t_1 + t_2)/2$. Accordingly, they adjusted the relative amounts of water and mercury so that the final temperature should be midway. Hypothesis 2 predicts that this situation should arise if the two fluids are of equal weights whereas hypothesis 2*a* calls for equal volumes. In the actual experiment the relative amounts were 1 part water and 20 parts by weight or 1.5 parts by volume of mercury. More precise experiments should have given 30 parts by weight or 2.2 parts by volume of mercury, an experimental error of nearly 50 per cent.

As mentioned already, Black's ideas rested originally on a very narrow experimental basis. The first repercussion of the new theory was to stimulate the required experimental work.

The measurement of the specific heat for various substances appeared now of obvious interest, and a whole new field opened up for experimental investigation. Since Equation 2*a* predicts that the distribution of heat is unaffected by material bodies, calorimetric experiments must have seemed pointless on the basis of the old theory.

In addition to the numerical values of the specific heats, the new calorimetric experiments also led to further general results. The latent heat (caloric) of phase transitions (freezing, boiling) was discovered by Black soon after the introduction of the specific heat (1762). This was an important step in extending the validity of the conservation law, Equation 3. Later Laplace and Lavoisier discovered that the specific heats are not constants but functions of the temperature (1780).

At this point the theoretical framework of the caloric theory attained a certain completion. Its results can be concisely summed up as follows:

The caloric content of a body is

$$Q(t) = Q(t_0) + \int_{t_0}^{t} C(t)\, dt \tag{5}$$

where t_0 is an arbitrary standard temperature

$$C(t) = C_0(t) + \sum_i L_i\, \delta(t - t_i) \tag{6}$$

Here $C_0(t)$ is the heat capacity and L_i is the latent heat associated with the phase transition taking place at the temperature $t = t_i$; the summation is taken over all phase transitions between t_0 and t; $\delta(t - t_i)$ is the well-known δ function. One has

$$\int f(t)\, \delta(t - t_i)\, dt = f(t_i)$$

if the interval of integration includes t_i; otherwise the integral vanishes.

If two bodies are brought into contact, they exchange caloric subject to the conservation law, Equation 3.

Equations 5 and 6 represent a preliminary version of the *caloric equation of state,* containing in a concise fashion the caloric properties of a body.

We may add at this point that the thermal properties of the body are given in the *thermal equation of state.* This equation expresses the experimental fact that the volume of a given quantity of a chemical element or pure compound in the fluid state is a well-defined function of the pressure and temperature:*

$$V = V(P, t) \tag{7}$$

The distinction between thermal and caloric equations of state clearly corresponds to the distinction between temperature and caloric.

The simple set of Equations 3, 5, 6, and 7 represent a conceptual framework that can be amplified with concrete experimental material, the extent of which is limited only by the available chemically well-defined substances. In particular, the identification of the common gases such as oxygen, nitrogen, hydrogen, and carbon dioxide opened the way for the quantitative study of the equations of state.

It is hardly a coincidence that the evolution of the caloric theory closely paralleled the clarification of the basic concepts of chemistry. In fact, the term "caloric" was advanced by Lavoisier in his *Traité élémentaire de chimie*, 1789, a landmark of the so-called chemical revolution.

The basic principles of the new chemistry are the conservation of mass and the invariance of the chemical elements. Together with the conservation of caloric, this was a powerful framework for the discussion of chemical reactions. The concept of heat of reaction was a natural extension of that of latent heat. With the addition of chemical reactions, the extent of experimental evidence supporting the caloric theory assumed truly impressive proportions. The contrast between free and latent caloric parallels that of kinetic and potential energy. Thus "caloric" can be conceived of in some ways, by no means superficial, as an ancestor of the energy concept.

The modern student might interject at this point that the caloric theory is not adequate to account for the energetics or chemical reactions, since it ignores the role of the compressional work performed automatically by the atmosphere. This is, of course, true and most significant in principle. Yet, from the purely empirical point of view, the contribution of mechanical work to heats of reaction is rather small and often within the accuracy of the early measurements. Thus the aforementioned experimental agreement could be claimed without unduly stretching the facts.

1.3 The Crisis of the Caloric Theory

In spite of its success described in the last section, the caloric theory was headed toward a crisis brought about by the dramatic phenomena of conversion discovered in the nineteenth century.

Best known to the modern student, even though of little influence for the actual development of ideas, were the experiments of Count Rum-

* We note that there are such exceptional situations as water near its density minimum at $t = 4°C$, where two values of t belong to the same P, V.

ford [1, 2, 3, and 4] performed with the definite intent to disprove the caloric theory. The boiling of water by means of the frictional heat created while boring a cannon is too well known to call for detailed description. Rumford's main point was that mechanical work was an inexhaustible source for the creation of caloric. Hence the latter is not conserved and cannot be of material nature. He suggested the revival of the mechanical concept abandoned half a century earlier.

The statement that heat is motion has a familiar ring to modern ears, and Rumford is frequently represented as a forerunner, if not the founder, of the energy theory of heat. In actual fact, Rumford had no conception of energy, provided we consider the principle of conservation an important aspect of this concept. Rejecting the conservation of caloric, Rumford did not envisage a more general conservation law to take the place of the old one. Also, the celebrated microscopic speculations of Rumford are seriously off the mark when scrutinized closely. He sought to establish a connection between temperature and the frequency of molecular vibrations. This may have been an anticipation of quantum theory, but it was not conducive to the crystallization of the concept of energy. It is interesting in this connection that Rumford believed in the "radiation of cold." This "frigorific radiation" he assumed to be different in kind from that of heat, but obeying the same laws [4].

Had Rumford's views been taken seriously, the consequences for the development of phenomenological thermodynamics would have been regressive. Black's distinction between the quantity and intensity of heat would have been given up for the naively empirical, in fact, prescientific dichotomy of "hot" and "cold."

Rumford has to be credited for pointing out that all was not well with the caloric theory and for emphasizing the idea of conversion (creation of caloric by mechanical action). However, his rejection of the principle of conservation of caloric without any hint of a more comprehensive principle of conservation was clearly not the way to initiate a constructive revolution. Rumford's influence was indeed much greater on late nineteenth-century historians in search of an ancestor of the mechanical theory of heat than on his contemporaries who went on to use the caloric for another half-century. This was most fortunate since the principles of classical thermodynamics arose from the ingenious attempts that stretched the caloric theory beyond its legitimate range. The problem ahead was to accept the idea of *conversion* between various forms of energy while at the same time recognizing the element of invariance, the principle of *conservation* amid the welter of transformations.

The scientific mind is bent on discovering the simple features of complexities, and the curious blend of the apparently contradictory principles of *transformation* and *conservation* was not easily accepted. And yet this difficulty was compounded by another even more subtle. Although diverse energy forms are convertible into each other according to definite conversion factors, they are still not on a completely equal footing: Mechanical energy can be converted into heat without reservations but the reverse conversion can occur only under certain additional conditions. This basic asymmetry of thermodynamic processes had to be incorporated into the new version of the theory.

The way this complex situation was mastered can be best described in terms of the following pattern. At first two conceptual schemes developed, each of which unified conversion and conservation in its characteristic manner. The first of these schemes referred only to the limiting case of ideal reversible processes and culminated in Carnot's principle (Section 1.5); the second found its general expression in the Mayer-Joule principle (Section 1.6). The two schemes are relatively simple when considered independently from each other, but they can be united in a single consistent scheme only at the price of accepting a considerable measure of abstraction. This unification of the principles was achieved by Clausius. It marks the beginning of classical thermodynamics (Section 2.1).

1.4 "Is Heat Matter or Motion?"*

Whatever is special, accidental and individual, will die, as it should; but that which is universal and essential should remain as an organic part of the whole intellectual acquisition. If that which is essential dies with the accidental, it must be because the accidental has been given the prominence which belongs to the essential.

J. W. Gibbs†

Mephistopheles. . . . Im ganzen — haltet
Euch an Worte!
Dann geht Ihr durch die sichre Pforte
Zum Tempel der Gewissheit ein.
Schüler. Doch ein Begriff muss bei dem
Worte sein.
Mephistopheles. Schon gut! Nur muss man
sich nicht allzu ängstlich quälen;
Denn eben, wo Begriffe fehlen,
Da stellt ein Wort zur rechten Zeit sich ein.
Mit Worten lässt sich trefflich streiten,
Mit Worten ein System bereiten,
An Worte lässt sich trefflich glauben,
Von einem Wort lässt sich kein Jota rauben.

J. W. Goethe: *Faust* I‡

* The contents of this section are not required for the continuity of the argument.
† *The Collected Works* (Yale University Press, New Haven, 1948), Vol. II, p. 170.
‡ Translation by Anna Sedgewick (Bell & Daldy, London, 1867).

The reader familiar with the conventional accounts of the history of thermodynamics may have missed in the foregoing discussion any concern with questions related to the "real nature of heat." Although this question smacks of "metaphysics" that physicists profess to avoid, it is nevertheless customary to reject the caloric theory as "utterly false, and, in some developments grotesque and absurd" [see page 40 of reference 14], presumably only because this theory adheres to the concept that "heat is matter." The attitude expressed in the present paper is very different. We have found a great deal of merit in the caloric theory and in the following sections we shall show it to be the natural, if not indispensable forerunner of classical thermodynamics.

We took "caloric" merely as an expression for a conservation law. The real issue is the probing of its range of validity and, eventually, its extension to more general situations. Such a critical evaluation and improvement of conservation laws is a perennial task of the physicist. Note that we do not speak of the "overthrow of classical thermodynamics" because the separate conservation laws for mass and energy proved to be incorrect in virtue of the relativistic relation $E = mc^2$. To be sure, the transition from caloric to classical thermodynamics is not a smooth and continuous affair, because it requires that several elements be added *simultaneously* to the more rudimentary theory. In actual fact, these various changes were introduced in installments extending over a stretch of decades. As a result, the transitional period was marked by considerable confusion, dispelled only as Clausius refocused and streamlined the accumulated conceptual material. Whether we call this sequence of events *a crisis of growth*, or a *revolution*, is immaterial, provided we realize that this revolution is constructive; it conserves the sound parts of the old theory even while some of its defects are eliminated. It is, however, misleading to speak of the "overthrow of a false doctrine by a correct one." The explanation for this divergence in interpretation is that evolution appears more organic if the theories are examined in terms of their logical-phenomenological or "operational" content, briefly their *core*, while the transition appears more abrupt with more backtracking involved, if attention is focused on the speculative fringe. Finally, the conflict between theories becomes entirely noncompromising, the transition cataclysmic if the argument is carried out on a purely verbal level. Theories have their structures which are interrelated, but these com-

---

Mephistopheles. . . . To sum up all — To words hold fast!/ Then the safe gate securely pass'd,/ You'll reach the fane of certainty at last.
Student. But then some meaning must the words convey.
Mephistopheles. Right! But o'er-anxious thought's of no avail;/ For there precisely where ideas fail,/ A word comes opportunely into play./ Most admirable weapons words are found,/ On words a system we securely ground,/ In words we can conveniently believe,/ Nor can we of one jot a word bereave.

plexities and nuances are not reflected in names, such as the "substantial" and the "material theories of heat," and in such slogans as "heat is matter" and "heat is motion." Yet, in spite of their fuzzy meaning, these expressions convey intuitive images of pedagogical and heuristic value. It is desirable to master the art of using them while they are helpful without holding on to them rigidly when they are getting in the way of a deeper understanding.

As an illustration, we shall digress here to survey the speculative ideas concerning the nature of heat if only to contrast the abrupt changes in their fortunes with the cumulative evolution of the phenomenological theory.

The idea that heat is a manifestation of microscopic motion was stated forcefully by Francis Bacon as far back as the beginning of the seventeenth century (page 43 of reference 2). A somewhat more quantitative interpretation of heat as molecular kinetic energy was advanced by Leibniz in 1695, while the corpuscular idea was embraced by Newton who extended it even to account for the nature of light.

The early molecular ideas failed to produce tangible phenomenological results. This was not for lack of trying but because of the inherent difficulty of the problem. To mention only one point, the casual use of the term "energy" might make us forget that the micromechanical energy concept is very different from the macroscopic energy of calorimetry. The former is specified in terms of innumerable molecular coordinates, the latter by a few macroscopic parameters. The two concepts can be connected with each other only in the case of thermodynamic equilibrium. This, however, is a most sophisticated *statistical* concept clarified only in recent years, while its rigorous reduction to deterministic mechanics is even now an unsolved problem.

The middle of the eighteenth century witnessed a series of developments which led within the span of a decade to a radical change of the intellectual climate. Where the molecular speculation seemed to fail, the idea of the continuum produced impressive results. In modern terminology, we can speak of the ascendancy of *field theory* in the perennial tug of war between theories favoring particles and fields, respectively. The turnabout started in hydrodynamics. The persistent efforts of the Bernoullis and of Euler to develop this discipline on a molecular basis had failed, and in 1753 Euler reluctantly formulated the alternative: One has to give up either the atomic hypothesis, or the validity of Newtonian mechanics on the atomic level [6]. Euler chose the former path and joined the postulate of continuous distribution of matter to Newtonian mechanics. The brilliant outcome was the mechanics of rigid bodies and the hydrodynamics of ideal (nonviscous) fluids, the basis of all future mathematical hydrodynamics. From our present vantage point, the alternatives discerned by Euler's extraordinary insight are not as incompatible with each other as they seemed to him. In principle, both of the above-mentioned approaches are important and practical in suitable approximations. We have an early manifestation of what nowadays is called complementarity. To be sure, for a long time after Euler there was not the least clue to the construction of a non-Newtonian atomic mechanics, and Euler's choice was the only fruitful one in his time.

The continuum conception of matter led in Euler's hands to a breakthrough in mathematical physics. Only a few years later a somewhat similar continuum conception of heat enabled Black to open up a new experimental field: calorimetry. Black's contributions, sketched out in Section 1.2, qualify him as the true founder of thermodynamics. For more detail we refer to pages 17 through 47 of reference 2. In particular, pages 42 through 46 deal with his critique of the mechanical theories. This critical attitude does not mean that Black was unaware of the facts of thermomechanical conversions. His student, coworker, and editor of his lecture notes, James Watt, is remembered for his momentous discoveries and inventions producing such conversions.

What Black realized was the necessity of a phenomenological framework for describing quantitative experiments pertaining to heat, and he noted that, at least for the time being, the mechanical speculations would not yield such a theory.

As the molecular theory of heat faded into temporary eclipse, the gap was filled with extensive microscopic speculations concerning the caloric, which was endowed with attractive and repulsive forces so as to account, in a qualitative fashion, for a wide variety of thermal and caloric phenomena [3]. The microscopic aspects of the caloric theory never received a quantitative mathematical elaboration. These aspects constitute an intuitive imagery rather than a real theory and, however ingenious, they are of no interest for our analysis concerned with the cumulative elements of the theory. In fact, the significant developments of the period 1760 to 1850 are all phenomenological that can be discussed without concern for the prevailing speculative ideas, whether right or wrong. The impressive rigor of classical thermodynamics was ensured by keeping clear of microscopic speculations.

A new element came into play, however, in 1859 with Clausius' discovery of the concept of molecular free path opening up the road to the kinetic theory and statistical mechanics of Maxwell, Boltzmann, and Gibbs. The molecular hypothesis finally proved its value; by long chains of mathematical reasoning, it could be quantitatively related to observations already described in the phenomenological theory.

Although the intuitive image that "heat is motion" provided the motivation for the kinetic theory and statistical mechanics, we should still not complacently expect that this model will serve us as well in the future. The model implies that all molecular motion stop at absolute zero, a prediction flagrantly contradicted by the experimentally verified facts of the quantum-mechanical zero-point motion. The distinction between thermal and quantal agitation is in their respective *incoherence* and *coherence*. Our microscopic interpretation of the nature of heat will not be satisfactory until these concepts are incorporated into the foundations of the theory.

1.5 The Carnot-Kelvin Principle

The first one to invent a theory simultaneously involving the ideas of conservation and conversion was the young French military engineer Sadi Carnot [7, 8]. The wealth of strikingly original ideas classes this work among the most brilliant new departures in theoretical physics.

Up to this time thermodynamics was developed primarily on an empirical basis provided by chemistry. Carnot approached an engineering problem, the efficiency of heat engines,* in terms of entirely new concepts which he developed for this purpose. These concepts turned out to be of even more fundamental importance in physics and chemistry.

Carnot's point of departure is the observation that the existence of temperature differences is a necessary condition for producing mechanical work by means of a heat engine. He simplifies the problem to its bare essentials and stipulates that his system, consisting essentially of a working substance, should exchange heat with its surroundings only at two fixed temperatures. In order to conceptualize such a situation, he introduces the idea of *heat reservoirs*. These are systems that are to provide the environment of the systems of interest. They are very much larger than the latter (the ratio of heat capacities $C_R/C \rightarrow \infty$), and they impose their temperature on the finite systems coupled to them. (Generally speaking, the contact of the bodies that were initially at different temperatures results in an intermediate final temperature.) It is assumed also that this coupling does not disturb the internal equilibrium of the reservoir. The aforementioned stipulations cannot be met for the direct coupling of two reservoirs with each other. Such coupling is to be established through the intermediary of a finite system, which is thus prevented from reaching equilibrium but which reaches in general a stationary state. This is, however, beyond the scope of the thermodynamics of equilibrium. Although heat reservoirs are idealized devices along with, say adiabatic walls, they can be realized to an excellent approximation. The atmospheric environment and thermostats regulated by automatic temperature control are such practical examples.

Carnot's engine consists then of a system, two heat reservoirs and a

* It was pointed out recently [9] that it may have been Cagnard's gas engine rather than the steam engine that served as the stimulus for Carnot's theory.

purely mechanical device, say a spring that is capable of storing or releasing mechanical energy without introducing any thermal effects into the situation, which is, of course, another idealization.

According to Carnot, the operation of this engine is analogous to the production of work through the lowering of a quantity of water between two levels in the gravitational field. More generally we can consider processes in which some kind of quantity X is transferred from a potential P_1 to P_2 while performing the amount of work W given by

$$X(P_1 - P_2) = W \tag{8}$$

The symbols in this relation can be given different interpretations. Thus X, P may be considered as mass and gravitational potential, or electric charge and electrostatic potential. Carnot proposes to use this analogy to describe the operation of the heat engine. He speaks of the descent of a conserved quantity from one temperature to a lower one. What is the nature of this quantity? The theory of heat contained up to this time only one conserved quantity, the caloric of Black and Lavoisier. It was natural that Carnot explicitly assumed that he was dealing with the same quantity. We know at present that this is not the case and we shall ease the reader's path by denoting Carnot's caloric as Q^*, to distinguish it from Black's caloric Q. This procedure saves us from logical error at the expense of historical accuracy. This star was invisible to Carnot and his contemporaries, and we have to ignore it in order to understand the malaise surrounding the last few years of preclassical thermodynamics.

In outlining Carnot's reasoning, we note first that the existence of temperature differences is a necessary but not a sufficient condition for the production of work. Heat may flow by conduction from the high- to the low-temperature reservoir without any production of work; the "work potential" is lost. Heat conduction and other dissipative factors will take their share of the "work potential" in case of the operation of any actual engine.

In order to give a quantitative expression to these distinctions, it is useful to introduce the concept of efficiency as the amount of work performed W divided by the amount of caloric Q^* transferred:

$$\eta^* = \frac{W}{Q^*} \tag{9}$$

We note that this quantity is not dimensionless. This defect cannot be removed until we clarify the meaning of Q^*.

Recognizing that only the ideal limiting case of maximum efficiency is suitable for a precise analysis, Carnot finds an ingenious way of characterizing such an engine. First, we suppose that the operation of the engine is cyclic. Each cycle consists of four stages: two isothermal exchanges of caloric with the two reservoirs and two so-called *adiabatic processes* in which the temperature of the system changes while thermally isolated due to the performance of work. After the completion of each cycle, the system returns to its initial state. The use of cycles is a simple but important artifice to eliminate effects arising from permanent changes in the system. It is also an effective implicit way of postulating that the system is always in equilibrium. Next we consider engines running backward: Caloric is lifted from low temperature to high temperature at the expense of performing work in much the same way as water is lifted from a lower to a higher level. A heat engine running backward is called a refrigerator, or heat pump.

The key concept of the theory is that of the *reversible engine.* An engine is reversible if the work obtained by it can be used to operate it as a heat pump in such a way that the initial distribution of caloric is entirely restored.

We consider now the class of all engines that operate between two fixed levels of temperature t_1 and t_2. Carnot's main theorem states that all reversible engines operating between these temperatures have the same efficiency regardless of the nature of the system constituting the working substance. Moreover, this efficiency is larger than that of any nonreversible engine.

The idea underlying the proof of Carnot's theorem is as follows. Let us consider a reversible engine run as a heat pump and couple it with another one that, contrary to the theorem, is supposed to be of higher efficiency. If the two engines are adjusted to work in each cycle with the same Q^*, the combined effect of the two engines is to restore the distribution of Q^* and to produce a net amount of work. The result would be a perpetual motion. Carnot declares this to be impossible.

Thus we obtained the remarkable result that the efficiency of the ideal reversible heat engine depends only on the temperatures t_1 and t_2 and not on the nature of the system. Comparison of Equations 8 and 9 suggests that the efficiency might be proportional to the difference of temperatures $t_1 - t_2$. However, Carnot points out that we have no right to assume that the linear relation 8 will apply also to the case of heat engines. In fact, it would be unclear which of the arbitrary empirical temperature scales should be chosen.

It was noticed in 1847 by William Thomson (the later Lord Kelvin) that this apparent defect of the theory can be turned into an advantage. Carnot's theorem can be used to eliminate the reliance of the temperature scale on the properties of individual substances. Since we are not yet committed to any particular scale, we can advance

$$W = Q^*(T_1 - T_2) \tag{10}$$

or

$$dW = Q^* \, dT \tag{10a}$$

as the definition of a temperature that is *absolute* in the sense that it is independent of the choice of thermometric substance.

The absolute temperature cannot be practically established with sufficient accuracy by directly measuring the quantities in Equation 10*a*, not even after the nature of Q^* will have been clarified (see Section 2.1). Nevertheless we shall see in Paper 2 that, indirectly, after appropriate elaboration of the theory, this definition can be made entirely "operational"; the result is known as the Kelvin temperature, or the Kelvin scale. We shall refer to Equation 10 as the *Carnot-Kelvin principle.*

The foregoing reasoning is most remarkable both from the practical and the theoretical points of view. Anticipating the operational character of the Kelvin temperature, we see that the study of *individual* substances allows us to arrive at universal conclusions relevant for *all* substances. This may be the first indication that thermodynamics has a significance going beyond the collection of empirical relations.

Carnot's results provided a new lease on life for the caloric theory. At least one type of energy conversion, the reversible conversion of heat into work and vice versa, was brought under its purview. Actually, this extension of caloric theory was tenuous and the source of new difficulties. The conversion of caloric in Carnot's theory was expected to hold in the reversible limit. How did this fit in with the conservation of caloric in the irreversible mixing process in the calorimeter discussed above? What about the production of heat through friction? Carnot was not unaware of these difficulties. He called for more experiments and also for the reconsideration of the foundation of the theory. His doubts and partial insight appear strengthened in his posthumously published notes [10]. However, his untimely death in 1832 prevented him from reaching full clarification.

The caloric thermodynamics described in Section 1.2 constitutes a consistent scheme within its range of validity. The same is true, of course, of classical thermodynamics within a greatly extended range. In contrast, Carnot's theory can be conceived of as a climb from one level to another. Its proper evaluation is a subtle question, and it is not surprising that this interpretation has been troubled by considerable controversy.

In the literature, Carnot was widely criticized for using the discredited caloric theory of heat instead of adhering to the mechanical point of view. Later, it was pointed out [11, 12] that Carnot's considerations are correct if his caloric Q^* descending from a higher to a lower temperature is interpreted as entropy. This is a valid point of view that we shall use later.

Going beyond this point, it was also claimed that Carnot should be considered as entirely correct, a modern writer on thermodynamics. This view is based on the fact that, in his notes, Carnot expressed himself to be in favor of the mechanical conception of heat. Moreover, he predominately used the term "calorique" for entropy and "chaleur" for heat quantity.

A different and rather more balanced interpretation of the historical situation is found in reference 13. We have to remember first of all that the caloric theory was by no means discredited in Carnot's time, and we need not perform an extensive search of the literature to prove this point. In Section 1.6, we quote a passage from Lord Kelvin, in which it is clear that this foremost authority had no hesitation in using the concept of conserved caloric 25 years after the appearance of Carnot's work. This fact may seem puzzling to those who believe that a theory has to be rejected as soon as found to be incorrect. In reality, an extended conceptual scheme that has proved its worth in organizing a significant range of experience is not rejected until there is something better to take its place. Carnot's stretching of the caloric theory beyond its then accepted limits is not unlike the technique pursued so successfully in the development of the modern atomic theory. Carnot's critics might just as well criticize Bohr for applying the discredited classical mechanics to the hydrogen atom. In contrast, Carnot's apologists would seem to ignore the difference between the old quantum theory and quantum mechanics.

Supposing even that Carnot had been consistent in using "calorique" for entropy and "chaleur" for heat quantity, this tacit method would have been entirely inadequate for initiating a major conceptual innovation: the splitting of the concept of caloric into two new concepts. In order to render such a conceptual bifurcation definite, we need clear instructions for the manipulation of these concepts, including their connection with experiment. The distinctive features of the new concepts have to be particularly emphasized in order to forestall their confusion in the future. While writing his posthumously published notes, Carnot had a clear conception of the irreversible creation of heat from work. He also raised the question concerning the reconciliation of this phenomenon with the reversible conversion that he had analyzed earlier, but this query remained unanswered at the time of his death at the age of 36.

Carnot and his contemporaries could conceive of a conservation law only in terms of an absolutely conserved fluid. At the time of the publication of his work, he did not fully appreciate the difficulties of using such a concept in the theory of heat engines. This is very fortunate, because otherwise his theory might never have come about. To state that Carnot discovered the concept of entropy implies an underestimation of the difficulty of "splitting" the concept of caloric into heat quantity, energy, and entropy. It means also the underestimation of the subtlety of the entropy concept with its peculiarly qualified law of conservation limited to reversible processes.

1.6 The Mayer-Joule Principle

In Carnot's principle, caloric is a quantity which is different in kind from work. Its dimension is "energy divided by temperature." This is, of course, quite obvious from Equation 10 of Section 1.5 but was much less clear in Carnot's time. Carnot was very sparing in the use of mathematics. Moreover, it was by no means obvious what is the mechanical quantity that is equivalent to caloric.

We turn now to the discussion of another principle in which heat appears as a quantity of the same kind as mechanical energy and obeys a joint conservation law. The idea of such a principle was "in the air" around 1840. Nevertheless, we shall confine the discussion to the achievements of Robert J. Mayer and James P. Joule, and we shall speak of the Mayer-Joule principle [1].

Mayer has the priority of first publication (1842) and the first theoretical derivation from existing experimental data of the conversion factor between heat and work, the so-called mechanical heat equivalent. Joule is to be credited with the most complete and thoroughgoing experimental investigation of the matter extending from 1843 for over three decades.

Mayer's belief in a principle of conservation of energy originated in physiological observations which, from our point of view, are more complicated than the principle to be obtained. However, his derivation of the mechanical heat equivalent is independent of the original motivation and starts with the analysis of a subtle difficulty of the caloric theory.

If an ideal gas is heated at constant volume, the amount of heat added results in an increase of the internal energy of the system. If the system expands on heating, an additional amount of heat is required to compensate for the expansion. This is a somewhat oversimplified argument* to show that the specific heat at a constant pressure C_p is larger than that at constant volume C_v. The difference $C_p - C_v$ is usually rather inconspicuous, but in gases it is quite considerable; the ratio $\gamma = C_p/C_v$ is 1.67 for monatomic and 1.4 for diatomic ideal gases.

These facts became known early in the nineteenth century. Their interpretation within the framework of the caloric theory was artificial.

* Some solids and liquids contract on heating within a small temperature range. The foregoing argument might suggest that in this case $C_p < C_v$. *This is wrong.* We shall see later that $C_p \geqq C_v$ regardless of the sign of thermal expansion. The energy balance is complicated in condensed systems because of the intermolecular potential energy.

The role of the performance of work was ignored, and it was assumed that the heat capacity of a given amount of gas increases as the density decreases. This assumption seemed to be confirmed by the cooling of a gas on adiabatic expansion (expansion of an adiabatically isolated system while doing work).

However, there were experiments which contradicted this interpretation. Gay-Lussac found that the free expansion of a gas into an evacuated chamber leaves the temperature unchanged (1807). The distinctive feature of this experiment compared to adiabatic expansion is that in the latter case the gas does work against a piston while no work is performed in the free expansion.

This difficulty was glossed over for a long time until the correct analysis of the experiments was carried out by Mayer in 1842. In modern notation, Mayer's argument can be formulated as follows.

All three types of experiment can be understood qualitatively by assuming that work performed on the gas has a similar effect as adding heat to it.

In the case of an ideal gas, the $C_p - C_v$ difference can be obtained quantitatively. Assuming that the temperature of one mole of the gas is raised by dT, we can write the heat-work balance as follows:

$$J(C_p - C_v)\, dT = P\, dV = R\, dT \tag{11}$$

where P is the pressure, dV is the increase in volume, and J is a proportionality constant. We have made use of the ideal gas law*

$$PV = RT \tag{12}$$

The left-hand side of Equation 11 is measured in calories, the right-hand side in mechanical units, such as liter-atmospheres, or joules; J is the mechanical heat equivalent. On the basis of available experimental data, Mayer obtained $J = 3.65$ joules/calorie as compared with the correct value, 4.18.

The calculation of this value of the mechanical heat equivalent amounted to an experimental prediction which could have been subjected to extensive experimental checks. Although Mayer's brilliant result did not get much attention from his contemporaries, the relevant experimental investigations were already under way for independent reasons.

We turn now to the discussion of the experimental investigations of Joule. On the face of it, these experiments seem to be the direct contin-

* This equation of state is the culmination of a huge body of experimental facts. Our cavalier handling of the matter can be excused, however, by the fact that the basic experiments leading to Equation 12 are well covered in elementary texts, while the truly subtle aspects of the concept of an ideal gas, and in particular the more searching queries concerning the temperature T, are preferably taken up only later.

uation of those of Rumford and the gap of some forty years between the two investigations has appeared puzzling to some authors. Thus Epstein in his well-known book on thermodynamics [5] tried to establish the continuity of the mechanical conception of heat by the intermediary of a tradition persisting among physiologists. Whether or not these speculations are correct, the ancestry of Joule's experiments is well documented, but the field is electrochemistry rather than physiology and the point of departure is Volta rather than Rumford [14].

In 1800, at about the same time when Rumford attempted to prove that work was an inexhaustible supply of caloric, Volta discovered the electric pile later named after him. He was impressed by the continued flow of electric current in a shorted pile, a striking contrast with the transient discharge of the Leiden jars and of other electrostatic devices. "This endless circulation or perpetual motion of the electric fluid may seem paradoxical and may prove inexplicable, but it is nonetheless real and we can, so to speak, touch and handle it." [Reference 14, page 73.]

Volta established the concept of tension originating in the contact of different metals. He proved that this tension is closely related to the electrostatic potential. However, it is interesting to note that Volta's idea of "tension" was a mingling of two ideas, namely electric charge and electric potential [reference 14, page 72]. This is another instance of confusion of an extensive quantity and its conjugate intensity, although this distinction was already understood in electrostatics.

Volta's challenge of the energy principle is interesting because it represents a rather unique attitude. There were, of course, inventors who attempted to construct perpetual motion machines, and, as we have seen, Rumford was rather cool toward the idea of a conservation of energy. Otherwise, however, there was among the important scientists of the period a surprisingly uniform adherence to some vague idea that the basic powers of nature are uncreatable and indestructible. The disagreements turned on the implementation of the general principle. Does the impossibility of perpetual motion require the conservation of caloric, as Carnot suggested, or is it rather incompatible with such a conservation as claimed by Mayer and Joule?

Although Volta's interpretation of his experiments turned out to be wrong, his attitude was not inherently absurd. Voltaic electricity was no less revolutionary at the time of its discovery than radioactivity about a century later. The two phenomena compare with respect to the immediate impact of their novelty and also in the theoretical and practical importance of their long-range implications. It is interesting that the breakdown of the energy principle was envisaged also in connection with radioactivity.

Whereas the re-examination of fundamental principles in the face of radically new phenomena is quite in order, the separation of the obsolete and the sound aspects of the inherited principles is a delicate task. At any rate, the energy principle has so far withstood every challenge although the concept of energy had to be redefined and extended more than once in the process. From our point of view this increased precision, scope, and articulation of the energy concept is of primary importance. It is this precise energy concept that is subject to mathematical description and is closely connected with a vast body of experimental material that is responsible for the difference between the modern idea of conservation and its rudimentary ancestor claiming the indestructible character of the "basic powers of nature."

The problem of the Voltaic cell was a first-rate puzzle amenable to large-scale experimental investigation. Its solution contributed substantially to the development of the precise energy concept. The technicalities of this development would involve us too deeply in chemistry, and we shall merely summarize some of the highlights of evolution [14].

The announcement of Volta's discovery gave rise at once to extensive experimental investigations. Among the first results obtained was the electrolytic decomposition of water. Since chemical effects were produced by electric currents, it was plausible to assume that the latter could be generated through chemical action. This was indeed the idea advanced by the famous chemist Humphry Davy only a few months after Volta's discovery.

Although Davy's chemical hypothesis is more nearly correct than Volta's contact hypothesis, the controversy between the supporters of these theories was carried out for a generation. Each hypothesis contained elements of truth. To unite these into a consistent scheme was a protracted task achieved only in small steps. The understanding of the complex experimental situation was slowed down by the fuzziness of the concepts in terms of which the phenomena were described. It was an important advance when the English physicist Peter M. Roget carefully distinguished between the quantity of electricity put into circulation by a cell and the *tension* at which the electricity is furnished (1832). Shortly thereafter, Faraday established his famous quantitative connection between the quantity of electricity circulated and the amount of chemicals consumed. Thus the basic idea of the chemical hypothesis was verified. On the other hand, De La Rive connected the *tension* of the cell with the *chemical affinities*. These are determined by the metals in contact, as Volta had claimed. Hence electrical *quantities* were connected with chemical *quantities* and electrical *intensities* with chemical *intensities*.

All the aforementioned workers were outspoken in rejecting Volta's attack on the principle of energy conservation. However, this acceptance of the conservation principle still lacked quantitative elaboration. This task was performed by Joule, who continued the researches of Roget and Faraday.

Joule's first discovery was that the flow of the current in a resistance R is accompanied by the development of heat proportional to Ri^2 (1840). To say that this amount of heat was *created* would have been jumping to conclusions. The current was produced by a chemical reaction; the latter might well be the source of caloric which is merely redistributed when appearing in the resistance. In order to decide the issue, Joule produced currents by electromagnetic induction. Finally, as the method of production turned out to be irrelevant, he concluded that caloric was indeed created rather than displaced in his experiments [15].

Joule was firmly convinced that there existed some conservation law of a general nature [16] and hence set out to investigate whether the conversion of the various forms of energy is governed by definite conversion factors. He considered the conversion of chemical, electric, caloric, and mechanical energy forms in all combinations. The determination of the mechanical heat equivalent forms the central part of these experiments, the results of which can be summed up in the general relation

$$W = QJ \tag{13}$$

where J is the mechanical heat equivalent introduced in Equation 11. Joule's experiments suggested that this relation may have universal validity with the same numerical value for J under all conditions.

Even while Joule was perfecting the experimental basis of the energy principle, Helmholtz published in 1847 his famous essay on the conservation of force [17]. In this work, he advanced the conservation of energy as a unifying principle extending over all branches of physics.

We might think that at this point the energy concept was ready to be generally accepted. Yet, in 1849, William Thomson published a paper on Carnot's theory [18] in which he expresses serious doubts that Joule actually created caloric. He considers the possibility that caloric might have leaked from the magnet into the calorimeter containing the induced current. His doubts are connected with the fact that he was impressed by the power and usefulness of Carnot's principle based on the conservation of caloric.

"The extremely important discoveries recently made by Mr. Joule of Manchester, that heat is evolved in every part of a closed electric conductor moving in the neighborhood of a magnet, and that heat is generated by friction of fluids in motion, seem to overturn the opinion commonly held that heat cannot be generated, but only produced from a source where it has previously existed either in a sensible or in a latent condition."

"In the present state of science, however, no operation is known by which heat can be absorbed into a body without either elevating its temperature, or becoming latent, and producing some alteration in its physical condition and the fundamental axiom adopted by Carnot may be considered as still the most probable basis for an investigation of the motive power of heat, although this, and with it every other branch of the theory of heat may ultimately require to be reconstructed upon another foundation when our experimental data are more complete. On this understanding, and to avoid a repetition of doubts, I shall refer to Carnot's fundamental principle, in all that follows, as if its truth were thoroughly established."

Although convinced of the validity of the conservation of energy, in particular of the impossibility of perpetual motion, Kelvin cannot quite reconcile this principle with either the Mayer-Joule or the Carnot principle.

"When thermal agency is thus spent in conducting heat through a solid, what becomes of the mechanical effect which it might produce? Nothing can be lost in the operations of nature — no energy can be destroyed. What effect then is produced in place of the mechanical effect which is lost? A perfect theory of heat imperatively demands an answer to this question; yet no answer can be given in the present state of science. A few years ago, a similar confession must have been made with reference to the mechanical effect lost in a fluid set in motion in the interior of a rigid closed vessel, and allowed to come to rest by its own internal friction, but in this case, the foundation of a solution of the difficulty has been actually found in Mr. Joule's discovery of the generation of heat by the internal friction of a fluid in motion. Encouraged by this example, we may hope that the very perplexing question in the theory of heat, by which we are at present arrested, will before long be cleared up."

"It might appear that the difficulty could be entirely avoided by abandoning Carnot's fundamental axiom, a view which is strongly urged by Mr. Joule (at the conclusion of his paper, 'On the Changes of Temperature Produced by the Rarefaction and Condensation of Air,' *Phil. Mag.*, May, 1845, Vol. XXVI). If we do so, however, we meet with innumerable other difficulties — insuperable without further experimental investigation — and an entire reconstruction of the theory of heat from its foundation. It is in reality to experiment that we must look, either for a verification of Carnot's axiom and an explanation of the difficulty we have been considering, or for an entirely new basis of the theory of Heat."

The confusion so vividly expressed by Kelvin was cleared up a year later, not by new experiments as he had expected, but by a critical reexamination of the fundamental concepts involved.

References*

1. E. Mach, *Principien der Wärmelehre* (J. A. Barth, Leipzig, 1896)
2. D. Roller, *The Early Development of the Concepts of Temperature and Heat* (Harvard University Press, Cambridge, Mass., 1950).
3. See S. C. Brown, *Am. J. Phys.* **18**, 367 (1950).
4. See S. C. Brown, *Proc. Am. Phil. Soc.* **93**, 316 (1949).
5. Paul S. Epstein, *Textbook of Thermodynamics* (John Wiley & Sons, Inc., New York, 1937).
6. L. Euler, *Opera omnia*, Ser. 2, Vol. 12, pp. 2–3. Also the Introduction by C. Truesdell, p. LXXXI.
7. S. Carnot, *Sur la Puissance Motrice du Feu* (Paris, 1824). Reimpression Paris, 1878, contains Carnot's unpublished notes. English translation by R. H. Thurston (John Wiley & Sons, Inc., New York, 1897).
8. W. F. Magie, *The Second Law of Thermodynamics* (Harper and Brothers, New York, 1899).
9. T. S. Kuhn, *Isis* **52**, 567 (1961).
10. E. Picard, *Sadi Carnot* (Gauthier-Villars, Paris, 1927).
11. H. C. Callender, *Proc. Phys. Soc. London* **23**, 153 (1911).
12. Victor K. LaMer, *Am. J. Phys.* **22**, 20 (1954); **23**, 95 (1955).
13. T. S. Kuhn, *Am. J. Phys.* **23**, 91 and 381 (1955). M. A. Hirschfeld, *Am. J. Phys.* **23**, 103 (1955).
14. E. Whittaker, *History of the Theories of Aether and Electricity* (Thomas Nelson and Sons, London, 1951), Second Edition, Vol. I, pp. 69–80, 177–184, 212–217.
15. J. P. Joule, "On the Caloric Effect of Magneto-Electricity and On the Mechanical Value of Heat," *Phil. Mag.*, Papers I, p. 123 (1843).
16. *Ibid.*, pp. 146, 268.
17. H. Helmholtz, *Ueber die Erhaltung der Kraft* (Berlin, 1847).
18. W. Thomson, *Edinburgh Trans.* **16**, Part V, p. 54 (1848–1849). See also E. Mach, reference 1, pp. 270–271, and R. Plank, *Naturwissenschaften* **37**, 361 (1950).

* This list contains the items that the author happened to use as historical source material.

CHAPTER 2. CLASSICAL THERMODYNAMICS

Clausius had this ability to bring order out of confusion, this breadth of view which could apprehend one truth without losing sight of another, this nice discrimination to separate truth from error . . .

J. W. GIBBS

2.1 Energy and Entropy

Classical thermodynamics was born out of the reconciliation of two apparently contradictory ingredients. The striking unification of the most diverse phenomena by the Mayer-Joule principle is to be contrasted with a peculiar new duality implicit in Carnot's principle: We have to draw the line between reversible and irreversible processes, a distinction characteristic of thermodynamics that is not predicted, or even allowed, by the fundamental equation of mechanics or electrodynamics. The precise and detailed description of this dichotomy is a most difficult problem, one of the central themes of this book.

The point of departure for this undertaking is the harmonization of the above-mentioned principles by Rudolph Clausius in 1850 and shortly thereafter by the equivalent but not identical contribution of Lord Kelvin [1].

Clausius and Kelvin accept Carnot's idea that the descent of heat from high to low temperature is a necessary feature of the conversion of heat into work. At the same time they give up the assumption that heat quantity is conserved: The heat quantity rejected to the cold reservoir is $Q_2 = Q_1 - W$, in agreement with the Mayer-Joule principle.

Clausius and Kelvin developed their theories at first by analyzing the properties of ideal gases. Instead of reproducing their arguments, we pursue a line of thought that is the continuation of our discussion of the Carnot-Kelvin principle in Section 1.5. Our attitude is frankly heuristic. The resulting concepts form the basis of an elaborate theoretical structure that is firmly rooted in experiment and that will enable us not only to "justify" our foundation in retrospect but also to gauge its limitations and to extend its scope.

We focus our attention on Equations 10*a* and 13 of Chapter 1:

$$W = (T_1 - T_2)Q^* \tag{1}$$

$$W = JQ \tag{2}$$

which are succinct expressions of the principles to be harmonized. These are in evident conflict if the quantities Q and Q^* are identified with each other. In order to establish their separate identity, we have to state the "operational rules" that govern their use.

First of all, we note that the mechanical heat equivalent J is a *universal constant* depending only on the units chosen. In particular, J does not depend on the nature of the system, on its state, e.g., its temperature, nor on the type of work involved. Following modern practice we shall henceforth express Q in terms of mechanical units and we set the constant $J = 1$. With this definition, Q becomes a quantity of the same kind as work, and hence mechanical *energy*.

The application of the Mayer-Joule principle to the Carnot cycle yields

$$W = Q_1 - Q_2 \tag{3}$$

Substituting into Equation 1 we obtain the connection between Q and Q^*

$$Q_1 - Q_2 = (T_1 - T_2)Q^* \tag{4}$$

It is plausible to sharpen this relation into a term by term equation and assume tentatively that the transfer of the heat quantity Q between two systems at practically the same temperature T is associated with the transfer of the quantity Q^* where

$$Q^* = \frac{Q}{T} \tag{5}$$

On the basis of this picture, we arrive at a transparent discussion of the reversible Carnot cycle. The hot reservoir, to be denoted $R(T_1)$, rejects the quantities Q_1 and Q_1^* where $Q_1 = T_1Q_1^*$, the cold reservoir $R(T_2)$ absorbs Q_2 and Q_2^* where $Q_2 = T_2Q_2^*$. The reversible cycle is characterized by the conservation law $Q_1^* = Q_2^* = Q^*$, whence

$$\frac{Q_1}{T_1} = \frac{Q_2}{T_2} = Q^* \tag{6}$$

We define now the efficiency of the cycle as the dimensionless parameter

$$\eta = \frac{W}{Q_1} = \frac{Q_1 - Q_2}{Q_1} = \frac{T_1 - T_2}{T_1} \tag{7}$$

which replaces Carnot's η^* defined in Equation 9 of Chapter 1.

The foregoing analysis, and in particular the conservation of Q^*, applies only to reversible cycles. We shall now extend our heuristic inquiry to nonreversible cases. It is convenient to start with the opposite extreme: We connect the two reservoirs by a conducting rod, which is insulated apart from its contact with the reservoirs. We know from experience that in due time a stationary situation will develop with a definite amount of heat Q flowing "downhill" from $R(T_1)$ to $R(T_2)$ during a fixed span of time. The reverse process, the "uphill" flow of heat, does not occur spontaneously, hence the conventional designation of this process as *irreversible*. Since in the stationary regime the state of the rod does not change,* the total variation of Q^* consists of the increase of $Q_2^* = Q/T_2$ in $R(T_2)$ and the decrease $Q_1^* = Q/T_1$ in $R(T_1)$. The net production of Q^* is

$$Q^* = Q\left(\frac{1}{T_2} - \frac{1}{T_1}\right) > 0 \tag{8}$$

Thus we conclude that the irreversible process is associated with a net production of Q^*. For the reverse of this process, the uphill flow of heat, the total Q^* would decrease. As a *tentative* generalization of this result we state the rule: *For real, irreversible processes there is a total increase of the quantity Q^*. For unnatural processes the total Q^* would decrease; for the limiting reversible processes the total Q^* is conserved.* This rule turns out to be correct over the entire range of our experience.

The foregoing results can be expressed also in terms of qualified conservation laws: The quantity Q is conserved for *no work* processes ($W = 0$), whereas Q^* is conserved for *reversible* processes.

The concept of irreversibility is of great theoretical and practical importance. It is worth while to ponder its meaning somewhat more carefully. First of all we note that the meaning of this term in thermodynamics does not conform to common usage. It *is* possible to transfer heat from the cold to the hot reservoir by reversing a Carnot engine to run as a heat pump (refrigerator), only this process is not spontaneous; we have to pay for it by dissipating in $R(T_1)$ the minimum amount of work $W = Q_1 - Q_2 = Q_2(T_1/T_2 - 1)$ that is in addition to the quantity Q_2 raised from $R(T_2)$ to $R(T_1)$. What is impossible is the complete re-establishment of the initial condition of all systems involved in the process.†

* Note that the use of stationary states constitutes a similar artifice to that of cycles. The influence of structural changes, hidden to the macroscopic observer, is eliminated from the analysis. We may add that the establishment of stationary states is practically feasible to a good approximation.

† It is a usual practice in physics to redefine words of the common language, such as "force," "inertia," not to mention "causality" and "determinism" into sharpened and also altered technical terms. This procedure is beset with pitfalls often not easily avoided. The term "irreversible" is particularly unfortunate since physics has

For irreversible processes, the total production ΔQ^* can be considered a quantitative measure of irreversibility, an indication that the quantity Q^* has an important meaning. However, the curiously qualified conservation law to which it is subjected is not good enough to convey an idea of just what it is we got. The key to a better understanding of the situation is a new concept known under the designation of *state function.*

Up to this point, our attention was focused on the auxiliary, or environment systems, such as heat reservoirs and purely mechanical devices ("work sources"). We turn now toward the system and we introduce the sign convention that the quantities Q and Q^* be taken with the positive sign if they correspond to an increment of the system at the expense of the reservoirs. Likewise, work is to have a positive sign when it is performed on the system by work sources. Equations 3 and 6 referring to reversible cycles appear with this convention as

$$Q_1 + Q_2 + W = 0 \tag{9}$$

$$\frac{Q_1}{T_1} + \frac{Q_2}{T_2} = Q_1{}^* + Q_2{}^* = 0 \tag{10}$$

Here $Q_1 + Q_2$ represents the total amount of heat transferred to the system during a cycle. We shall denote this quantity symbolically as $\oint đQ$, and similarly the work per cycle as $\oint đW$. The bar across the differential sign indicates that we are dealing with an inexact differential. This is in contrast with the exact differential of a function, say dQ^*, the integral of which taken along a closed path vanishes. Equations 9 and 10 are written now as

$$\oint đQ + \oint đW = 0 \tag{11}$$

$$\oint \frac{đQ}{T} = \oint dQ^* = 0 \tag{12}$$

---

no technical terms to distinguish between situations in which the initial state of a system *can,* respectively *cannot* be re-established, even with the performance of work. We propose to call states of the second kind *irretrievable.* Thus the Koh-i-noor diamond once oxidized would be irretrievable. The same is not true of small diamond crystallites that can be now produced by means of available technology although at an energy input orders of magnitude larger than the one computed from thermodynamic considerations of irreversibility. More flagrant examples could be invoked concerning biological organisms and species. The concept of irretrievability is much too complicated for a fruitful discussion at this point. This note is merely a warning that the very simple consideration of *irreversibility* should not be thoughtlessly applied to situations which in fact involve the question of *retrieval.* In particular, considerations of entropy have no bearing whatever on the last-mentioned problem.

These integrals can be easily visualized for special systems. Let us consider, say, fluids where the relevant work is compressional:

$$đW = -P\,dV \tag{13}$$

In the P–V diagram, Carnot cycles are represented as closed curves consisting of two isotherms and two adiabatics each. The integrals in Equations 11 and 12 correspond to closed paths in this thermodynamic diagram. The work per cycle is numerically the area of the closed curve.

These results can be generalized to other coordinates corresponding to, say, electric or magnetic work. It is also possible to interpret Equations 11 and 12 in terms of arbitrary closed paths, since these can be built up from infinitesimal Carnot cycles. [See page 168 of reference 2.]

The significance of the vanishing of loop integrals is well known from mechanics where

$$\oint \mathbf{f}\cdot\mathbf{d}s = \oint đW = 0 \tag{14}$$

is the condition that the force f be conservative and derivable from a potential, or in other words that the integral

$$\int_i^f đW = \phi_f - \phi_i \tag{15}$$

be the same for all paths that connect the same initial and final points. As a consequence, ϕ is defined as a function of position.

The formal analogy with Equations 11 and 12 is perfect, but there is an important difference in interpretation. The mechanical path integral is in ordinary space; the thermodynamic path integral refers to a thermodynamic phase space, the points of which correspond to states of equilibrium. Accordingly, the "potentials" that arise from these relations are functions of equilibrium states; briefly, they are the *state functions* mentioned previously.*

First we have the *internal energy*

$$U_f - U_i = \int_i^f (đQ + đW) \tag{16}$$

or in differential form

$$dU = đQ + đW \tag{16a}$$

We say also that dU is a perfect differential in contrast to $đQ$ and $đW$ which are infinitesimals, but not the differentials of functions; their integrals are path-dependent.

The quantity Q^* was recognized as a state function by Clausius, who called it the *entropy*. We change our preliminary notation to the more

* Note that this similarity and contrast of the geometries of mechanics and thermodynamics will reappear throughout this volume.

conventional one

$$\int_i^f \frac{đQ}{T} = \int_i^f đQ^* = \int_i^f dS = S_f - S_i \tag{17}$$

In differential form

$$dS = \frac{đQ}{T} \tag{17a}$$

Relations 16*a* and 17*a* are summed up by saying that the incomplete differential $đQ$ is supplemented either by addition or by multiplication to yield the complete differential of a state function. The reciprocal temperature $1/T$ is called in this context an integrating factor.

For processes that are partially or completely irreversible, Equation 17 is replaced by an inequality

$$dS \geqq \frac{đQ}{T} \tag{18}$$

It is time now to make a preliminary remark on the possibility of measuring our new quantities. The key problem is the measurement of the temperature, which means the calibration of actual thermometers according to the Kelvin scale. Actually, Carnot cycles can seldom be performed sufficiently reversibly to afford us a practical approach to the Kelvin scale. Fortunately for the evolution of thermodynamics, the ideal gas temperature provides a partial answer to our quest.

An ideal gas is defined by means of the relations

$$PV = k\theta \tag{19a}$$

$$\left(\frac{\partial U}{\partial V}\right)_\theta = 0 \tag{19b}$$

where θ is the temperature and k is a constant.*

Carrying this system through a reversible Carnot cycle (in a thought experiment!), we can compute each stage to show that the temperature θ is identical to the Kelvin scale. [See page 165 of reference 2.] This result is of great practical importance since θ can be established for a number of real gases, and in the limit of small densities all of these scales tend to a common limit which is acceptable as an excellent approximation to the Kelvin scale.

In spite of this practical success, the procedure just outlined is not entirely satisfying. It seems unfair to lean so heavily on the fact that nature happens to be kind enough to provide us with real systems exhibiting nearly ideal behavior. Moreover, at low temperature where

* Note that Equation 19*a* by itself does not define the ideal gas. We shall see, however, in Paper 2 that Equation 19*a* implies 19*b* if θ is interpreted as the Kelvin temperature.

gases condense, the method no longer works. Fortunately, the question can be handled in a fashion that is immune to these objections; but we have to have much of the thermodynamic formalism at our disposal for this discussion. We shall return to this point in Section 2.5 of Paper 2 in a context devoted to the systematic discussion of the problem of measurement. Meanwhile, we shall consider the Kelvin scale as measurable.

With the measurement of temperature ensured, the determination of U and S reduces to measurements of macroscopic work. To be more precise, experiment yields only increments over a standard state: $U - U_0$, $S - S_0$. We return later to the question of integration constants along with other subtleties surrounding the connection of theory and experiment. The intuitive meaning of energy and entropy will emerge gradually from these discussions.

The unification of principles can be formally expressed by eliminating $đQ$ from Equations 16a and 18:

$$T\,dS \geqq dU - đW \tag{20}$$

or, in the special case in which only compressional work is relevant,

$$T\,dS \geqq dU + P\,dV \tag{21}$$

These relations play a central role in thermodynamics and are the origin for many ramifications. In order to develop the formalism best suited for particular applications, we have to enrich further the conceptual material and we shall proceed with our heuristic considerations in Section 2.3.

2.2 The Thermodynamics of Cycles

In contrast to our fast-moving survey, the actual historical evolution paused more extensively at this juncture. The relatively simple results obtained indeed cry out for consolidation. In hindsight, our lengthy heuristic argument can be considerably streamlined,* a task actually accomplished by Clausius and Kelvin. They built in the process an impressive theoretical structure, to which we shall refer as CK thermodynamics. This theory is of great historical importance and decisively influenced modern thermodynamics. Nevertheless, we survey these ideas only briefly, since we shall find it advantageous to postpone the consolidation of the theory.

It is a curious aspect of the crisis of preclassical thermodynamics that all of the protagonists rested their case on the impossibility of a perpetual

* Why not eliminate the heuristic argument altogether? We hope that it is worth our while to present classical thermodynamics as a stage in the evolutionary process; an antidote against the attitude that considers preclassical thermodynamics mildly ridiculous and postclassical thermodynamics all but inconceivable.

motion machine. The trouble stemmed from the fact that, unwittingly, they used this term (anyway, a misnomer!) to designate two distinct types of processes. This distinction is clearly recognized in CK thermodynamics, and the theory is based on two postulates stating the impossibility of both processes. A standard formulation of these postulates follows:

First Law of Thermodynamics. It is impossible to construct a device that, operating in a cycle, will produce no effect other than the performance of work on or by an auxiliary mechanical system.

Second Law of Thermodynamics. It is impossible to construct a device that, operating in a cycle, will produce no effect other than the extraction of heat from a reservoir and the performance of work on a mechanical system (conversion of heat into work).

This statement of the second law is usually designated as the Kelvin-Planck formulation. It can be shown to be equivalent with Clausius' formulation:

Second Law of Thermodynamics. It is impossible to construct a device that, operating in a cycle, will produce no effect other than the transfer of heat from a colder to a hotter body.

On the basis of these postulates one can derive the relation 20 and establish the concepts of internal energy, entropy, and Kelvin temperature much more rigorously than we have done in our heuristic considerations. We shall not reproduce here the classical arguments, partly because these are available in most texts on thermodynamics [see references 2 and 3], and partly because our program calls for a further heuristic extension of the conceptual material before turning to more rigorous deductions.

The establishment of the concepts of energy and entropy opened the way for innumerable applications. Naturally enough, these applications were handled with the method of Carnot cycles, the only technique that was already available. Although this "thermodynamics of cycles" yielded abundant results, it seems heavyhanded compared with more elegant modern methods. It is not hard to discern the reasons which called for a retooling of the theory. The cycles constitute a most ingenious device that enables us to arrive at the concepts of energy and entropy without making unwarranted assumptions about the structure of matter. In this artifice the unknown properties of the systems are automatically eliminated and all statements are in terms of reservoirs and other idealized macroscopic devices. Since energy and entropy are intimately connected with the structural properties of the system, it is a real *tour de force* that these concepts could be precisely defined in a procedure in which the system is hardly more than a "black box" that we are not supposed to open. However, once this task is achieved we are confronted with a new situation. We can bring to bear our new conceptual tools to a direct study of the system and the black box attitude becomes more a hindrance than a help. The adequate method for this structural problem was developed primarily by Gibbs [4].

2.3 Gibbsian Thermostatics

For the sake of simplicity we start with the case of a homogeneous chemically inert fluid and we write the relation 20 of Section 2.1 as an equation for reversible processes as

$$dS = \frac{1}{T}\,dU + \frac{P}{T}\,dV \tag{22}$$

This equation contains no explicit reference to a specific substance, and has, in this sense, a universal character. (The relation is, of course, specialized, inasmuch as work is assumed to be compressional. A generalization will be considered in Section 4 of Paper 2.)

However, the same relation can be brought to bear also on specific systems. We start from the observation that systems of well-defined chemical identity tend to assume spatially homogeneous forms called phases.*

Let us consider a fixed amount of a phase measured in grams, or preferably moles, and interpret Equation 22 as a process performed on this system.

The choice of U and V as independent variables closely corresponds to actual experimental procedures. Thus in the experiments of the type performed by Joule controlled amounts of energy are added to otherwise isolated systems. Equipped with a pressure gauge and a thermometer we may establish the empirical functions

$$\frac{1}{T} = f_1(U, V) \tag{23a}$$

$$\frac{P}{T} = f_2(U, V) \tag{23b}$$

The use of the absolute energy U rather than of dU implies that we have taken care of the energy constant by convention as stipulated in Section 2.1.

The functions 23 express the characteristic equilibrium properties of the substance under study; we refer to them as *equations of state*. If the

* We assume that the terms *chemical identity* and *homogeneous phase* convey a clear enough meaning for this preliminary analysis. The subtler aspects of these concepts will call for our attention later.

role of the dependent and independent variables is inverted,* these equations appear in a more familiar form and are identified as the *thermal* and the *caloric equations of state:*

$$V = V(P, T) \tag{24a}$$

$$U = U(P, T) \tag{24b}$$

The two equations of state are on an equal footing: Both are needed for the specification of the system, but they are not independent from each other. The existence of the entropy as a state function requires that the integrability condition

$$\frac{\partial 1/T}{\partial V} = \frac{\partial P/T}{\partial U} \tag{25}$$

be satisfied. This relation is the first instance of those thermodynamic identities that we shall consider systematically in Paper 2.

The redundancy of the equations of state is not really objectionable. Nevertheless, it is a gain in economy and elegance to remove this feature, which is easily achieved by integrating Equation 22 to yield

$$S = S(U, V) \tag{26}$$

the so-called *fundamental equation* of the system. The equations of state are, of course, recovered through differentiation:

$$\frac{\partial S}{\partial U} = \frac{1}{T} \tag{27a}$$

$$\frac{\partial S}{\partial V} = \frac{P}{T} \tag{27b}$$

Equations 22 and 26 are not entirely equivalent, since the latter involves an integration constant. The integrated equation contains more than the original differential form, and the addition would seem to be entirely unwanted. We shall see, however, that this is not so. The integration constant can be determined from quantum statistics (Section 2.4), and we shall find that the concept of equilibrium and its description in terms of the entropy is significantly enriched by the shift to the integrated form. (See Section 5 of Paper 2.) This is one of many instances in which Gibbsian thermodynamics is readily adapted to modern quantum-mechanical developments and an indication, later confirmed, that the fundamental equation truly deserves this name. Its most remarkable property is that it contains a complete statement of the thermodynamic information concerning all equilibrium states of the system in question. We express this important fact by saying that the fundamental equation

* The conditions of solubility are usually satisfied for these equations, although there are interesting exceptions in the so-called critical points. (See Paper 3.)

with its ingredients, the energy and the entropy, provide a *parametrization of all equilibrium states.* The correctness of this assertion is by no means obvious; the demonstration consists of the explicit construction of MTE from such a basis* (Papers 2 and 3).

Equation 26 can be represented as a surface in the space spanned by the variables S, U, and V. This space plays a particularly important role among all thermodynamic phase spaces. We propose to call it *Gibbs space.* More generally, we shall refer thus to the space spanned by all extensive variables (see the following).

Gibbs himself generalized the scope of Equation 26 by introducing additional independent variables. Before turning to these matters, however, we shall discuss a number of problems that arise in connection with the transition from the *differential inequality* 21 to the *integrated equality* 26. From the purely formal point of view this transition is simple enough. Nevertheless the logical structures that are built around these expressions as their points of departure differ in a surprising number of ways.

Centering thermodynamics around the integrated relation 26 means that we consider *equilibrium states* rather than *processes* as the basic conceptual element of the theory. It is now widely recognized that this is indeed an appropriate attitude in the classical theory which is often designated as *thermostatics.* However, the consistent development of this idea is contingent on the solution of a number of difficulties.

The first problem is to reintroduce the concept of processes into the basically static theory. A partial answer to this question is simply to differentiate relation 26 and regain Equation 22. This relation appears now clearly associated with a transition between infinitesimally different equilibrium states. It is usual to speak of a *quasi-static process.* We recall that we had established Equation 22 for *reversible processes* and conclude now that these concepts with ostensibly different meanings designate really the same idealized limiting situation. If we have a quasi-static path, a sequence of equilibrium states, then we can make the system run through these states in either sense by adjusting the infinitesimal differences of temperature or pressure (see page 129). Conversely, we established differences of entropy by integrating along *reversible paths.* This conforms to the notion that entropy, a state function, is associated with *equilibrium states.*

The foregoing arguments do not help us to account for irreversible processes. Were we to accept this as a limitation of thermostatics, we

* Massieu [5] seems to have been the first to advance the concept of fundamental equation. (He is quoted by Gibbs on page 86 of reference 4.) His choice of variables was, however, different from the foregoing. It was Gibbs who laid the foundation for a systematic transformation theory that ensures a complete flexibility in the choice of variables. See pages 63 and 237.

should be unable to formulate properly the extremal property of entropy, which is an indispensable element of MTE.

Gibbs stated the extremum principle in two versions: *In an isolated system the entropy tends to a maximum at constant energy, or alternatively, the energy tends to a minimum at constant entropy.* Although these statements undoubtedly express important truths, they lack precision to the point of being paradoxical. If an isolated system is not in equilibrium, we can associate no entropy with it, and if it is in equilibrium, its entropy can no longer increase. Many authors have grappled with this dilemma until a satisfactory solution was found in terms of the *composite system.* Consider a system of two or more spatially disjoint parts separated by adiabatic partitions. After reaching equilibrium each part has a definite entropy, the sum of which is associated with the composite system. The relaxation of an internal constraint will, in general, trigger a process, namely the redistribution, say, of energy at constant volume,

$$(DU')_{V'} + (DU'')_{V''} = 0 \tag{28}$$

leading to a new equilibrium with higher entropy. Here D denotes a finite change that actually takes place in the system and for which all the conservation laws and boundary conditions are satisfied. Thus the increase of entropy is perfectly well defined since it is associated with a transition from a more restricted to a less restricted equilibrium. This very important matter is discussed in detail in Papers 2 and 3.

The concept of composite systems seems to have been first used by Carathéodory [6], but in a rather different context (see Section 2.5). The role of composite systems in dispelling conceptual confusion of the type mentioned was particularly emphasized by Planck [7]. At present this concept is in general use and appears to be so simple that its slow acceptance might seem surprising. Actually, however, the idea of composite systems is a bold innovation that was to start a "logical chain reaction." First of all, we have to accept the distinction between two kinds of processes: those occurring in an isolated system under constant external conditions from those that involve the manipulation of constraints, e.g., the removal of the partition in a composite system. We propose to call the latter *parametric processes or thermodynamic operations.**

When fully exploited, the use of composite systems opens up a great extension of the scope of thermodynamics. It enables us to describe

* Strictly speaking, only the processes of the first kind can be considered as micromechanical processes described as trajectories in a fixed mechanical phase space, while parametric processes correspond to the transitions from one phase space to another, and these do not fit into the scheme of Hamiltonian mechanics. This was presumably the reason for the early hesitation of making use of this simple conceptual device.

complex space-time structures and spells the definite abandonment of the restrictive "black box" attitude of the thermodynamics of cycles.

Whereas the artifice of composite systems enables us to extend the range of processes that are accounted for in the formalism, this extension still leaves much to be desired. Thus we could consider the analogue of process 28 in which the volume changes at constant energy

$$(dV')_{U'} + (dV'')_{U''} = 0 \tag{29}$$

This is a somewhat artificial process, much less important than the so called adiabatic or isentropic compression:

$$(dV')_{S'} + (dV'')_{S''} = 0 \tag{30}$$

The constancy of the entropy means that there is neither irreversibility to produce a net entropy increase, nor is there a heat exchange between the systems. In such a way we preclude also the exchange of conserved entropy. There are many situations in which this is a reasonable approximation for a real process.*

In order to arrive at the process 30 in a systematic fashion we solve Equation 26 for the energy

$$U = U(S, V) \tag{31}$$

Or, in differential form

$$dU = T\,dS - P\,dV \tag{32}$$

with

$$T = \left(\frac{\partial U}{\partial S}\right)_V \tag{33a}$$

$$-P = \left(\frac{\partial U}{\partial V}\right)_S \tag{33b}$$

Evidently Equations 26 and 31 are equivalent† since they are transformed into each other by a rotation in Gibbs space. Therefore we have to accept Equation 31 as a fundamental equation of equal standing to Equation 26. We shall refer to the formalism derived from Equations 26 and 31 as the *entropy* and the *energy* schemes, respectively. They are expressed in terms of different independent variables, and this is significant because differentiation of the fundamental equation is used to generate the "natural processes" of the scheme. Thus the analogues of Equations 28 and 29 are now

$$(dS')_{V'} + (dS'')_{V''} = 0 \tag{34a}$$

$$(dV')_{S'} + (dV'')_{S''} = 0 \tag{34b}$$

* Surprisingly, this static description goes a long way to describe even the propagation of mechanical waves (sound and ultrasound). Of course, absorption and dispersion are beyond the scope of thermostatics.

† For a slight limitation of this equivalence at $T = 0$ and ∞ see page 130.

Equation 34*b* is identical to Equation 30, but instead of being advanced ad hoc it emerges as the "natural" process of the energy scheme. Equation 34*a* corresponds to the reversible transfer of entropy. We had encountered such processes, say, in the Carnot cycle.

The fact already established that the total entropy of all systems involved in a process must increase is expressed in our composite system as follows:

$$DS' + DS'' > 0 \tag{35}$$

The corresponding relation in the energy scheme is

$$DU' + DU'' = W \tag{36}$$

where the work W performed by the external work source can be of either sign. The conservation laws include now the conservation of total entropy; hence the process is reversible, in contrast with the irreversible process 35. This important duality is firmly built into our formalism.

It is worth noting that both types of processes can lead to the decrease of entropy of a particular system: $DS' < 0$, provided that a compensating entropy increase takes place in the environment $DS'' \geqq -DS'$.

The entropy maximum principle can be expressed also in a very useful indirect form. Since the phenomenologically known trend toward equilibrium corresponds formally to the increase of entropy, the state of equilibrium will be attained if no further entropy increase can be brought about by any process consistent with the conservation laws and the constraints of the problem. Processes which satisfy these conditions are often called *virtual*, the corresponding virtual changes of the entropy are designated by ΔS (or δS if the process is infinitesimal). The condition of equilibrium is that

$$\Delta S \leqq 0 \tag{37}$$

for all *virtual* processes, since none of these can *actually* occur. The precise interpretation of this principle has many fine points which are discussed in detail in Paper 3. It will be shown also that an equivalent extremum (minimum) principle can be stated in the energy scheme:

$$\Delta U \geqq 0 \tag{38}$$

The principle expressed in Equations 37 or 38 plays a key role in the derivation of the formalism of MTE, a role comparable to that of Carnot cycles in CK thermodynamics.

At this point some readers might ask whether the formalism of thermodynamics is enriched by the addition of the energy scheme. This question is so much more relevant because a number of other schemes will be introduced later (see particularly Paper 2). On the basis of the discussion leading up to Equations 33 and 34 we might be inclined to answer in the affirmative. Yet we have noted earlier that the two fundamental equations are equivalent and correspond only to a different selection of the dependent and independent variables in Gibbs space. Accordingly, it is occasionally stated that thermodynamics is complete and self-contained in each of its schemes.

The resolution of this paradox is not particularly hard, but it forces us to render explicit another of those hidden dualities inherent in thermodynamics, namely the one connected with the aims and objectives of this discipline. First of all, as mentioned before, the fundamental equation contains a succinct summary of all thermostatic information concerning each specific substance. We still have to demonstrate what the exact nature of this information is, but it has certainly structural taxonomical features in the same sense as chemistry has. Second, however, MTE enables us to account for the gross aspects of space-time processes conceived as redistributions of conserved quantities in a composite system. This description is at first very sketchy, but there are various devices to widen its scope, and adding new schemes to the formalism is one of them. This procedure however, adds nothing to the taxonomical part of MTE, to the storage of thermostatic information. In fact, we shall see that the transformation theory of MTE hinges on the requirement that each scheme have the same information content.

Gibbs himself contributed in essential ways to the extension of the formalism by introducing additional variables into the fundamental equation and by developing a transformation theory enabling one to switch to the intensities as independent variables. These developments are integral parts of MTE, and we shall discuss them in detail in Papers 2 and 3. There is only one item in this general context that we wish to take up as a conclusion of this section.

Up to this point we have considered only *closed systems* consisting of a fixed amount of matter. This is quite adequate so long as we are dealing with the structural properties of homogeneous phases, since such properties can be scaled from one size to another, which is merely another way of expressing what homogeneity is.* The situation is different, however, when we turn to problems of the second type and wish to account for actual processes involving the passage of matter between phases or subsystems. It is then necessary to deal explicitly with variables specifying the total amount of matter in the system. We introduce a variable N

* This "scale invariance" of thermodynamic systems of laboratory size is so important that there is no theory that can do without it, although few authors care to formulate it explicitly. While some take this property as *self-evident*, others might reject it as *incorrect*, since homogeneity breaks down in the very small and the very large. (See page 123.) The tradition that favors the formulation of the postulates of thermodynamics as "impotence axioms" is not conducive to the formulation of postulates of limited validity: It is clearly *not* impossible to find nonhomogeneous systems. Yet we cannot do without postulates of limited validity, but we should be particularly careful in making them explicit so that the limitations should be observed and eventually overcome.

to denote either the number of moles or molecules. As long as we are not solving any numerical problem we can afford to be vague on this point, since all results can be scaled by Avogadro's number.

Within the validity of the homogeneity postulate the variables S, U, V, N are proportional to the size of the system; in a conventional terminology they are called *extensive*. (For the entropy this follows from the definition $dS = dQ/T$, since the heat quantity is extensive and the temperature is independent of the size of the system.) This fact can be used in various ways to introduce other variables and to express the homogeneity of the system in the formalism.

First of all we may single out one of the variables as a "scale factor"; its main role is to define the size of the system whereas other variables correspond to its intrinsic properties. Thus, using the mole number N in this capacity we introduce the intrinsic *molar quantities:*

$$s = \frac{S}{N}, \qquad u = \frac{U}{N}, \qquad v = \frac{V}{N} \tag{39}$$

An interesting alternative is to use the volume as a scale factor and the *densities*

$$s = \frac{S}{V}, \qquad u = \frac{U}{V}, \qquad \rho = \frac{N}{V} \tag{40}$$

as intrinsic variables. There is, of course, a conflict of notation between Equations 39 and 40 that has to be taken care of in each context. Unfortunately, there are not enough symbols to go around to have a fixed set for each thermodynamic scheme. However, this is only a small price to pay for the flexibility gained by using these different schemes. Thus Equations 39 and 40 refer to entirely different situations: They correspond to *closed* and *open systems*, respectively. It is remarkable that these cases appear here on an equal footing, and we can choose either one to suit the problem on hand.

The traditional approach favors closed systems and accordingly in the fundamental Equation 26 we have assumed the mole numbers to be fixed. Therefore, this equation can be interpreted in terms of the variables 39 as

$$s = s(u, v) \tag{41}$$

The alternative form in terms of the densities of an open system is*

$$s = s(u, \rho) \tag{42}$$

There are also situations, however, in which neither N nor V is fixed. Take, for example, a system consisting of water and ice. In the case of

* There are indications that this scheme has potentialities of practical application not yet exploited (see Paper 5).

melting or freezing the N and V of both systems are variable. This situation we handle as follows: From Equations 39 and 41 we have

$$S = Ns\left(\frac{U}{N}, \frac{V}{N}\right) = S(U, V, N) \tag{43}$$

The function of three independent variables appearing here has special properties due to the homogeneity of the system. We must have for any positive scale factor λ

$$S(\lambda U, \lambda V, \lambda N) = \lambda S(U, V, N) \tag{44}$$

Functions of this sort are known in mathematics as *homogeneous functions* of the first order, a neat parallel between physical and mathematical homogeneity.

We are now in a position to express the entropy maximum principle for the above-mentioned case of melting ice, i.e., for any two-phase system:

$$S(U, V, N) = \max \{S'(U', V', N') + S''(U'', V'', N'')\} \tag{45}$$

where the comparison states for this extremum problem are selected from the virtual states satisfying the additivity conditions:

$$U' + U'' = U \tag{46a}$$

$$V' + V'' = V \tag{46b}$$

$$N' + N'' = N \tag{46c}$$

The two functions in the bracket in Equation 45 have different functional forms corresponding to the two phases. We shall see that this extremal problem has a unique solution that specifies the amounts of the two phases N_0' and N_0'' under equilibrium conditions and also the intrinsic properties of the coexisting phases:

$$u_0' = \frac{U_0'}{N_0'} \qquad u_0'' = \frac{U_0''}{N_0''}, \cdots$$

In Papers 2 and 3, Equation 45 is extended in an obvious fashion to several phases, and the generalized extremum principle is justified by the fact that it implies the entire theory of phase equilibrium.*

As a further heuristic step it is plausible to extend the theory to systems consisting of several chemical species by postulating a fundamental equation of the form

$$S = S(U, V, N_1, N_2, \cdots) \tag{47}$$

* Some adjustments are needed, however, to take into account symmetry properties and surface effects.

The processes corresponding to this equation include diffusion of matter between subsystems. Our generalization is meaningful only if these processes are also governed by the entropy maximum principle, which indeed turns out to be the case.

By solving Equation 47 for the energy we obtain another useful form of the fundamental equation:

$$U = U(S, V, N_1, N_2, \cdots) \tag{48}$$

Since these equations are used in conjunction with appropriate extremum principles, it is not surprising that their derivatives turn out to be important for the formulation of the conditions of equilibrium. Say, for Equation 48 we have

$$T = \left(\frac{\partial U}{\partial S}\right)_{V,N_j} \tag{49a}$$

$$-P = \left(\frac{\partial U}{\partial V}\right)_{S,N_j} \tag{49b}$$

$$\mu_j = \left(\frac{\partial U}{\partial N_j}\right)_{S,N_{j'}} \qquad j' \neq j \tag{49c}$$

where $j = 1, 2, \cdots$ runs through all chemical species.

While the first two equations are identical to Equations 33, the last equation is new; it serves to define the parameters μ_j called chemical potentials. The derivatives of the fundamental Equations 47 or 48 are called *intensive variables* or *intensities.* They are also intrinsic properties of systems yet quite distinct from densities and molar quantities considered earlier. We may recall that the temperature is the same for all subsystems of a system in thermodynamic equilibrium, regardless of the assumption of homogeneity. The new parameters μ_j play a quite similar role in governing the direction of the diffusion of matter as the temperature does for the diffusion of heat (energy).

It is an important aspect of the formalism under discussion that processes are primarily redistributions of conserved quantities. Here we seem to be running into a snag, since chemical species are not, in general, conserved, but are subject to chemical reactions. This difficulty is entirely eliminated, however, by another Gibbsian discovery, the concept of independent chemical components. As we shall see in Section 3 of Paper 2, the validity of fundamental Equations 47 and 48 can be saved even in the presence of chemical reactions by the simple device of reinterpreting the numbers N_j to represent invariants of the relevant chemical reactions.

Sketchy as the discussion of this section may be, it ought to suffice to justify the fundamental Equations 47 and 48 as a tentative basis for the development of MTE. A further support to this viewpoint is provided in the next section devoted to a digression into quantum mechanics.

2.4 Digression on Entropy

It is a remarkable fact that Gibbsian thermostatics harmonizes very well with quantum mechanics developed a half a century later. The relation of these disciplines has many subtle aspects. At this point, we confine ourselves to the discussion of a few simple ideas to show that the aura of mystery that shrouds the concept of entropy is reduced when viewed from a quantum-mechanical context.

Let us consider an isolated system of $N_1, N_2, \cdots$ particles of the species $A_1, A_2, \cdots$ in a box of volume V. Let

$$g = g(U, V, N_1, N_2, \cdots) \tag{50}$$

be the number of linearly independent eigenfunctions belonging to the energy eigenvalue U. We shall refer to this function as the *structure function* of the system.

For a composite system consisting of isolated parts (adiabatic partitions inhibit the transfer of heat and mass), the structure function is evidently multiplicative, and since it depends on the same set of independent variables as the fundamental Equation 47, it is plausible to postulate a relation of the form

$$S = k \ln g(U, V, N_1, N_2, \cdots) \tag{51}$$

This relation is the quantum-mechanical version of Boltzmann's principle, and k is Boltzmann's constant.*

* Reversing the historical process it is easy to change the mechanical interpretation of this formula from quantum to classical mechanics. However, this switch involves important conceptual changes which are, of course, all in favor of the quantum-mechanical interpretation. (i) The classical value of g involves an arbitrary multiplicative factor, depending on the arbitrary cell size in mechanical phase space. This leads to an arbitrary constant, in apparent agreement with the fact that the experimental definition in terms of heat quantities yields only *differences* of entropy. In contrast, the quantum-mechanical interpretation assigns definite numerical value to g. We shall see that this number is meaningful and is an important element in the operational definition of thermodynamic equilibrium. (ii) According to quantum mechanics the basic constituents of matter form classes of indistinguishable entities the permutation of which does not lead to new microstates. This is in complete agreement with the point of view of thermodynamics and also with experiment. In contrast, the description in classical-mechanical phase space leads to a spurious increase in the number of microstates that seemed paradoxical and had to be removed by an ad hoc correction, a replacement of g by $g/N_1!\, N_2! \cdots$ (Gibbs paradox). (iii) In contrast to Boltzmann's interpretation of g as a "thermodynamic probability," the structure function has no probabilistic character. This removes the difficulty that this alleged probability does not have the properties associated with the notion of probability (see Paper 7). We note that relation 51 has a precise meaning only for isolated systems. For systems that exchange energy with each other (thermal equilibrium), Equation 51 has to be replaced by a statistical expression (see Paper 7).

The interpretation of the entropy maximum principle is now formally very simple. The Schrödinger equation can be solved for different boundary conditions, and it is evident from the well-known extremal properties of eigenvalues that the number of solutions belonging to a fixed value of the energy, i.e., the structure function, cannot decrease when internal constraints are relaxed.

We have mentioned before that MTE consists of a taxonomic structural part and a dynamic part describing space-time processes. The quantum-mechanical foundation of the latter is a difficult problem that we are not prepared to discuss at this point. In contrast, the structural part of MTE can be considered just a subdivision of quantum mechanics. We consider only situations governed by the time-independent Schrödinger equation, and instead of solving this equation, we use little more than the fact that there exists a structure function.

A more sophisticated integration of thermodynamics and quantum mechanics requires the use of the concept of density matrix, the adaptation of which into the present context requires deeper methods than those considered in this volume.

2.5 The Axiomatics of Carathéodory

With the discussion of Section 2.4 we have accumulated already the conceptual elements needed for the deductive reconstruction of the theory. However, before proceeding to develop this axiomatic neo-Gibbsian theory to which we shall refer as MTE, it seems appropriate to clarify the relation of our approach to the well-known axiomatization of thermodynamics by Carathéodory [6]. Our discussion is directed to the reader who is to some extent familiar with the Carathéodory theory, and no introduction to the subject is intended.

In comparing different axiomatizations of the same discipline it is convenient to consider a deductive system as a logical device that processes the postulational basis, the "logical input" into the main body of the theory, the "logical output."

We have seen previously that classical thermodynamics is concerned with two principal problems. First, the establishment of the concepts of energy and entropy by means of macroscopic measurements along the lines initiated by Clausius and Kelvin (Section 2.2). Second, the *parametrization of equilibrium;* that is, the development of a mathematical formalism that accounts for the equilibrium properties of thermodynamic systems, following the ideas of Gibbs (Section 2.3). Although Carathéodory used some Gibbsian ideas concerning the specification of the chemical and phase structure of thermodynamic systems, his theory can be considered as the axiomatization of the first of the above-mentioned problems; we shall briefly refer to this structure as the CKC theory. The macroscopic thermodynamics of equilibrium, briefly MTE, is in a similar relation to the second problem. Although the two theories are consistent with each other, they differ in their aims, their scope, and their style.

The central idea of MTE is that the satisfactory parametrization of equilibrium makes use of the concepts of energy and entropy; hence these have to be available as primitive concepts of the theory. Conversely, the reduction of the energy and entropy to macroscopic measurable quantities within the CKC theory depends on the assumption that the equilibrium states can be parametrized in terms of such empirical parameters as pressure and volume.* Clearly, the transition from CKC theory to MTE implies the interchange of the roles of the logical input and output. There are no *a priori* reasons to prefer one of these theories over the other, and both are to be appreciated by their results.

The main achievement of the CKC theory is the constructive definition of the thermal quantities: heat quantity, internal energy, entropy, and temperature in terms of macromechanical measurable parameters. In this endeavor Carathéodory introduced the concept of composite systems which is an important ingredient of MTE as well. The theory was worked out in terms of differential geometry, a welcome progress over the peculiar technique of Carnot cycles that cannot be considered as an instance of any sort of "normal mathematics." We have to remember also that the theory dates from the beginning of the century. The atomic theory passed through a crisis, and it appeared important to establish concepts related to molecular structure in purely "operational" terms. This task was accomplished for the simplest systems (nearly ideal gases) for which the postulate of parametrization seemed harmless enough.

The whole approach was undoubtedly inspired by the positivist "operational" program, according to which the basis of a theory should contain only directly observable quantities. It is also among the most successful products of that program.

Granted, however, that the CKC theory has its merits, it is in order to point out its shortcomings. First of all, the Carathéodory axiomatics is not an appropriate basis for the construction of a reasonably complete thermodynamic theory. For example, the entropy does not have all the properties that we consider at present essential and the third law of thermodynamics specifying the behavior of entropy under the limiting

* In contrast to the importance here attributed to the problem of parametrization, this aspect is ignored or underplayed in most of the traditional formulations. The relevance of the question may have been first recognized by Carathéodory, as it appears from the study of the original reference. The results of this subtle and difficult paper were taken over very slowly in the literature. While at present there are easily available accounts of the theory, these are still rather incomplete. The tendency is to emphasize the purely mathematical aspects, particularly Carathéodory's theorem, at the expense of the full statement of the postulates.

conditions of absolute zero of temperature, and small density (Section 5 of Paper 2) still has to be postulated in an ad hoc fashion. More generally, the theory pretends to be absolutely "safe," since it is supposedly based only on empirical facts. On closer inspection such claims generally turn out to be illusory since a mathematical formalism is never identical with its actual experimental basis. Thus Carathéodory has to postulate that the equilibrium of one component fluids be uniquely parametrized in terms of the pressure and volume. It has been pointed out recently that substances with negative expansion coefficients, such as water, are exceptional in this respect [8]. Another point is that Carathéodory has to assume the possibility of nonsingular transformations from the set of empirical to "deformation" variables. This means that the Jacobian of the transformation should nowhere vanish. However, the empirical variables are too vaguely defined to allow us to draw any conclusions as to the validity of this assumption. Needless to say, these difficulties are inherent also in the Clausius-Kelvin theory, but they were not made explicit until Carathéodory.

We conclude this section by indicating our attitude concerning the aforementioned difficulties within the forthcoming development of MTE. First, and most important, we claim no absolute validity for our postulational basis. The validity of the postulates and the usefulness of the primitive concepts are only tentative and have to be justified by the experimental verification of the implications of the theory. We shall often prefer sharply stated postulates having many precise implications even if we have to pay for this precision by a narrower range of validity. Let us consider, in particular, the aforementioned difficulties of parametrization in the CKC theory. The first of these, connected with the change of sign of the expansion coefficient of certain substances, does not appear in MTE; the parametrization in terms of the entropy is not singular under these conditions. In contrast, the nonvanishing of the Jacobian of the thermodynamic transformations will be required also in MTE. However, within this theory we shall be in a position to provide practical criteria of validity. The fact that these conditions break down for certain states (critical points and absolute zero) offers no methodological difficulties. In fact, this breakdown will be the point of departure for the extension of the theory into the microscopic statistical direction.

References

1. R. Clausius, *Pogg. Ann. Phys.* **79**, 378 and 500 (1850). See also references 1, 8, and 18 to Chapter 1.
2. M. W. Zemansky, *Heat and Thermodynamics* (McGraw-Hill Book Company, Inc., New York, 1957), 4th edition.
3. E. Fermi, *Thermodynamics* (Dover Publications, New York, 1956).

4. J. W. Gibbs, *Collected Works* (C. Scribner's Sons, New York, 1902; Yale University Press, New Haven, 1948), Vol. I.
5. M. Massieu, *Compt. Rend.* **69**, 858 and 1057 (1869).
6. C. Carathéodory, *Math. Ann.* **67**, 355 (1909); *Sitzber. Preuss. Akad. Wiss.* 39 (1925). M. Born, *Physik. Z.* **22**, 218, 249, and 282 (1921). M. Born, *Natural Philosophy of Cause and Chance* (The Clarendon Press, Oxford, 1949).
7. M. Planck, *Physica* **2**, 1029 (1935); see also R. Ehrenfest-Afanassjewa and G. L. de Haas-Lorentz, *ibid.* **2**, 743 (1935), and M. Planck, *Ann. Physik* **19**, 759 (1934).
8. J. S. Thomsen and T. J. Hartka, *Am. J. Phys.* **30**, 26 and 388 (1962).

PAPER 2

AN OUTLINE OF MTE

1. THE POSTULATIONAL BASIS OF MTE

The basic structure of the macroscopic thermodynamics of equilibrium (MTE) is developed in Paper 3. The present contribution is partly a commentary and partly a supplement to that paper.

The postulational basis of the theory is contained in Section I of Paper 3. The somewhat cut-and-dried presentation should not cover up the fact that we are dealing with a basically simple situation: We describe the distribution of energy and matter over regions of space in terms of additive conserved quantities. We assume that the system is put together of homogeneous parts (phases) in which the additive quantities are proportional to the size of the system. Such variables are said to be extensive; we propose to call the abstract space spanned by them the *Gibbs space.* The domains, also called cells, are bounded by walls that are permeable to some of the quantities, while they restrain others. Thus some of the extensive variables are *fixed* while others are *free* to change within certain limits insofar as the constraints (conservation laws and boundary conditions) are concerned.* The actual variation of the free variables is determined by the microscopic dynamics of the system. In general, this is a formidable theoretical problem that can be approached, however, by successive approximations. The first step exploits the simplifications brought about by the trend toward thermodynamics equilibrium: In isolated systems the free variables tend to constant values, and the basic assumption of MTE is that these are determined as solutions of a constrained extremum (maximum) problem; the extremal function is the entropy, expressed as a function of the extensive variables, the so-called fundamental equation of the system.

It is not surprising that a maximum principle leads in a very direct fashion to conditions of equilibrium and stability involving the first and

* The simplicity of the formulation of MTE depends largely on the sharpness of the distinction between free and fixed variables. It is good to remember, however, that in reality what we can say is only that the relaxation time of the so-called fixed variable is large, and that of the free variables is small compared to the duration of the experiments envisaged. We shall consider this matter in somewhat more detail in Section 3 in connection with the problem of chemical composition where a similarly dynamic situation prevails. Even the statistical theory STE is based on this assumption about the distribution of the relaxation times of the system. The occurrence of processes, the relaxation time of which is of the order of magnitude of the duration of the experiment, calls for the use of much more profound theories.

second derivatives of the fundamental equation. The result is the theory of phase equilibrium; its essential aspects are contained in Paper 3. There remains, however, a searching question to answer: How do we determine the entropy that appears as an undefined concept by postulation fiat in MTE? In CKC thermodynamics, great care is taken to *construct* the entropy from empirical data. The *postulation* of this concept might seem cavalier to the traditionalist student of the subject, and he may even invoke Bertrand Russell's famous remark contrasting honest work and theft in an apparently similar constellation.

Yet such an attitude misses a most important point. Different axiomatizations of the same empirical discipline lead to different theories which enable us to ask and answer different types of questions. In CKC thermodynamics one admits only operational elements into the postulational basis and proceeds from empirical data to the entropy. This concept appears as the culmination of the whole effort, but the theory has not much to tell us about the practical use of the newly won concept.

In contrast, the entropy is a conceptual tool, a means to an end in MTE. The postulational basis is formulated in terms of variables that bring out the intrinsic symmetry of the theory in an elegant form, even though not all of these variables are strictly speaking operational. The apparent conflict between the tastes of the theorist and the experimentalist in choosing their variables is resolved by means of a transformation theory that ensures a perfect flexibility of choice without any need of compromise. The result is a richer theory and a correspondence with experiment that ranges wider and deeper. In particular the use of the entropy enables us to answer elusive questions about the microscopic aspects of equilibrium that are altogether outside the scope of the classical theory. Since these matters are not treated in sufficient detail in Paper 3, we take them up in the next section.

Some of the subtle aspects of the conceptual and empirical definition of entropy are closely tied up with a sharpened formulation of Nernst's law. The question has been beset by considerable controversy and we defer its discussion to Section 5. This discussion will also involve some quantum statistics. Actually, the modern student who is more familiar with quantum mechanics than with the intricacies of the Carnot cycle might best look at MTE as a rudimentary branch of quantum mechanics. The postulates of MTE are then merely reminders of the rules that are used in this analysis. These rules are very simple compared to those used in more genuinely quantum-mechanical contexts. The reasons are several. Primarily, we ask only a restricted type of question, which is very modest from the microscopic point of view but which happens to be quite important for the macroscopic observer. Also, a fuller use of quantum mechanics and of statistics would enable us to prove some of our postulates as theorems. A first step is made in this direction in Paper 7.

Although the quantum-mechanical substructure is, in general, well camouflaged in MTE, it shows unmistakably in the marginal, yet important argument of Section 5.

2. MTE AND EXPERIMENT

2.1 Intensities

MTE is centered around the fundamental equation expressed uniquely in terms of extensive variables. The intensive variables are introduced in a systematic manner as derivatives of the fundamental equation. The transformation theory discussed in the next section will enable us to use these derivatives as independent variables whenever experimental or theoretical considerations warrant.

The situation is somewhat complicated by the fact that we have the alternative formulations of the entropy and the energy schemes:

$$S = S(X_0, X_1, X_2, \cdots) \tag{1}$$

$$U = U(X_0, X_1, X_2, \cdots) \tag{2}$$

where $X_1 = V$, $X_2 = N_1, \cdots$, and X_0 is interpreted as U and S in the entropy and the energy schemes, respectively.

The transition between Equations 1 and 2 is also a sort of transformation that is discussed in Section 2.3 of Paper 1; see also Section IV of Paper 3 and Section 5.

The first derivatives of Equations 1 and 2 are called intensities in the respective scheme. Their proper interpretation is important for establishing the empirical meaning of the formalism.

We start with the energy scheme and define the intensity P_i conjugate to X_i as

$$P_i = \left(\frac{\partial U}{\partial X_i}\right)_{X_j} \tag{3}$$

or also

$$dU = \sum P_i \, dX_i \tag{4}$$

The conjugate intensities in the entropy scheme are defined as

$$\pi_i = \left(\frac{\partial S}{\partial X_i}\right)_{X_j} \tag{5}$$

or

$$dS = \sum \pi_i \, dX_i \tag{6}$$

We can easily forego the use of these entropic intensities in MTE, but they are useful in STE (see Paper 7). We note for the record that from

Equations 4 and 6

$$\pi_0 = \frac{1}{P_0} = \frac{1}{T} \tag{7}$$

$$\pi_i = -\frac{P_i}{P_0} = -\frac{P_i}{T} \tag{8}$$

We have also made use of Equation 9.

It is evident from Sections III and V of Paper 3 that the conditions of equilibrium with respect to the transfer of the quantity X_i are expressed in terms of the constancy of the conjugate intensities.

Let us consider now some of the problems that arise in connection with identification of the individual intensities.

It is apparent from Section III of Paper 3 that

$$T = P_0 = \left(\frac{\partial U}{\partial S}\right)_{X_i} \tag{9}$$

has all the qualitative features associated with the empirically defined temperature. It is also identical to the absolute Kelvin temperature as it was defined in Section 2.2 of Paper 1. The problem of actual measurement will be taken up in Section 2.5. For the present it is sufficient to anticipate that, except for a trivial scale factor, the absolute temperature scale can be uniquely established.

Next we consider the chemical potentials

$$\mu_j = \left(\frac{\partial U}{\partial N_j}\right)_X \qquad j = 1, 2, \cdots \tag{10}$$

that are conjugate to the mole numbers and perform a function in the transfer of matter which is the direct analog of the function of temperature in heat (energy) transfer. In spite of this analogy there are objective and subjective reasons why temperature should be more familiar than chemical potentials. There is a chemical potential to each of the innumerable chemical components, whereas temperature derives its unique character from the uniqueness of energy. Energy can be transferred without any matter being displaced, yet every transfer of matter is associated with a transfer of energy. There is even a thermodynamic system consisting only of energy, a zero-component system, namely blackbody radiation.

On the subjective side we have a physiological conception of temperature which, in a qualitative sort of way, parallels the thermodynamic concept. In the case of chemical potentials we are neither helped nor hindered by our prescientific experience, although chemical potentials are connected with chemical affinity (Section 4), a concept not without some intuitive appeal.

The method of measuring chemical potentials will be considered in Section 5.

In addition to the transfer of energy and matter, MTE is also concerned with the performance of work. The most common kind of work is compressional. Its precise formulation takes on a somewhat different guise in the entropy and in the energy schemes. In the former, the displacement of a partition (piston) between two subsystems can be conceived as an exchange of volume which is completely on a par with the exchange of energy and matter.

In the energy scheme we should rather speak of a change of volume as due to a variation in the boundary conditions of the system produced by coupling with an auxiliary mechanical system, a work source. (See also Section 4.)

We claim that the intensity that is conjugate to the volume is the negative pressure:

$$-P = \left(\frac{\partial U}{\partial V}\right)_{S,N} \tag{11}$$

This follows from a sort of correspondence principle that coordinates the relation of MTE to neighboring disciplines, in particular with the macromechanics of the system as a whole. (Whenever we say macromechanics we could also add macroscopic electromagnetism.*)

Equation 11 evidently has the form of the well-known expression of hydrostatics. The only question concerns the stipulation that the variables S and the N's be constant on differentiation. This is indeed the only possibility to ensure consistency with macromechanics, since a variation of S would mean either irreversibility, or the reversible transfer of heat, or the combination of both, and both of these effects are inconsistent with macromechanics. The constancy of the N's means that the system is closed, a natural requirement under the conditions considered.†

* We have to be on our guard against a confusion in terminology. Thermodynamics is usually said to be macroscopic, which is correct to the extent that, say, MTE has only a single parameter the entropy to account for the microstructure of the system. Yet MTE is not as macroscopic as macromechanics, which ignores the microstructure altogether. In particular, this macromechanics should be nondissipative and hence reversible. Irreversible dissipative terms would place the theory already within the domain of thermodynamics. We do not stress the macroscopic aspects of MTE because relatively small changes in the postulational basis suffice to deepen MTE into STE and thus render it more microscopic (see Papers 7 through 12).

† Some time ago E. A. Guggenheim [*Proc. Phys. Soc.* (*London*) **79**, 1079 (1962)] criticized Callen's method of introducing the entropy that parallels the present approach. Guggenheim points out that the entropy maximum principle does not fully determine the entropy and might lead to an incorrect expression for the pressure. He suggests the additional postulate that adiabatic work be equivalent to a process at constant entropy. The foregoing remarks show, however, that such a specific postulate is unnecessary, provided one invokes the "correspondence principle," a methodological postulate that cannot be ignored without serious trouble. This matter was

Compressional work is only one of the many types that can be performed on the system. Of course, this is in many ways of a very special kind because the atmosphere acts as a "volume reservoir" and maintains systems in contact at a fixed pressure. Hereby the atmosphere performs an amount of work which has a different character from the directly measurable work performed by a work source consisting, say, of weights and springs.

We introduce the notation

$$đW = -P\,dV + đW' \tag{12}$$

where $đW'$ denotes the additional infinitesimal work. We have now

$$dU = T\,dS - P\,dV + \sum\mu\,dN + đW' \tag{13}$$

and the special case

$$đW = -P\,dV + đW' = (dU)_{S,N} \tag{14}$$

where the variables used as subscripts are to be held constant. This relation is reminiscent of that of mechanics according to which the work performed by conservative forces on an otherwise isolated system is equal to the increase of the potential energy. We see that in MTE the internal energy assumes this role of the potential energy, provided the system is closed to the transfer of matter and the work is reversible adiabatic.

It would be tempting to introduce additional macroscopic variables and write

$$đW' = \sum P\,dX \tag{15}$$

Yet we defer the discussion of this matter to Section 4 since it involves more complications than one might expect at first blush. A short statement of these difficulties is found on pages 123–124. The additional intensities do not have all the properties of the "true" intensities T, $-P$, and μ. In particular, the following property is interesting and important.

Let us consider a system of fixed size, interacting with a large "environment system," a reservoir. We let the size of the latter tend to infinity in such a way that the intensities remain constant. The question of whether the coupling of the two systems will lead to a transfer of the quantities U, N, V into the system or out of it, depends only on whether the intensities T, μ, $-P$ are smaller or larger than the corresponding intensities of the environment. Therefore, the limiting systems of infinite size are adequately characterized by their intensities, while the extensive quantities associated with them become infinite and therefore useless.

---

treated somewhat casually on page 134 and deserves more emphasis. We shall see in Section 5 that Nernst's law also has a bearing on the correspondence principle.

2.2 Transformation Theory

Two of the intensities, namely the temperature and pressure, lend themselves to direct measurement, and these are favorite independent variables of experimental thermodynamics. Mathematical physics possesses in the Legendre transformation a device to achieve an appropriate switch of independent variables with ease and precision. The best-known example for the use of this transformation is the transition from the Lagrangian to the Hamiltonian function in analytical mechanics. The transformation is simple enough to be accepted as an ad hoc formal device, yet its meaning is best understood in terms of a geometrical interpretation:* A sufficiently smooth curve can be equally well represented in terms of its points as its tangents. The duality of points and lines, so important in some branches of geometry, appears in MTE as the duality of extensive and intensive variables. As a result, Gibbs space is endowed with a fairly elaborate geometry that is discussed in Paper 6. Here we confine ourselves to summarizing the formalism and discuss at first a single independent variable.

Let us consider the function

$$y = y(x) \tag{16}$$

with its derivative

$$p = \frac{dy}{dx} \tag{17}$$

or

$$dy = p\,dx \tag{18}$$

We define the Legendre transform of $y(x)$ as

$$\psi(p) = px - y(x) \tag{19}$$

We assume that Equation 17 can be solved for $x = x(p)$ which is possible if

$$\frac{d^2y}{dx^2} \neq 0 \tag{20}$$

Thus x can be eliminated from the right-hand side of Equation 19. This relation can also be put in the form

$$\psi(p) = x\frac{dy}{dx} - y = x^2\frac{d(y/x)}{dx} \tag{21}$$

The equivalence of $\psi = \psi(p)$ with Equation 16 is evident from the fact that the latter can be regained from the former. The relation is dual

* See R. Courant and D. Hilbert, *Methods of Mathematical Physics* (Interscience Publishers, Inc., New York, 1962), Vol. 2, p. 32, also Paper 6.

in the sense that the inverse and direct relations have the same form. Indeed, we have from Equation 19

$$d\psi = x\,dp + p\,dx - dy = x\,dp \tag{22}$$

and hence

$$x = \frac{d\psi}{dp} \tag{23}$$

This equation can be solved for $p = p(x)$ provided

$$\frac{d^2\psi}{dp^2} \neq 0 \tag{24}$$

Therefore p can be eliminated from

$$y = px - \psi = p^2\,\frac{d(\psi/p)}{dp} \tag{25}$$

and we recover Equation 16. Thus in the domain where the conditions 20 and 24 are satisfied the transformation $y, x \leftrightarrow \psi, p$ is information-preserving and entirely symmetric. In order to appreciate this fact, we ought to compare the Legendre transformation with the "naive" attempt to replace $y = y(x)$ by

$$y = y(p) = y\left(\frac{dy}{dx}\right)$$

This procedure would save us the trouble of introducing the new function ψ. Unfortunately, it does not work. The last relation is a differential equation which yields on integration $y = y(x, c)$ with the arbitrary integration constant c, i.e., a family of curves rather than the original $y = y(x)$. Hence information is lost.

The generalization of the foregoing discussion to functions of, say, r variables offers no difficulties except that of notation. We can replace any set of extensive variables by their conjugates and thus we have to contend with as many as 2^r transformed functions. Although not all of these are of practical interest, it is convenient to have a systematic notation for any of them. We introduce

$$\begin{aligned}\psi(p_1, p_2, \cdots, p_k | x_{k+1} \cdots x_r) &\equiv -y[p_1, p_2, \cdots, p_k | X_{k+1} \cdots X_r] \\ &\equiv -y[p_1, p_2, \cdots, p_k] \\ &\equiv \sum_i^k x_i p_i - y \end{aligned} \tag{26}$$

Here the variables $x_1, x_2, \cdots, x_k$ are to be eliminated from the right-hand side by using the relations

$$P_i = \frac{\partial y}{\partial x_i} \qquad i = 1, 2, \cdots, k \tag{27}$$

We can indeed solve for $x_1, x_2, \cdots, x_k$ provided

$$\frac{\partial(P_1, P_2, \cdots, P_k)}{\partial(x_1, x_2, \cdots, x_k)} \neq 0 \tag{28}$$

We shall use the more compact notation in Equation 26 if the set of variables x_i can be assumed to be known. The negative sign in Equation 26 is used because the tradition in thermodynamics is to use the Legendre transforms with the opposite sign.

We shall apply these results to the most common thermodynamic situation in which we start from the energy fundamental equation and replace one or both of the variables S, V by their conjugates. We have thus altogether four fundamental equations which, because of their importance, rate standard notations and designations: We have in their order of listing the energy, the Helmholtz free energy, the Gibbs free energy (or function), and the enthalpy:

$$U = U(S, V, N) \tag{29a}$$

$$U[T] = U - TS \equiv F(T, S, N) \tag{29b}$$

$$U[-P] = U + PV \equiv H(P, S, N) \tag{29c}$$

$$U[T, -P] = U - TS + PV = G(T, P, N) \tag{29d}$$

The physical meaning of these functions is apparent from the differential expressions obtained on inserting them into Equation 13:

$$dU = T\,dS - P\,dV + \textstyle\sum\mu\,dN + đW' \tag{30a}$$

$$dF = -S\,dT - P\,dV + \textstyle\sum\mu\,dN + đW' \tag{30b}$$

$$dH = T\,dS + V\,dP + \textstyle\sum\mu\,dN + đW' \tag{30c}$$

$$dG = -S\,dT + V\,dP + \textstyle\sum\mu\,dN + đW' \tag{30d}$$

We obtain from here

$$đW' = (dU)_{S,V,N} = (dF)_{T,V,N} = (dH)_{S,P,N} = (dG)_{T,P,N} \tag{31}$$

where the variables used as subscripts are to be held constant (see Equation 29). We see that the functions U, F, H, and G play the role of the mechanical potential energy, each for its own type of work: (a) adiabatic-isochoric, (b) isothermal-isochoric, (c) adiabatic-isobaric, and (d) isothermal-isobaric. Accordingly, the functions are called collectively thermodynamic potentials, the independent variables appearing in Equations 29, 30, and 31 are their *natural variables.* We can realize the various situations by coupling the system with devices that keep the indicated variables constant. The extensive variables are held constant by means of passive devices (walls) whereas the intensities are controlled by coupling to reservoirs that are active in the sense of performing positive or negative work or transferring positive or negative amounts of heat to the system. The work $\text{đ}W'$ is the amount in excess of that performed while enforcing the boundary conditions.

If $\text{đ}W' > 0$ for two neighboring equilibrium states, then this quantity represents the minimum amount of work required to "lift" the system from the initial to the final state. If $\text{đ}W' < 0$, then $|\text{đ}W'|$ is the maximum amount of work that can be extracted from the process.*

The conditions of stable equilibrium for systems coupled to constraining devices are

$$\delta W' = 0, \qquad \delta^2 W' > 0 \tag{32}$$

for all virtual processes originating in the state under scrutiny. Moreover, we see from Equation 30 that for each type of constraint there is a potential that takes on its minimal value at equilibrium. The variables held constant by the constraints are the natural variables of the potential in question.†

Equations 30 contain a great deal of useful information, namely the derivatives of the potentials with respect to their natural variables, which are constantly needed in thermodynamic manipulations. This information is contained in a compact form in the diagram of Figure 1, apparently

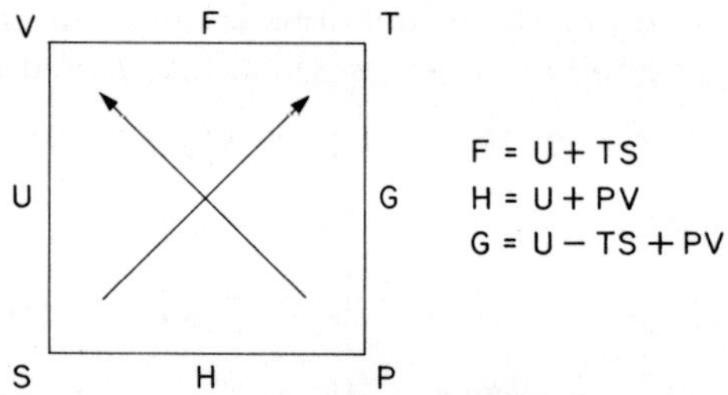

Fig. 1. The Born diagram.

first used by Max Born in his Göttingen lectures on thermodynamics in the late 1920's. Here each potential is between its "natural" independent variables. Its derivatives are obtained by following the diagonal of the square, and the sign is determined in obvious fashion by the direction of the arrow.

* In an actual process the work performed by the source will be larger and the one absorbed by it smaller than indicated above. The balance is absorbed by a heat reservoir, say the atmosphere.

† The situation is actually somewhat more complicated, because a state of equilibrium can have different types of stability as described on page 13. These results can be immediately generalized to the present case of constrained systems by the simple device of substituting W' for W.

We leave it to the reader to set up other transformation schemes, e.g., involving the conjugate pairs N, μ and magnetic variables (after these have been introduced in Section 4). See also H. Callen, quoted in the Preface.

The various elementary derivatives occurring in the diagram of Figure 1 are not all independent, since every expression must be reducible to the scheme based on any of the potentials. This fact leads to a large number of thermodynamic identities which we shall discuss in Section 2.4. Moreover, within each scheme there is a so-called Maxwell relation following from the equality of the mixed second derivatives:

$$\frac{\partial y}{\partial x_i \partial x_k} = \frac{\partial y}{\partial x_k \partial x_i} \tag{33}$$

Two of the relations involving the derivatives of F and G are particularly useful, and we note them for future reference:

$$\left(\frac{\partial S}{\partial V}\right)_T = \left(\frac{\partial P}{\partial T}\right)_V \tag{34a}$$

and

$$\left(\frac{\partial S}{dP}\right)_T = -\left(\frac{\partial V}{\partial T}\right)_P \tag{34b}$$

So far the thermodynamic potentials have been discussed in relation to the work $đW'$ performed on the system. This enabled us to handle all of the potentials in a unified manner.*

It is worth noting, however, that the energy and the enthalpy are related also to the heat absorbed by the system. The definition of heat quantity is again the classical expression $đQ = T\,dS$. Assuming further $đW' = 0$, we obtain from Equations 30

$$đW' = 0, \qquad đQ = T\,dS = (dU)_{V,N} = (dH)_{P,N} \tag{35}$$

This means that heat added to the system at constant volume (isochoric heat denoted as $đQ_V$) appears as an increase of the internal energy U, whereas heat added at constant pressure (isobaric heat denoted by $đQ_P$) appears as an increase in enthalpy H. In most practical cases, heating is carried out in systems open to the atmosphere and thus at constant pressure. For this reason enthalpy is frequently called the "heat content" of the system, although this terminology is not strictly correct, since heat refers to a *process* rather than to a state.

We define now in the usual fashion the isochoric and isobaric heat

* This is true of the potentials based on the energy scheme. The entire procedure can be repeated for the entropy scheme. Of not much use in MTE, some of these transforms, such as $S[1/T]$ and $S[1/T, -\mu/T]$ are very important in STE, where they occur in the canonical and grand canonical distributions. (See Paper 7.)

capacities (or specific, or molar heats, depending on the scaling of the system).

$$C_V = \left(\frac{đQ_V}{dT}\right) = T\left(\frac{\partial S}{\partial T}\right)_V = \left(\frac{\partial U}{\partial T}\right)_V \tag{36a}$$

$$C_P = \frac{đQ_P}{dT} = T\left(\frac{\partial S}{\partial T}\right)_P = \left(\frac{\partial H}{\partial T}\right)_P \tag{36b}$$

2.3 Empirical Construction of the Fundamental Equation of One-Component Systems

By merely postulating the existence of the fundamental equation we have been living on credit. It is time to start paying off by establishing the connection with experiment. Thanks to the transformation theory established in the previous section we are free to choose the fundamental equation best suited for the purpose. Objective convenience and tradition are in agreement to favor the Gibbs function.* We confine ourselves to one-component systems, but in Section 4 we shall consider some of the fundamental aspects of the many-component theory. We assume that the relevant work is compressional ($đW' = 0$). Also, to begin with we consider a single phase; the inclusion of several phases will offer no difficulties.

Because of the homogeneous character of the phase, the size of the system is an unimportant scale factor insofar as the intrinsic properties of the system are concerned. Therefore, we consider the molar Gibbs function, which, for one-component systems, happens to be identical to the chemical potential:

$$g(T, P) = \frac{G(T, P, N)}{N} = \mu(T, P) \tag{37}$$

Instead of measuring the (molar) Gibbs function directly, it is more convenient to measure the derivatives and determine the function by integration, with proper attention to the constants of integration.

The first derivatives are

$$\left(\frac{\partial g}{\partial T}\right)_P = -s \tag{38a}$$

$$\left(\frac{\partial g}{\partial P}\right)_T = v \tag{38b}$$

These relations express the entropy and the volume as functions of temperature and pressure. They are called the caloric and the thermal equations of state, respectively. The second of these equations is di-

* However, see Section 6 for a somewhat different situation.

rectly measurable but the former is not, and we proceed to the second derivatives:

$$\left(\frac{\partial^2 g}{\partial T^2}\right)_P = -\left(\frac{\partial s}{\partial T}\right)_P = -\frac{c_p}{T} \tag{39a}$$

$$\frac{\partial^2 g}{\partial P \partial T} = \frac{\partial^2 g}{\partial T \partial P} = -\left(\frac{\partial s}{\partial P}\right)_T = \left(\frac{\partial v}{\partial T}\right)_P = v\alpha \tag{39b}$$

$$\frac{\partial^2 g}{\partial P^2} = -\left(\frac{\partial v}{\partial P}\right)_T = v\kappa_T \tag{39c}$$

We have introduced here three coefficients, the molar specific heat at constant pressure c_p, the expansion coefficient α, and the isothermal compressibility κ_T. In the terminology of Section VII of Paper 3 these are elements of the compliance matrix or *compliance coefficients*. They are all measurable and listed in physical-chemical tables. We retrace the foregoing procedure and obtain the equations of state and the Gibbs function by successive integrations. The thermal equation of state is

$$v(T, P) = v_0 + \int_{T_0}^{T} \left(\frac{\partial v}{\partial T}\right)_{P=P_0} dT + \int_{P_0}^{P} \left(\frac{\partial v}{\partial P}\right)_T dP \tag{40}$$

where the integration constant $v_0 = v(T_0, P_0)$ can be obtained by measuring the molar volume (or the density) at the standard state P_0, T_0. The first integral is carried out at constant pressure P_0, the second at the constant temperature T. Other paths can be chosen, of course, without affecting the value v and of other state functions.

The caloric equation of state is handled similarly:

$$s(T, P) = s_0 + \int_{T_0}^{T} \frac{c_p}{T} dT - \int_{P_0}^{P} \left(\frac{\partial v}{\partial T}\right)_P dP \tag{41}$$

where we have made use of Maxwell relation 34*b*. The only difference is that the integration constant is in this case not accessible to measurement. Yet this constant is meaningful and will be discussed in Section 5.

In general, it will happen that the path of integration crosses a line of phase equilibrium. Here the compliance coefficients are infinite, their integrals contribute to Equations 40 and 41 finite changes of volume Δv and entropy $\Delta s = l/T_1$, where l is the latent heat and T_1 the temperature of the transition.

A further integration leads to the Gibbs function

$$g(T, P) = g_0 - \int_{T_0}^{T} s(T, P_0)\, dT + \int_{P_0}^{P} v(T, P)\, dP \tag{42}$$

Inserting Equation 41 into the first integral and integrating by parts we obtain

$$g = g_0 - s_0(T - T_0) - T\int_{T_0}^{T}\left[\int_{T_0}^{T'} c_P\, dT\right]\frac{dT'}{T'^2} + \int_{P_0}^{P} v\, dP$$

Integrating by parts we obtain

$$g = h_0 - s_0 T - T\int_{T_0}^{T} \frac{c_p}{T}\, dT + \int_{T_0}^{T} c_p\, dT + \int_{P_0}^{P} v\, dP \qquad (43)$$

We have set here

$$h_0 = g_0 + s_0 T_0 \qquad (44)$$

The same result can be obtained also from

$$h = h_0 + \int_{T_0}^{T} c_p\, dT \qquad (44a)$$

and the so-called Gibbs-Helmholtz relation

$$h = -T^2 \frac{\partial}{\partial T}\left(\frac{g}{T}\right) \qquad (45)$$

which is a special case of Equation 25.

Summing up, we may say that the construction of the fundamental equation is completed with the following reservations: The measurement of temperature remains to be explained and the meaning of the constants s_0 and h_0 clarified.

2.4 The Thermodynamic Identities

The multiplicity of schemes that give equivalent descriptions of a thermodynamic system leads to a high degree of overdetermination and thus to a large number of identities. Although most of these are uninteresting, there is a considerable residue of important relations which form the raw material of thermodynamic routine calculations. Mathematically the derivation of these identities involves only the simplest rules of differentiation; nevertheless the novice might have considerable trouble in finding the sequence of steps that yields his desired formula. It has been a long-felt need to formulate an unambiguous recipe to serve as a guide in these operations. There does not seem to be a simple recipe that is entirely satisfactory, but we shall advance two sets of simple instructions. The use of one or the other enables us to cope with most, if not all situations.

The identities involve largely derivatives of the form $(\partial y/\partial x)_z$ where x, y, z are various combinations of the potentials and variables, as they appear, in the diagram of Figure 1.

First Method

This method centers around the use of Jacobians. We summarize their definition and relevant properties in terms of two pairs of variables. The generalization to several variables is obvious.

Definition:

$$\frac{\partial(u, v)}{\partial(x, y)} = \begin{vmatrix} \dfrac{\partial u}{\partial x} & \dfrac{\partial u}{\partial y} \\[2ex] \dfrac{\partial v}{\partial x} & \dfrac{\partial v}{\partial y} \end{vmatrix} \tag{46}$$

As a natural generalization of the derivative the Jacobian obeys similar rules:

$$\frac{\partial(u, v)}{\partial(x, y)} \cdot \frac{\partial(x, y)}{\partial(r, s)} = \frac{\partial(u, v)}{\partial(r, s)} \tag{47}$$

As a special case, one gets for $r = x, s = y$

$$\frac{\partial(u, v)}{\partial(x, y)} = \left[\frac{\partial(x, y)}{\partial(u, v)}\right]^{-1} \tag{48}$$

moreover

$$\frac{\partial(u, v)}{\partial(x, y)} = -\frac{\partial(v, u)}{\partial(x, y)} = -\frac{\partial(u, v)}{\partial(y, x)} \tag{49}$$

All of this follows at once from the definition in terms of determinants. The connection with the thermodynamic formalism hinges on

$$\frac{\partial(u, y)}{\partial(x, y)} = \begin{vmatrix} \dfrac{\partial u}{\partial x} & \dfrac{\partial u}{\partial y} \\[2ex] 0 & 1 \end{vmatrix} = \left(\frac{\partial u}{\partial x}\right)_y \tag{50}$$

Suppose some physical consideration yields an expression 50 and we want to switch from the independent variables x, y to z, t. This is achieved by writing

$$\frac{\partial(u, y)}{\partial(x, y)} = \frac{\partial(u, y)}{\partial(z, t)} \left[\frac{\partial(x, y)}{\partial(z, t)}\right]^{-1} \tag{51}$$

Example: Express the specific heat at constant volume C_v in terms of the independent variables T, P.

$$\begin{aligned} c_v &= T\left(\frac{\partial s}{\partial T}\right)_v = T\,\frac{\partial(s, v)}{\partial(T, v)} \\ &= T\,\frac{\partial(s, v)}{\partial(T, P)}\left[\frac{\partial(T, v)}{\partial(T, P)}\right]^{-1} \\ &= T\left[\left(\frac{\partial v}{\partial P}\right)_T\right]^{-1} \frac{\partial(s, v)}{\partial(T, P)} \end{aligned}$$

For the last factor we write explicitly

$$\frac{\partial(s, v)}{\partial(T, P)} = \begin{vmatrix} \left(\frac{\partial s}{\partial T}\right)_P & \left(\frac{\partial s}{\partial P}\right)_T \\ \left(\frac{\partial v}{\partial T}\right)_P & \left(\frac{\partial v}{\partial T}\right)_T \end{vmatrix}$$

By making use of the Maxwell relation 35 and the symbols defined in Equations 39, we obtain finally

$$\begin{aligned} c_v &= T\left(\frac{\partial S}{\partial T}\right)_P - T\,\frac{(\partial v/\partial T)_p^2}{(\partial v/\partial P)_T} \\ &= c_P + \frac{Tv\alpha^2}{\kappa_T} \end{aligned} \tag{52}$$

Although we can go a long way deriving identities with this technique,* it is usually simpler to follow another method.

Second Method

We take our cue from the step in the foregoing example, in which we eliminated the entropy by means of the Maxwell relation. Very often the motivation behind searching for a thermodynamic identity is the fact that we have arrived at an expression that has certain features we consider undesirable. The method provides instructions for eliminating the typical features that occur in practice. The nice thing about it is that the repeated application of these elementary rules leads usually to the simple derivations of the experienced student. We refer to the entire procedures as "reducing the formula."†

(i) Change of an independent variable into a dependent one. ("Pulling up into the numerator"):

$$\left(\frac{\partial x}{\partial y}\right)_z = \left(\frac{\partial y}{\partial x}\right)_z^{-1} \tag{53}$$

and

$$\left(\frac{dx}{dy}\right)_z = -\,\frac{(\partial z/\partial y)_x}{(\partial z/\partial y)_y} \tag{54}$$

The second relation follows from

$$dz = \left(\frac{\partial z}{\partial x}\right)_y dx + \left(\frac{\partial z}{\partial y}\right)_x dy \tag{55}$$

* H. Margenau and G. M. Murphy, *The Mathematics of Physics and Chemistry* (D. van Nostrand Company, Inc., New York, 1943).

† The meaning of this term is somewhat flexible, and intentionally so, since we chose not to define in any rigid fashion what is "undesirable" in the original formula. Otherwise we could not encompass such a wide variety of thermodynamic identities as we actually do.

by setting $dz = 0$ (z = constant), or can be obtained by means of Jacobians.

(ii) We want to eliminate a number of variables, usually one of the potentials and/or the entropy. We start with the elimination of the former.*

Suppose, for definiteness' sake, that our formula contains the enthalpy H that we want to eliminate. We recall that

$$dH = T\,dS + V\,dP \tag{56}$$

This follows from Equation 30 with $dN_j = 0$ and $đW' = 0$,† and can be inferred from the diagram of Figure 1 along with the analogous relations for the three other potentials.

According to (i) we pull up H into the numerator and have

$$\left(\frac{\partial H}{\partial x}\right)_y = T\left(\frac{\partial S}{\partial x}\right)_y + V\left(\frac{\partial P}{\partial x}\right)_y \tag{57}$$

(iii) In order to eliminate the entropy we first pull it into the numerator: $(\partial S/\partial x)_y$. If $x = T$, this expression is proportional to one of the specific heats. (See Equation 36.) If $y = T$, the entropy is eliminated by one of the two Maxwell relations (Equation 34). If neither x nor y is the temperature T, one writes

$$\left(\frac{\partial S}{\partial x}\right)_y = \left(\frac{\partial S}{\partial T}\right)_y\left(\frac{\partial T}{\partial x}\right)_y \tag{58}$$

in which the right-hand side is reduced.

(iv) The preceding steps leave us with an expression in terms of C_v, C_p, and $(\partial x/\partial y)_z$ where the variables are permutations of P, v, T. We may wish to eliminate v (or alternatively P) as independent variables. Thus, according to (i) we have

$$\left(\frac{\partial P}{\partial T}\right)_v = -\frac{(\partial v/\partial T)_P}{(\partial v/\partial P)_T} = \frac{\alpha}{\kappa_T} \tag{59}$$

and we have everything in terms of c_v, c_P, α, and κ_T. The reduction of c_v to the independent variables P, T is given by Equation 52. The simple method of (i) fails in this case, but we can simplify the derivation of

* The sequence is important; otherwise the result of the first step might be undone by the second. This "going around in circles" is well known to those attempting their reduction by random walk.

† The procedure can be extended to these additional variables, provided $đW'$ is expressed in terms of pairs of conjugate variables (Section 4; see also Callen, as quoted in the Preface).

Equation 52 as follows:

$$ds = \left(\frac{\partial s}{\partial T}\right)_P dT + \left(\frac{\partial s}{\partial P}\right)_T dP$$
$$= \frac{c_P}{T} dT - \left(\frac{\partial v}{\partial T}\right)_P dP$$

Differentiating both sides with $dv = 0$:

$$\left(\frac{\partial s}{\partial T}\right)_v = \frac{c_v}{T} = \frac{c_P}{T} - \left(\frac{\partial v}{\partial T}\right)_P \left(\frac{\partial P}{\partial T}\right)_v$$

By using Equation 59 we obtain Equation 52.

In contrast to the experimentalist's preference for the variables T, P, theoretical (statistical) calculations are more convenient in terms of the variables T, v, or T, μ. We leave it to the reader to work out the corresponding reductions.

Finally we note that the entropy and the potentials can be eliminated only from formulas that contain these functions in derivative relations. The functions themselves have to be expressed in terms of integral relations such as Equations 41 and 43.

Applications. Instead of demonstrating the rules of this section in terms of arbitrary examples, we consider a number of processes having practical importance in thermodynamics, we indicate the quantities which naturally arise in connection with these processes, and show how their reduction leads to useful information about the system. The reductions will be carried out according to the rules laid down in the preceding section. Detailed references to the thermodynamic diagram and to the basic reduction formulas already derived will be omitted.

(1) *Isothermal compression:* This process is accompanied by a rejection of heat whose amount is

$$đQ = -T\,dS = -T\left(\frac{\partial S}{\partial P}\right)_T dP$$

The reduction of this expression gives

$$-đQ = -T\,dS = T\left(\frac{\partial V}{\partial T}\right)_P dP \tag{60}$$

For finite (reversible) compressions, we find on integration

$$-Q = \int_{P_1}^{P_2} T\left(\frac{\partial V}{\partial T}\right)_P dP = \int_{P_1}^{P_2} TV\alpha\, dP \tag{61}$$

(2) *Adiabatic compression:* In this process, the quantity of interest is the change in temperature; hence, we consider

$$\left(\frac{\partial T}{\partial P}\right)_S = \frac{(\partial S/\partial P)_T}{(\partial S/\partial T)_P} = \frac{T}{C_P}\left(\frac{\partial V}{\partial T}\right)_P = \frac{Tv\alpha}{C_P} \tag{62}$$

or, integrating,

$$T_2 - T_1 = \int_{P_1}^{P_2} \frac{TV\alpha}{C_P}\, dP \tag{63}$$

If $\alpha > 0$ (the usual case), then an adiabatic compression increases the temperature and expansion decreases it. The combination of isothermal compressions and adiabatic expansions of a gas constitutes an important cooling method with many practical applications.

(3) *Free expansion:* In this process, the system, usually a gas, is allowed to change its volume at constant energy. The whole system is adiabatically insulated to preclude any heat interchange with its surroundings and no external work is done by the system. The problem is to find the change of temperature with volume:

$$\left(\frac{\partial T}{\partial V}\right)_U = -\frac{(\partial U/\partial V)_T}{(\partial U/\partial T)_V}$$

Since $dU = -P\,dV + T\,dS$, we have

$$\left(\frac{\partial T}{\partial V}\right)_U = \frac{P - T(\partial S/\partial V)_T}{T(\partial S/\partial T)_V} = \frac{1}{C_V}\left[P - T\left(\frac{\partial P}{\partial T}\right)_V\right] \tag{64}$$

This expression belongs to the $V - T$ scheme but can be transformed to the $P - T$ scheme by the methods of the previous section. However, if, in an actual case, the equation of state of the gas is more easily solved for P than for V, then Equation 64 is more satisfactory for carrying out the analysis of the system.*

(4) *Joule-Thomson (Kelvin) effect (porous plug experiment):* A gas is passed through a throttle (a porous plug) so that pressures P_1 and P_2 are maintained on opposite sides. The friction in the throttle is so large that the kinetic energy of the flow is completely dissipated and the two sides are assumed to be adiabatically insulated. Suppose a certain amount of gas occupies volume V_1 before passing the throttle and volume V_2 after, then, the energy change under the given conditions will be equal to the net work done on the system:

$$W = P_1V_1 - P_2V_2 = U_2 - U_1$$

hence

$$U_1 + P_1V_1 = U_2 + P_2V_2$$

* *Note:* It should be remarked that free expansion is an irreversible process, so that the validity of Equation 64 requires some clarification. Actually, it can be considered as giving the temperature change with volume for a reversible process in which the gas does work and absorbs heat with the total energy remaining constant. In the final state of this reversible process, V and U have the same value as for the irreversible process. Since the system depends on only two independent variables, the final state is thus completely determined. This situation is typical of those irreversible processes for which the final state is determined by the condition of the process.

and we conclude that in such a throttling process the enthalpy is constant. The pressure difference will, in general, be associated with a temperature difference which can be found from:

$$\left(\frac{\partial T}{\partial P}\right)_H = -\frac{(\partial H/\partial P)_T}{(\partial H/\partial T)_P}$$

Since $dH = V\,dP + T\,dS$, we have

$$\left(\frac{\partial T}{\partial P}\right)_H = -\frac{V + T(\partial S/\partial P)_T}{T(\partial S/\partial T)_P}$$

$$= \frac{1}{C_p}\left[T\left(\frac{\partial V}{\partial T}\right)_P - V\right] = \frac{V}{C_p}(\alpha T - 1) \tag{65}$$

This expression is frequently called the Joule-Thomson coefficient. It can be positive or negative. If it is positive, a pressure drop leads to a temperature drop. This effect has been used industrially for cooling and the liquefaction of gases in the well-known Linde process. The throttling process is an irreversible one, and the footnote on free expansion applies to this case as well.

(5) *Pressure dependence of heat capacity:* The relevant expression is

$$\left(\frac{\partial C_p}{\partial p}\right)_T = T\frac{\partial^2 S}{\partial P \partial T} = T\frac{\partial^2 S}{\partial T \partial P} = -T\left(\frac{\partial^2 V}{\partial T^2}\right)_p \tag{66}$$

(6) *Sound propagation:* From hydrodynamics, the velocity of sound propagation in a fluid is known to be

$$C_{\text{sound}} = (\rho\kappa_s)^{-1/2} \tag{67}$$

where κ_s is the adiabatic compressibility

$$-v\kappa_s = \left(\frac{\partial v}{\partial p}\right)_s = \frac{(\partial s/\partial p)_v}{(\partial s/\partial v)_p}$$

$$= -\frac{(\partial s/\partial T)_v(\partial T/\partial p)_v}{(\partial s/\partial T)_p(\partial T/\partial v)_p} = \frac{c_v}{c_p}\left(\frac{\partial v}{\partial p}\right)_T \tag{68}$$

Using the standard notation for the ratio of the specific heats

$$\gamma = \frac{c_p}{c_v} \tag{69}$$

we have

$$\kappa_s = \frac{\kappa_T}{\gamma} \tag{70}$$

A combination with Equation 52 yields another useful relation:

$$\kappa_T - \kappa_s = \frac{c_p - c_v}{c_p}\kappa_T = \frac{vT\alpha^2}{c_p} \tag{71}$$

The relations 70 and 71 are important for the comparison of dynamic (adiabatic) elastic coefficients and moduli as measured, for example, in experiments involving ultrasonics with the corresponding static (isothermal) quantities commonly measured in cases where the changes in the system take place comparatively slowly.

Similar relations are obtained in solids; only the compressibility is replaced by other elastic coefficients.

It is sometimes convenient to express the sound velocity in terms of the variables T, v. This is the case, e.g., in the neighborhood of the critical point where α, κ_T, $c_p \to \infty$.

The reduction to the variables T, v yields easily

$$-\left(\frac{\rho}{v}\right)C^2_{\text{sound}} = \frac{c_p}{c_v}\left(\frac{\partial P}{\partial v}\right)_T = \left(\frac{\partial P}{\partial v}\right)_T - \frac{T}{c_v}\left(\frac{\partial P}{\partial T}\right)^2_v \tag{72}$$

2.5 Absolute Temperature

In the course of their evolution some concepts fare better than others. While many factors have conspired to make entropy appear unduly mysterious, there were as many reasons to induce us to gloss over the intricacies of the temperature concept. Actually, absolute thermodynamic temperature and entropy are inextricably connected with each other. If someone claims to understand temperature while being mystified by entropy, then his statements may be presumed to be 50 per cent accurate.

Temperature got off to a good start because of our familiarity with a somewhat related notion going under the same name. It should be evident to the reader of Paper 1 that it is a long way from our sense experience to the thermodynamic concept.

The first stage of the evolution of the objective concept is to consider temperature as an ordered sequence of points marking common levels of thermal equilibrium (zeroth law of thermodynamics). This concept is precise and intuitive at the same time.*

The problem starts to be subtle as we wish to establish a quantitative temperature scale without relying on accidental instrumental properties.

As luck would have it, this problem was solved to a considerable extent by the simple concept of the ideal gas temperature. On the one hand, this concept turned out to be identical to the thermodynamic temperature; on the other hand, it is "operational" in a wide range and can be established with excellent accuracy by extrapolation from data obtained on real gases (see Zemansky, Chapter 2, page 13, reference 2, Paper 1).

* This simplicity prevails as long as we do not pry into the intricacies of the thermodynamic equilibrium. This we can do only by using the entropy concept to the utmost. See Section 5.

Under these conditions it is not surprising that low-temperature physicists are almost alone concerned about the subtler aspects of the temperature concept. For them this is a necessity, since the ideal gas thermometer reaches the limits of its usefulness around 1°K.

We do not wish to enter here into the special problems arising in the low-temperature domain, partly because thermometry relies there mostly on the properties of magnetic systems which we have not yet discussed. What we do emphasize is the interesting aspect of thermodynamics that enables us to use any odd substance to establish the absolute scale in the region in which the system exhibits a temperature-sensitive property.*

The decisive point is that thermodynamic identities are valid only in terms of the correct temperature. The most frequent use of the identities is to reduce physically significant, but not easily measured expressions to the basic compliance coefficients that are measured and tabulated. In the present context, however, we consider identities, all constituents of which are measurable in terms of our *empirical* temperature, and the identity is considered a differential equation for the *thermodynamic* temperature.

We demonstrate this procedure by considering the expression for the Joule-Thomson (Kelvin) effect of Example 5 of the last section. We denote the effect by λ and write according to Equation 65:

$$\lambda = \left(\frac{\partial T}{\partial P}\right)_H = \frac{V}{C_p}(T\alpha - 1) \tag{73}$$

In contrast to our procedure followed thus far we consider the temperature T as unknown. We introduce, rather, an empirical temperature: e.g., we use the same gas on which Equation 73 is to be tested as a constant pressure gas thermometer and define the "temperature" θ

$$v = A\theta \tag{74}$$

where A is a constant so long as the pressure of the gas is constant.†

* We note that the term "absolute" is used in a double sense. (i) To indicate that the lowest conceivable choice of temperature is $T = 0$, rather than some conventional zero such as the ice point. Note that the so-called negative temperatures are higher than $T = \infty$ and not lower than $T = 0$. The absolute zero will be further discussed in Section 5. For a comparison of the various scales, see G. N. Hatsopoulos and J. H. Keenan, *Principles of General Thermodynamics* (John Wiley & Sons, Inc., New York, 1965), p. 147. (ii) The scale is absolute also in the sense that it is not dependent on the choice of the thermometric substance. However, our lucky sequence persists and confusion is avoided because both interpretations are correct both for the ideal gas temperature, for the thermodynamic temperature, and even for the temperature that appears as a parameter in the canonical distribution (see Paper 7). This situation is to be contrasted with the confusion surrounding the concept of absolute entropy (see Section 5).

† If the gas were ideal, we would have $T \sim \theta$. The proportionality can be made into an identity by assigning a conventional value to a single fixed point. (See later discussion.) It is evident from Equation 73 that in this case $\lambda = 0$. However, our

We denote by a star the coefficients in Equation 73 measured in terms of the empirical scale θ. We have

$$\lambda^* = \left(\frac{\partial \theta}{\partial P}\right)_H, \qquad C_p^* = \left(\frac{\partial H}{\partial \theta}\right)_P, \qquad \alpha^* = \frac{1}{v}\left(\frac{\partial v}{\partial \theta}\right)_P$$

and

$$\lambda = \lambda^* \frac{dT}{d\theta}, \qquad C_p = C_p^* \frac{d\theta}{dT}, \qquad \alpha = \alpha^* \frac{d\theta}{dT}$$

Inserting into Equation 73 we have

$$\frac{1}{T}\frac{dT}{d\theta} = \alpha^* \left(\frac{\lambda^* C_p^*}{V} + 1\right)^{-1}$$

and hence

$$\ln \frac{T}{T_0} = \int_{\theta_0}^{\theta} \frac{\alpha^*\, d\theta}{(\lambda^* C_p^*/V) + 1} \tag{75}$$

It is evident from Equation 75 that experiment yields only the ratio of temperatures. In order to go beyond this we have to assign a conventional value to an identifiable state of constant temperature. Nature actually provides us with states that lend themselves to such a procedure. It is a consequence of the phase rule (see Section VI of Paper 3), and is borne out by experiment, that one-component systems exhibit triple points in which three phases in equilibrium coexist only at a fixed value of temperature and pressure.

It was indeed decided at the Tenth Conference on Weights and Measures in Paris in 1954 to select the triple point of water as the fixed point, and to assign it the value $T_0 = 273.16°\text{K}$. In Equation 75 we have then $T_0 = \theta_0$.

There is, of course, a long story behind this decision. For some hundred years the thermodynamic temperature was defined by the requirement $T_s - T_i = 100°$, where T_s and T_i are the steam point and the ice point, respectively (the temperatures of boiling and freezing of water at the pressure of one atmosphere). The old convention was dictated by the desire of maintaining agreement with the traditional Celsius scale. The advantages of the one fixed-point over the two fixed-point scheme were pointed out by Giauque in a spirited paper that is highly recommended reading.*

The above-mentioned value of T_0 was determined so as to ensure the best possible agreement with the $T_s - T_i = 100$ requirement. It implies $T_i \approx 273.15°\text{K}$, $T_s \approx 373.15°\text{K}$. These values, particularly the latter,

---

present argument does not rely on the notion of an ideal gas. In particular, we do not assume that our test system is nearly ideal.

* W. F. Giauque, "A Proposal to Redefine the Thermodynamic Temperature Scale with a Parable of Measures to Improve Weights," *Nature* **143**, 623 (1939).

may be slightly changed with future improvements in the accuracy of measurements.

The remarkable fact about Equation 75 is that it can be applied to an unlimited number of substances, and all of them yield the same temperature scale. Moreover, there are many other relations that can be put to a similar use. We may sum up by saying that the existence of a unique absolute temperature is among the impressive experimental verifications of thermodynamics.

We close this section by discussing an objection that might be brought up by a skeptical reader. What keeps us from introducing a new temperature as some function $f(T)$, say, $\log T$, $1/T$, and the like. Such possibilities cannot be excluded, of course, so long as we merely wish to identify the different levels of equilibrium. However, the formalism of MTE and STE requires that the temperature be an intensity defined as the derivative of the fundamental equation with respect to an extensive quantity. (The importance of this requirement will be further explained in Paper 5.) This leaves us with an intensity $\sim T$ in the energy scheme and $\backsim 1/T$ in the entropy scheme. It would be indeed very practical to have a term and permanent symbol for the latter. This need is met to some extent by $\beta = 1/kT$ where k is Boltzmann's constant.

3. CHEMICAL THERMODYNAMICS

3.1 Between Physics and Chemistry

Every fundamental equation of MTE corresponds to a well-defined chemical substance. Every investigation of experimental thermodynamics presupposes that substances of well-defined identity were made available by chemists.*

The historical connection between chemistry and thermodynamics is very close. It is no exaggeration to say that thermodynamics is a theoretical discipline distilled out of empirical chemistry, much as classical mechanics was distilled out of observational astronomy. The closeness to chemistry may transpire to some extent from the historical Paper 1 (see Section 1.6 for locating the "missing link" in the evolution of the first law within electrochemistry).

In spite of these facts, physicists tend to present thermodynamics with only minimal ties to chemistry. The present work is not very much different in this respect. It is indeed natural for specialists to keep away from the complexities of neighboring disciplines. Notwithstanding these subjective considerations, the objective relation of physics and

* There is one important exception: Blackbody radiation is a system consisting only of energy.

chemistry has changed a great deal in recent years. The classical physicist had no choice but to keep away from chemistry because he was helpless in the face of the phenomena of the chemical bond and the discontinuous transitions between states of definite chemical identity. All this changed with the advent of quantum theory. Einstein* stated, as far back as 1916, that within quantum theory with its states and transitions it is no longer even possible to distinguish sharply between physics and chemistry. This trend became more pronounced within quantum mechanics. To be sure, it is commonly interpreted in the sense that chemical phenomena have been at long last subjected to a mechanical interpretation. What may not be quite realized is that the conceptual makeup of mechanics has changed beyond recognition due to the assimilation of chemical concepts. The purpose of this section is to point out the nature of some of the chemical concepts that become relevant within the limited range of MTE. This can be achieved without getting involved with the characteristic complexities of chemistry, a restraint easily ensured by the author's ignorance.

3.2 Chemical Species and Components

Let us consider thermodynamic systems made up of a set of chemical species $A_1, A_2, \cdots, A_s$. The composition of a particular system can be specified in terms of the masses $m_1, m_2, \cdots, m_s$ or preferably the mole numbers $\mathfrak{N}_1, \mathfrak{N}_2, \cdots, \mathfrak{N}_s$ of these species.

There is, of course, an enormous amount of information contained in the knowledge that the subscript in $\mathfrak{N}_i$ refers to a well-defined substance, say, to H_2, H_2O, etc. It is not *a priori* obvious that the materials offered by nature in such bewildering abundance can be purified into a manageable set of chemical species having reproducible, invariant properties. That this is indeed so deserves to be stated, we shall speak of the *principle of chemical invariance.*

We note that the following formulas can all be scaled by Avogadro's number, thus passing from the molar to the molecular level. Insofar as our calculations are not numerical, we can afford to be quite casual about interpreting the $\mathfrak{N}_i$ as mole numbers or as number of molecules.

The introduction of the variables $\mathfrak{N}_i$ into the fundamental equation offers no difficulties if the corresponding species are inert. Yet this assumption is too strong, and it is within the range of MTE to handle a much more general situation. We have to assume only that the species considered are subject to reactions that can be classified into slow and fast ones with respect to the time scale of observation. The former are slow enough to be ignored, the latter fast enough to have run their

* A. Einstein, *Deutsch. Physik. Ges. Verhandl.* **18**, 318 (1916).

course to equilibrium. The reader will notice the analogy with the classification of walls into adiabatic and diathermic, and variables into fixed and free.

We consider at first the case that there is a single relevant fast reaction that we represent as follows:

$$\sum_{i=1}^{s} \nu_i A_i = 0 \tag{1}$$

where the ν_i are the so-called stoichiometric coefficients defined by the relations

$$\Delta\mathfrak{N}_i = \nu_i \, \Delta\xi \qquad i = 1, 2, \cdots, s \tag{2}$$

Here $\Delta\xi$ is a proportionally constant called the *advancement of the reaction.* The $\Delta\mathfrak{N}_i$ are the increments of the $\mathfrak{N}_i$ owing to this reaction. Positive and negative ν_i correspond to the creation or annihilation of the species A_i. Evidently the ν_i are defined only up to an arbitrary multiplicative constant. One often selects the set of smallest possible integers, or alternatively one sets one of the $\nu_i = 1$.

The reaction 1 can be used to express one of the species, say, A_s in terms of the others:

$$A_s = \sum_{i=1}^{s-1} \lambda_i A_i \tag{3}$$

where $\lambda_i = -\nu_i/\nu_s$. We say that the $s - 1$ species on the right are "*independent components*" or briefly *components,* whereas A_s is a *derived species.*

As an example consider the reaction

$$2H_2O - 2H_2 - O_2 = 0 \tag{4}$$

where we may choose any two of the three species as independent components.* While the mole numbers of the three species are subject to change due to the foregoing reaction, the total amount of the components, say, H_2 and O_2, is unchanged if the mole numbers of the "free" H_2 and O_2 are augmented with the amounts bound in H_2O. Thus we define the mole number N_{H_2} and N_{O_2} of the two components connected with the mole numbers of the species (designated by script capitals) as follows:

$$N_{H_2} = \mathfrak{N}_{H_2} + \mathfrak{N}_{H_2O}$$

$$N_{O_2} = \mathfrak{N}_{O_2} + \tfrac{1}{2}\mathfrak{N}_{H_2O}$$

Evidently N_{H_2} and N_{O_2} are conserved, they are invariant with respect to the reaction 4.

* Actually the choice is even freer, and any of the following will do: H, O; OH, O_2; H_2O, H, etc. A general statement concerning this situation is given in the following.

Returning to the relation 3, we define more generally the mole numbers of components

$$N_i = \mathfrak{N}_i + \lambda_i \mathfrak{N}_s \tag{5}$$

The formation of $\Delta\mathfrak{N}_s$ moles of A_s is associated with the disappearance of $-\Delta\mathfrak{N}_i$ moles of A_i where

$$\Delta\mathfrak{N}_i = -\lambda_i \, \Delta\mathfrak{N}_s \tag{6}$$

Consequently $\Delta N_i = 0$, the mole numbers of the components are indeed invariant.

These results are readily generalized for the case of several fast chemical reactions. Suppose we have ρ independent reactions

$$\sum_{i=1}^{s} \nu_{\kappa i} A_i = 0, \qquad \kappa = 1, 2, \cdots, \rho \tag{7}$$

The independence of the reactions is reflected in the linear independence of the forms 7. This is ensured if the rank of the matrix $\|\nu_{\kappa i}\|$ is ρ. Evidently $\rho > s$. We divide now the s species into $c = s - \rho$ components, say, $C_1, C_2, \cdots, C_c$ and ρ derived species $D_{c+1}, D_{c+2}, \cdots, D_s$. Equations 7 can be solved for the latter

$$D_\kappa = \sum \lambda_{\kappa i} C_i \tag{8}$$

The creation of $\Delta\mathfrak{N}_\kappa$ moles of D_κ is associated with the annihilation of a proportional amount of the species A_i:

$$\Delta\mathfrak{N}_i = -\lambda_{\kappa i} \, \Delta\mathfrak{N}_\kappa, \qquad i = 1, 2, \cdots, c \tag{9}$$

We define the mole numbers of components as

$$N_i = \mathfrak{N}_i + \sum_{\kappa=c+1}^{s} \lambda_{\kappa i} \mathfrak{N}_\kappa \tag{10}$$

It follows from Equations 9 and 10 that the numbers N_i are invariant even while the $\mathfrak{N}_i$ vary because of the chemical reactions.

We have noticed already that, given the set of species and reactions, the selection of the components is not unique. Let us assume now quite generally that a new set of components is introduced by means of the transformation

$$C_i = \sum_{j=1}^{c} a_{ij} C_j' \qquad i = 1, 2, \cdots, c \tag{11}$$

The transformation should be nonsingular, its determinant $|a_{ij}| \neq 0$.

We write now the composition of the system symbolically as

$$\sum_{i=1}^{c} N_i C_i = \sum_{i,j=1}^{c} N_i a_{ij} C_j' = \sum_{j=1}^{c} N_j' C_j' \tag{12}$$

where the new mole numbers are defined as

$$N_j' = \sum N_i a_{ij} \tag{13}$$

These formulas can be concisely written in terms of matrices, a form which is also convenient for numerical calculations. Interpreting N and C as column vectors and using $A \equiv \|a_{ij}\|$ we have

$$C = AC' \tag{14}$$

The composition of system is written symbolically as

$$\tilde{N}C = \tilde{N}AC' = \tilde{N}'C' \tag{15}$$

where the tilde means "transpose" and

$$\tilde{N}' = \tilde{N}A \tag{16}$$

or also

$$N' = \tilde{A}N \tag{17}$$

As an example let us consider the transformation from the components H_2, O_2 to H, OH. We have

$$H_2 = 2H$$

$$O_2 = -2H + 2OH$$

with the matrix

$$A = \begin{Vmatrix} 2 & 0 \\ -2 & 2 \end{Vmatrix}$$

and

$$N_1 H_2 + N_2 O_2 = N_1' H + N_2' OH$$

with

$$N_1' = 2N_1 - 2N_2$$

$$N_2' = 2N_2$$

We note that the transformed numbers are not mole numbers of stable entities, but merely invariants of the relevant reactions. It will be apparent from the discussion of the following section that this distinction is not as trifling as it might seem at first.

3.3 The Classification of Reactions*

The discussion of the previous section is based only on the fact that the reaction equation 1 is linear, without any restriction on the nature of the species in question.

However, from the experimental point of view, reactions fall into broad classes such as molecular and nuclear reactions. To these we may

* The contents of this section will not be further used in the present volume.

add transitions between excited states associated with the emission or absorption of photons and the so-called fundamental processes occurring primarily in high-energy physics. That these are conventionally not classified under chemistry is, of course, explained by the radical differences in the experimental and theoretical techniques in their study. Although for the most general purposes all of these processes fall within the scope of the formalism, it is possible to explain the above-mentioned classification of reactions in conceptual terms, rather than by just naming the objects involved.

We claim that molecular reactions are singled out from all reactions by the validity of two criteria:

(i) Conservation of mass
(ii) Existence of structure formulas

The meaning of the second condition calls for amplification. Let us assume that out of the set of all species it is possible to single out a finite subset of "elements," say, $E_1, E_2, \cdots, E_t$ in such a way that all the other species are represented as "compounds":

$$A_r = \sum_{i=1}^{t} \kappa_{ri} E_i \tag{18}$$

where the stoichiometric numbers κ_{ri} are integers.

Superficially, the formation from components (Equation 3) and the formation from elements (Equation 18) seem alike; actually, there are essential differences.* The elements E_i are themselves species, whereas the components C_i may be more abstract invariants that are not observable in isolation. Since the E_i are indestructible, they are "contained" in the compounds A_m. If decomposed, the A_m lead always to the same set of elements regardless of the chains of reactions. The reaction equations refer to processes and are directly inferred from observations. The structure formulas refer to states and are inferred only indirectly.

Evidently, condition (ii) is satisfied for classical molecular chemistry and also for nuclear reactions consisting of the rearrangement of nucleons, as in spontaneous and induced α decay. The formal distinction between the two types of reactions can be achieved by means of the conservation of mass (condition (i)) that we formulate as follows: It is possible to associate with each element E_i a mass M_i, such that the mass of the compound A_r is

$$M_r = \sum_{i=1}^{t} \kappa_{ri} M_i \tag{19}$$

* These were overlooked, e.g., in the otherwise sophisticated paper by S. R. Brinkley, *J. Chem. Phys.* **14**, 563 (1946).

In accordance with our remark made before, it is a simple matter of scaling whether we speak of *molar* or *molecular* (and *atomic*) masses. Strictly speaking, the relation is never satisfied, since the binding energy (heat of formation from the elements) of the compound is reflected in a relativistic mass defect. However, the binding energies of molecules are an order of magnitude lower than those of nuclei, and the corresponding mass defect is negligible for the former but not for the latter.

The logical-historical status of condition (ii) is very curious and warrants some closer scrutiny. Within the present context the general validity of elementary structural formulas is somewhat artificial and hardly self-evident. It certainly did not seem evident to many eighteenth-century chemists who believed that the observed substances might be built out of components that are themselves not observable (say, "metallic" and "sulfuric principles"). The idea that some of the existing substances have an entirely special status as "elements" was slow to crystallize. Clearly, no "crucial experiment" could be at first invoked to decide the issue. Substances were declared elements after attempts at their decomposition had consistently failed. A more positive attitude became possible only with the establishment of the periodic table of elements. However, it fits the usual pattern of the evolution of concepts that the unlikely hypothesis of elementary structural formulas gradually hardened into a firmly held and unshakable axiom. It seemed indeed good riddance to abandon unobservable "principles" in favor of ponderable substances always closely observed with the balance. The massive chemical evidence was reinforced by the appealing mechanistic picture according to which molecules contain indestructible and uncreatable atoms. Hence every chemical change is represented as rearrangement in a fixed mechanical phase space.

As atoms proved to be subject to change, their original role of ultimate elements was transferred to *elementary particles.* We notice only recently a growing skepticism concerning the "elementary" nature of the structures so designated. Such skepticism is certainly reinforced if the concept of element is formally introduced as stated above. Already the β decay is inconsistent with the elementary structural concept, since it is inadmissible to picture a neutron as "containing" a proton, electron, and neutrino.* At present it would seem that the additive invariants of fundamental processes are electric charge, baryonic, and leptonic numbers, which in isolation do not appear as observable entities. Such a situation is entirely within the spirit of thermodynamic theory. Of course, this

* Looking at molecules in terms of wave functions and electron charge distributions, we realize that not even molecules "contain" atoms in the strict sense of the word since the latter are strongly deformed within the molecule, and their electron clouds are intimately mixed.

fact is of little, if any, practical consequence within MTE and even in the more microscopic equilibrium theories. However, this extended view should be kept in mind in the formulation of a deeper dynamics.

3.4 Chemical Equilibrium

The distinction between the mole numbers of (chemical) *species* and (independent) *components* significantly enriches the theory and lends it a flexibility which is much needed in view of the extreme complexity of the problem of chemical composition. Describing a closed system in terms of its fixed amount of components, we can forget about the chemical reactions that may or may not take place within the system. However, if we *are* interested in these reactions we simply switch to chemical species. There is not even a need to be too pedantic in making our choice explicit if it is clear from the context which of the two interpretations of the term "mole number" happens to be relevant.

In this section attention is focused on chemical processes, and the term *mole number* is to refer to species rather than to components.

The conditions of chemical equilibrium are derived most conveniently in terms of the Gibbs function, and we consider Equation 51 of Section 2 with $dT = 0$, $dP = 0$, $đW' = 0$. The variation of G in terms of the variation of mole numbers is

$$\delta G = \sum \mu_i \, \delta \mathfrak{N}_i = \delta \xi \sum \nu_i \mu_i = 0 \tag{20}$$

We have made use of Equation 2. The last relation yields

$$\sum \nu_i \mu_i = 0 \tag{21}$$

This important relation bears out the designation of the intensities μ as chemical potentials. The expression 21 can be considered a measure of chemical affinity. If the condition 21 is not satisfied in an isolated system, this expression can only decrease. An increase can be brought about, however, at the expense of a positive work input W'.

Substituting in Equation 21 $\mu = h - Ts$, we obtain the heat of reaction

$$Q = \sum \nu_i \, \Delta h_i = T \sum \nu_i \, \Delta s_i \tag{22}$$

In particular the integration constants h_0 in Equation 44 of Section 2 are related so as to give the heat of reaction at the standard state T_0, P_0.

In classical molecular chemistry it seemed possible to assign arbitrary h_{i0} to each element, whereas the heat of formation from the elements determines the h_{i0} for all compounds.

From the point of view of low-energy nuclear physics all elements can be built up from protons, neutrons, and electrons with a reduction of the number of arbitrary constants to three.

Finally, taking into account relativistic quantum effects, all particles can transform into radiation and the $E = mc^2$ relation reduces the arbitrariness to a single constant that can be interpreted as the local potential energy.

Particularly simple and important implications follow from Equation 21 for perfect gases* defined by the equation of state

$$P = \frac{RT}{v} = \frac{\mathfrak{N}RT}{V} \tag{23}$$

Inserting this relation into Equation 43 of Section 2 we obtain

$$g = \mu = RT[\ln P + \phi(T)] \tag{24}$$

where the dimensionless function $\phi(T)$ is determined from the caloric equation of state which also includes the entropy constant s_0. We shall repeatedly return to this point, in particular in Section 5.

We consider next a mixture of inert perfect gases. The Gibbs function is

$$G = \sum \mathfrak{N}_i \mu_i \tag{25}$$

Note that this is a general thermodynamic relation, valid even for condensed systems; G is a first-order homogeneous function of the mole numbers (T and P being intensities), and Equation 25 is the well-known Euler relation for homogeneous functions. (See also Equations 5.23 and 5.24 of Paper 3.)

One arrives at the thermodynamic description of perfect gas mixtures by the simple stipulation that Equation 24 be still valid for each species provided that P is replaced by the partial pressure P_i of that species:

$$\mu_i = RT[\ln P_i + \phi_i(T)] \tag{26}$$

with

$$P_i = \frac{\mathfrak{N}_i RT}{V} \tag{27}$$

Moreover the total pressure is

$$P = \sum P_i = \frac{RT}{V} \sum \mathfrak{N}_i \tag{28}$$

The last statement is known as Dalton's law.

* Although a stock-in-trade of elementary texts, the concept of a perfect gas is not as simple as all that. First of all, Equation 23 does not define a bona fide thermodynamic system, since it fails to satisfy Nernst's law (see Section 5). We have indeed $(\partial s/\partial v)_T = (\partial P/\partial T)_v = R/v$, and this expression does not tend to zero with $T \to 0$. It is possible to define the perfect gas in a more sophisticated fashion (page 267). This definition leads to the Fermi-Dirac and Bose-Einstein gases that do satisfy Nernst's law, a striking instance of the perfect consistency of thermodynamics and quantum mechanics. However, these refinements do not concern us in the present context in which we confine ourselves to highly dispersed, low-density states of matter. In this case the use of the classical gas of Equation 23 is justified as the common limit of the two quantum gases.

The remarkably simple Equations 26 through 28 are approximately verified by experiment on real systems. A theoretical derivation is possible only in terms of a statistical model.

Introducing the mole fraction

$$x_i = \frac{\mathfrak{N}_i}{\sum \mathfrak{N}_i} = \frac{\mathfrak{N}_i}{\mathfrak{N}} \tag{29}$$

we obtain from Equations 27 and 28

$$P_i = x_i P \tag{30}$$

From Equations 25 through 28 we have

$$\begin{aligned} G &= RT \sum \mathfrak{N}_i[\ln P_i + \phi_i(T)] \\ &= RT \sum \mathfrak{N}_i[\ln P + \phi_i(T) + \ln x_i] \end{aligned} \tag{31}$$

We obtain from here the entropy

$$\begin{aligned} S = -\left(\frac{\partial G}{\partial T}\right)_P &= -R \frac{d}{dT} \sum \mathfrak{N}_i T \phi_i \\ &-R\mathfrak{N} \ln P - R \sum x_i \ln x_i \end{aligned} \tag{32}$$

The last term in this expression is known as the entropy of mixing; it is positive, since $0 < x_i < 1$. We can also write the entropy as

$$S = -R \frac{d}{aT} \sum \mathfrak{N}_i T \phi_i - R \sum \mathfrak{N}_i \ln P_i \tag{33}$$

The entropy is simply the sum of the entropies that the constituent gases would have were each of them to occupy the total volume V in the absence of the other constituents. The meaning of the entropy of mixing becomes apparent in another situation.

Suppose that the different components, all at the same temperature T, are separated and each occupies an appropriate volume $V_i = x_i V$ so that each exerts the same pressure P. Now let the components mix without doing any work, in an adiabatic diffusion process. The energy stays constant, and so does the temperature. (This is Joule's law. It follows by inserting Equation 23 into Equation 64 of Section 2.) According to Equations 27 and 28, the final pressure of the system is P, so that the total pressure is unchanged, although the partial pressures necessarily drop from P to P_i. From Equation 32, the entropy change connected with this mixing process is

$$\Delta S_{\text{mix}} = -nR \sum x_i \ln x_i > 0 \tag{34}$$

in accordance with the irreversible character of the diffusion.

So far, it has been assumed that the gases of the mixture were inert. If this is not the case and a chemical process, Equation 1, takes place, we have to insert Equation 26 into the condition 21. Hence

$$\sum \nu_i[\ln P_i + \phi_i(T)] = 0 \tag{35}$$

and

$$\prod P_i^{\nu_i} = e^{-\Sigma \nu_i \phi_i(T)} = K_p(T) \tag{36}$$

where $K_p(T)$ is the equilibrium constant, a function only of the temperature.

By using Equation 31 we find that the Gibbs-Helmholtz relation 45, of Section 2, yields the relation for the heat of reaction

$$\begin{aligned} \frac{\Delta H}{R} &= -\frac{\partial}{\partial 1/T} \sum \nu_i \phi_i \\ &= -\frac{\partial}{\partial 1/T} \ln K_p(T) \end{aligned} \tag{37}$$

Equation 36 is the mass-action law and is of great practical and theoretical importance. It has been subjected to extensive experimental tests and justifies our assumptions implicit in Equations 26 through 28.

Actually it does even more. We pointed out earlier that the entropy constants of the various species s_i enter into the definition of $\phi_i(T)$ and hence of $K_p(T)$. If the entropy constants were arbitrary — and there is no reason to expect otherwise on the basis of the traditional definition of entropy — then $K_p(T)$ would be likewise arbitrary and we would fail to get a definite distribution of the components over the available species. In reality we do have such a definite distribution, an important aspect of the principle of thermostatic determinism. This circumstance motivated Nernst to take the entropy constants to be meaningful entities (Section 5).

4. GENERALIZED WORK

The role of work in MTE is of fundamental importance. This apparently simple concept also has many ramifications, some of which are very complex.

We start from the basic fact emphasized in Paper 3 that for processes in the energy scheme the total entropy is conserved $\Delta S + \Delta S' = 0$, whereas the total energy varies and has to be balanced by an amount of work $W = \Delta U + \Delta U'$ performed by a "work source." This is any idealized device of weights or springs that does not produce entropy and does not upset the conservation of the latter.

In this context, W is analogous to the entropy production $\Delta S + \Delta S'$

associated with energy conserving processes $\Delta U + \Delta U' = 0$ in the entropy scheme.

All this is simple and entirely rigorous provided the necessary details are spelled out.

The complication starts if the work source is directly coupled to a simple system and if we wish to express the infinitesimal work in terms of the variation of the extensive parameters of the system:

$$đW = \sum P\, dX = (dU)_{S,N} \tag{1}$$

At first sight it seems encouraging that macroscopic mechanics and the electromagnetic theory indeed leads to work-energy relations of this sort. Closer inspection, however, indicates a number of difficulties.

The processes for which the formalism of MTE works most smoothly are exchanges of conserved quantities between spatially separated systems. The displacement of a piston in a cylinder can be indeed conceived of formally as an exchange of volume, and accordingly we obtain the corresponding condition of equilibrium: the constancy of the pressure throughout the system.*

There are, nevertheless, differences compared to the analogous situation in the case of thermal equilibrium.

Although a piston released from an initial position of pressure unbalance will ultimately come to an equilibrium, the exact position of this cannot be predicted since it depends on the details of the dissipation process. (The dissipated energy appears as heat, but it is undetermined how this is shared by the subsystems.) Thus we have an infringement of the *principle of thermostatic determinism* (see page 163). For another limitation of the analogy of pressure and temperature see page 138.

The foregoing difficulties are aggravated in the case of elastic, electric, and magnetic work. It is more or less artificial to consider these processes as exchanges of conserved quantities, and we have to be careful in applying to these situations the entropy scheme. This limitation has important implications. Thus the components of elastic stress tensor, the electric and magnetic intensities are *not* constant throughout a heterogeneous system. For this reason we propose to call these variables *pseudointensities*, the conjugates of the pseudothermodynamic variables of page 123. It calls for a special study to sift the results of MTE that hold or do not hold for them.

The situation can be better handled in the energy scheme, but even here we must proceed with care. We shall discuss the important case of magnetic work, more to indicate the type of problems involved than aiming at any measure of completeness.

* It is assumed that the variation of an external field, say, gravitation, is not significant over the system.

According to the electromagnetic theory the magnetic work performed on a system is

$$đW = \int dV\, \mathbf{H} \cdot d\mathbf{B} \tag{2}$$

where the integral is over all space. On such a basis it is often assumed that $V\mathbf{B}$ and $\mathbf{H}$ (or some other variant obtained by splitting off the vacuum field) can be considered as a conjugate pair of variables. (This was assumed in particular in Paper 4.) Although there are cases of particularly simple geometry where this is correct, it is a matter of principle that this is not so in general.

It was pointed out in an important paper by V. Heine* that $\mathbf{H}$ in Equation 2 is the actual field within the system, rather than an externally applied field that has an experimental meaning. He indicated a transformation that leads from Equation 2 to one having the desired form. Heine's point of view agrees entirely with the one of Papers 3 and 7: One of the main objectives of thermodynamics is to carve out a system from its environment, to describe the former in detail primarily in terms of extensive variables and the latter in a schematic fashion in terms of intensities, which is adequate to deal with the coupling of system and environment. This approach is very natural for processes involving the transfer of energy and matter. The case of the performance of work can be fitted into this scheme with some difficulty and only for well-chosen special cases.

We give a brief outline of Heine's transformation for the case of magnetic work. For the electric case and for further details we refer to the original paper. We note, however, that the same result can be obtained by a more intuitive but more special method of Pippard.†

We consider a static magnetic field generated in a coil in which the current is maintained constant. Let the magnetic vectors be $\mathbf{H}_0$ and $\mathbf{B}_0$ with $\mathbf{B}_0 = \mu_0 \mathbf{H}_0$. The insertion of the magnetic material modifies the field to $\mathbf{H}$, $\mathbf{B}$ with

$$\mathbf{B} = \mu_0(\mathbf{H} + \mathbf{M}) \tag{3}$$

where $\mathbf{M}$ is the magnetization in the sample. We define

$$\mathbf{B}_i = \mathbf{B} - \mathbf{B}_0 \tag{4}$$

and

$$\mathbf{H}_i = \frac{1}{\mu_0}\mathbf{B}_i - \mathbf{M} = \mathbf{H} - \mathbf{H}_0 \tag{5}$$

* V. Heine, *Proc. Cambridge Phil. Soc.* **52**, 546 (1956).

† A. B. Pippard, *Elements of Classical Thermodynamics* (Cambridge University Press, Cambridge, 1957), p. 23, also reproduced by H. B. Callen, *Thermodynamics*, quoted in the Preface.

We have

$$\nabla \cdot \mathbf{B}_0 = 0, \qquad \nabla \cdot \mathbf{B} = 0, \qquad \nabla \cdot \mathbf{B}_i = 0 \tag{6}$$

and

$$\nabla \times \mathbf{H}_i = 0 \tag{7}$$

because $\mathbf{H}_i$ is produced by molecular currents only. Heine starts from the identity

$$\mathbf{H} \cdot d\mathbf{B} = (\mathbf{H} - \mathbf{H}_0) \cdot d\mathbf{B} + (\mathbf{H}_0 \cdot d\mathbf{B} - \mathbf{B}_0 \cdot d\mathbf{H}) + \mathbf{B}_0 \cdot (d\mathbf{H} - d\mathbf{H}_0) + (\mathbf{B}_0 \cdot d\mathbf{H}_0 - \mathbf{H}_0 \cdot d\mathbf{B}_0) + \mathbf{H}_0 \cdot d\mathbf{B}_0 \tag{8}$$

On integrating over all space the first and the third terms vanish because of Equations 6 and 7. (The integral over the scalar product of an irrotational and a solenoidal vector vanishes.*) The fourth term vanishes identically, and the second term is

$$\mathbf{H}_0 \cdot d\mathbf{B} - \mathbf{B}_0 \cdot d\mathbf{H} = \mathbf{B}_0 \, d\left(\frac{\mathbf{B}}{\mu_0} - \mathbf{H}\right) = \mathbf{B}_0 \, d\mathbf{M}$$

Thus

$$đW = \int dV \, \mathbf{H} \cdot d\mathbf{B} = \int dV \, \mathbf{B}_0 \cdot d\mathbf{M} + \int dV \, \mathbf{H}_0 \cdot d\mathbf{B}_0 \tag{9}$$

which is of the desired form; in particular, the important first integral is taken only over the system rather than over all space.

At this point another characteristic difficulty may be mentioned. In spite of appearances, it is not justified to identify the first term in Equation 9 with the system and the second with the environment or to the vacuum. In general no assignment of the energy or free energy to localized systems is possible without ambiguity, because the interaction energy can be assigned to either system. Most authors make this assignment in arbitrary but consistent ways and arrive at correct answers for, say, the magnetocaloric effect. Heine suggests that such an assignment of the energy can be dispensed with, since what one really wants is the entropy obtained as the negative temperature derivative of the free energy. The last term in Equation 9 is temperature-independent, and the entropy resulting from the preceding one is assigned without ambiguity to the system.

For further details we refer to the literature. Evidently, a great deal remains to be worked out in this direction, since different types of work processes and also systems with different geometrical properties give rise to specific problems.

I have a good rationale for not pursuing these matters any further. My principal aim in this volume is to give justice to the specific features of the thermodynamic method. The complications of the problem under

* See J. A. Stratton, *Electromagnetic Theory* (McGraw-Hill Book Company, Inc., New York, 1941), p. 111.

discussion stem from the fact that we have a blend of the coupling method of thermodynamics in terms of the exchange of additive invariants (including the energy) and the mechanical method in which the coupling is due to the nonadditive interaction energy. It is natural to leave these matters for later consideration and to the attention of those interested in specific problems.

5. NERNST'S POSTULATE AND THE CLINCHING OF THE EQUILIBRIUM PROBLEM

In the previous few sections we have discussed the subtler aspects of such concepts as temperature and work that we have encountered in a less sophisticated form already in the most elementary parts of thermodynamics. We now turn to the discussion of the concept of equilibrium that exhibits in a most extreme form the double aspects of simplicity and complexity.

Virtually every thermodynamic phenomenon involves the concept of equilibrium in a more or less hidden fashion. For example, when we measure the temperature of an object we trust that thermal equilibrium is being established between the system and the instrument. Yet we have to face the embarrassing fact that there is no satisfactory empirical definition of equilibrium. The usual requirement that visible intrinsic processes should cease is only a necessary criterion that serves us well in simple situations, but we lack a practical sufficient criterion to reassure us that equilibrium is actually established on any particular level of molecular organization.

In MTE the assignment of a definite entropy value can be considered as a sort of definition of equilibrium. Of course, the cynic might interject that we explain the ill-understood concept of equilibrium in terms of a concept that we grasp even less. I propose to show that the cynic is wrong.

Purely empirical methods allow us to determine the entropy difference between two states along any particular path. The entropy difference is path-independent, and any deviation from this rule is evidence that we do not measure the true entropy because of lack of equilibrium somewhere along the line.*

This procedure is superior to the merely visual observation of the damping out of internal processes, but it still provides only necessary conditions. A criterion of equilibrium that yields a quantitative estimate for the extent of equilibrium reached is provided by the use of Nernst's

* Workers in the field usually pose the question whether or not thermodynamics is valid for their system under study, a curious contradiction to the claim that thermodynamics is generally valid because it is implied by impotence axioms.

postulate. At least, this is how this postulate appears from the point of view of MTE.*

The idea is to use the quantum-mechanical interpretation of the entropy (see page 259) but only in such limiting cases in which the result can be reliably established. First of all the entropy tends to zero in the limit of absolute zero. (See pages 177 and 275.) The practical application of this statement is at first impeded by two big *if's*. Is the lowest temperature reached close enough to $T = 0$, and is the system in thermodynamic equilibrium? Neither of these questions can be decided on a purely empirical basis, but a decision can be reached if caloric measurements are supplemented by the theoretical knowledge of the entropy at another fixed point: in the limit of the low-density highly dispersed gaseous state. The entropy of the perfect gas of N molecules is simply N times the entropy of a single molecule. There are many substances for which this quantity is easily calculated since the moment of inertia, the vibrational frequencies, and other data, such as electronic and nuclear spin multiplicities, are obtained with great precision from spectroscopy.

This low-density high-temperature entropy is chosen as the initial state for our empirical integration down to the lowest temperature T_0 for which the specific heat of the system has been actually measured. By using Equation 41 of Section 2, one arrives at the low-temperature entropy $s_0 = s(T_0, P_0)$. This quantity has been found in many cases to vanish within experimental error. This is not only a brilliant verification of Nernst's postulate, but it allows us to infer that for practical purposes T_0 is close enough to absolute zero and also that the system is in thermodynamic equilibrium in the entire range of investigation. If s_0 is significantly larger than zero, at least one of these statements is incorrect. The clarification of the situation often requires a considerable ingenuity combining experimental and theoretical techniques. The method is far from trivial, but it works, and under no conditions are we led to doubt Nernst's postulate.

The discrepancy is sometimes eliminated by further lowering the temperature. Otherwise the finite zero-point entropy s_0 can be interpreted in terms of frozen-in disorder (nonequilibrium state) of one sort or another.† In this case a further lowering of the temperature is of no avail.

* Strictly speaking we have to invoke also some simple results of quantum statistics. This blurring of the limits between phenomenological and statistical methods is not as shocking in 1965 as it was a few years ago. It may be even in order to recall that the classical thermodynamicist was justified in objecting to the uncritical use of oversimplified models. Within the limitations of equilibrium theories STE (Paper 7) goes a long way toward meeting thermodynamic standards. A statistical theory coping in similar generality with microscopic dynamics is yet to be developed.

† The reader is encouraged to get acquainted with some of the rich material that arose from pursuing this program. See, e.g., the monograph: J. Wilks, *The Third Law of Thermodynamics* (Oxford University Press, London, 1961), which is excellent for orientation and contains an ample bibliography. We note that nonequilibrium states

The relevant arguments are clearly microscopic, but they need not involve complicated calculations. The essential point is that $s = 0$ implies a coherence developing within the system, and the task is to find out about the nature of those variables (orientation of electronic and nuclear spins, of molecular groups, isotopic distribution, and the like) that do develop the coherent pattern and those that "freeze-in" before developing it.

It is interesting to ask in this connection: What are we to make of it if we obtain $s_0 < 0$? This, of course, should not happen since s_0/k is the logarithm of an integer. Yet we might get such a result if we ignore certain degrees of freedom in the computation of the statistical entropy. Suppose we omit to include nuclear spin states which in most cases are irrelevant in the sense that they are entirely random in both comparison states. If, however, there is a partial ordering of the spin states, this will appear as a spurious negative entropy, a situation corrected by including the appropriate degrees of freedom into the calculation of the statistical entropy.

The procedures outlined represent an extraordinary extension of the scope of the thermodynamic method. The naive equilibrium concept of the phenomenologist depending on the cessation of visible processes is deepened to provide us with significant insight into the microstructure of the system. This is achieved by careful dovetailing of the experimental and theoretical procedures where each is assigned an important task that it is best suited to solve. *This careful dovetailing of procedures takes the place of the traditional exclusion of microscopic considerations from thermodynamics.*

The great significance of Nernst's law was, of course, duly recognized. In addition to Planck, it was Nernst's former student, the late Sir Francis Simon, who demonstrated the usefulness of the new heat theorem to the physicist, particularly for unifying the great variety of low-temperature phenomena.* This point of view, or at least the emphasis, differed from Nernst's chemical orientation,† and of that of Giauque and his coworkers.

Although Nernst's postulate has been duly "celebrated," this was not without dissent. An unusually blunt, but quite characteristic opinion is that of Fowler and Sterne:‡ "We reach therefore the rather ruthless

arise only in the low-temperature region, with no difficulties expected in the gas phase. Exceptional cases such as the lack of equilibrium between ortho- and parahydrogen will be spotted, however, as anomalies in the course of the standard analysis.

* His brilliant review article (reference 15 of Paper 3) makes interesting reading even today; although the most striking cryogenic developments were still to come, they fit perfectly into the early framework.

† See K. Mendelssohn, "Walther Nernst, An Appreciation," *Cryogenics* 4, 129 (1964).

‡ R. H. Fowler and T. E. Sterne, *Rev. Mod. Phys.* 4, 537 (1932). Italics in original paper.

conclusion that *Nernst's Heat Theorem strictly applied may or may not be true, but is always irrelevant and useless — applied to 'ideal solid states' at the absolute zero which are physically useful concepts the theorem though often true is sometimes false, and failing in generality must be rejected altogether.*"

We do not fully understand the meaning of Nernst's postulate unless we go to the root of this negative view. It is apparent even from this short quotation and is confirmed by the study of the extensive literature that much of the argument turns around a basic misunderstanding.

Whereas Nernst formulated his statements in terms of entropy differences, Planck recognized that a considerable simplicity could be achieved by speaking of an absolute entropy and claiming the vanishing of this quantity with $T \rightarrow 0$. At this point trouble developed, however, since there are obvious cases in which the entropy does not go to zero, such as supercooled liquids and most mixtures. Planck attempted to separate the sheep from the goats, and for many years it was usual to accept his claim that the entropy would exhibit the proper limiting behavior for *pure crystalline* substances. It is now evident that this wording is too restrictive and insufficiently restrictive all at the same time. On the one hand, there are pure crystalline substances, such as ice, that have finite zero-point entropies due to frozen-in disorder. On the other hand, the entropy tends to zero for *liquid* helium, for alloys that make a transition into an ordered state, and even for supercooled liquids if in a metastable equilibrium. All apparent exceptions are traceable to frozen-in nonequilibrium states in which the trend to equilibrium is impeded by potential barriers or selection rules. The inference suggested by Simon and now all but generally accepted is that Nernst's postulate is always valid, provided the system is in equilibrium, or more precisely it is valid for those degrees of freedom for which the equilibrium is established.

At this point we should stop to realize that there is a fundamental methodological difference between the Planck and the Simon statements of Nernst's postulate. The former has the form of a prediction; we know at the outset whether or not a system on hand is within the purview of the law. In contrast, the Simon statement can hardly be considered a bona fide prediction, since we have no way of knowing whether a given system is or is not in equilibrium. Hence this version of Nernst's postulate does not conform to the widely held methodological rule that the statements of physics have taken the form of verifiable predictions. Under these conditions one may take one of three major attitudes: (i) Reject Nernst's postulate because of its lack of conformity with the accepted methodology; (ii) Hold on to Nernst's postulate and also to the old methodology, while closing one's eye to their conflict; (iii) Point out the limitation of the old methodology and extend its scope in order to fit the new situation.

Needless to say, only the last-mentioned attitude harmonizes with the spirit of this volume. We shall constantly point out that the theorems of generalized thermodynamics are more general than the tacitly accepted methodology distilled from Newtonian mechanics. In particular in MTE we deal with the properties of the *normal equilibrium* state and arrive at results that are so quantitative as to serve as *norms*, to be compared with actual measurements and thus to make inferences about the structure of the system in question (see also the next section).

Although differing in emphasis, the foregoing discussion is consistent with the currently accepted views inasmuch as the validity of Nernst's postulate is tied to the establishment of equilibrium. There is, however, a significant discrepancy inasmuch as there is a widespread tendency to avoid the use of absolute entropy, and accordingly Nernst's postulate is often formulated as follows: It is impossible to reach absolute zero as a result of a finite sequence of operations.

The unattainability postulate and the emphasis on entropy differences satisfy the methodological rule that the statements of physics have to be unconditional predictions. All this fits in well with CKC thermodynamics. The statement becomes too weak, however, to serve as a criterion of equilibrium. For this we need the conditional formulation of Nernst's postulate in terms of absolute entropy that in turn harmonizes with MTE.

Generally speaking, it would seem that the distrust of the notion of absolute entropy has been overdone. Few objections are voiced against the use of energy, although operationally we never measure anything but energy differences. The arguments against absolute entropy remind one of a spectroscopist protesting against the use of *term* values because (i) spectral lines involve only differences of terms and (ii) term values can be changed by a common additive constant. It is clearly preferable to use term values, while keeping in mind that absolute values are not altogether meaningful, rather than forbid their use categorically. The issue is also beclouded by semantic difficulties, for a discussion of which we refer to pages 120 and 275.

Finally we note that a reasonable definition of the entropy constant renders the chemical potential as definite as the constants u_0 and s_0 themselves. Certainly the combination 25 of Section 3 has a good operational meaning. If the addition of catalysts changes the definition of components and of s_0, then the chemical equilibrium changes as well.

6. CRITICAL POINTS

The theory of phase equilibrium and, in particular, of critical points, constitutes the central part of MTE that leads us to an interesting confrontation with experiment. In contrast to the quantitative verification involved, say, in the temperature concept, this comparison is only qualitative in the same sense as is topology. Yet the experimental check is so exacting that even while confirming the theoretical framework in its general aspects it strongly suggests its deepening as well.

These matters are to a large extent covered in Paper 3 and we discuss them in this section mainly because of major experimental advances since the appearance of that paper. It was found, almost exactly a hundred years after the discovery of the critical point, that the specific heat $c_v(T)$ measured at the critical volume has a sharp peak that seems to be best represented as a logarithmic singularity.*

This discovery caught us theorists by surprise. To be sure such a result was predicted by the model of lattice gas based on a close analogy between the critical points of a fluid and the Ising model.† There was little apparent justification, however, to take this model that seriously.

In order to appraise the situation from the point of view of MTE, we have to remember our recurrent theme on the complementary nature of this theory to particle mechanics and hence also to statistical mechanics. Whereas in this discipline one computes the properties of a definite model and thus predicts its phenomenological behavior, in MTE one specifies certain general requirements a model has to satisfy in order to exhibit a particular phenomenological behavior.‡ Such criteria are indeed provided by the phase rules.

The first phase rule specifies the maximum dimensionality of the set of points in the space of intensities in which a given number of modifications (phases) can coexist in a system of a given number of components.§

The second phase rule specifies the dimensionality of critical points. In order to explain the meaning of the new experiments, we shall sum up the situation for the critical points of one-component fluid systems. For proofs and other details we refer to Paper 3.

The stability of the system in question depends on a quadratic form

* M. I. Bagatskii, A. V. Voronel, and V. G. Gusak, *Zhur. Eksperim. i Teor. Fiz.* **43**, 728 (1962) [translation: *Soviet Phys.—JETP* **16**, 517 (1963)]; A. V. Voronel, Yu. R. Chashkin, V. A. Popov, and V. G. Simkin, *Zhur. Eksperim. i Teor. Fiz.* **45**, 828 (1963) [translation: *Soviet Phys.—JETP* **18**, 568 (1964)]; M. R. Moldover and W. A. Little, *Proceedings of the Ninth International Conference on Low Temperature Physics* (Plenum Press, New York, 1965), p. 653, and *Conference on Phenomena in the Neighborhood of Critical Points*, National Bureau of Standards, Washington, D. C., April 5–9, 1956 (to be published).

† T. D. Lee and C. N. Yang, *Phys. Rev.* **87**, 410 (1952); C. N. Yang and C. P. Yang, *Phys. Rev. Letters* **13**, 303 (1964).

‡ Compare with the similar methodological contrast described in connection with Nernst's postulate in the previous section.

§ The classical theory expresses the complexity of the system only in terms of the number of distinct independent chemical components. The main point of Paper 3 is to show that the intrinsic symmetry should be taken into account as well.

the matrix of which, briefly the stiffness matrix, is written in the energy scheme $u = u(s, v)$ as

$$\begin{bmatrix} \left(\frac{\partial T}{\partial s}\right)_v & \left(\frac{\partial T}{\partial v}\right)_s \\ -\left(\frac{\partial P}{\partial s}\right)_v & -\left(\frac{\partial P}{\partial v}\right)_s \end{bmatrix} = \begin{bmatrix} \frac{T}{c_v} & -\frac{T}{c_v}\left(\frac{\partial P}{\partial T}\right)_v \\ -\frac{T}{c_v}\left(\frac{\partial P}{\partial T}\right)_v & \frac{1}{v\kappa_s} \end{bmatrix} \tag{1}$$

Here κ_s is the adiabatic compressibility. This matrix can be diagonalized by a nonsingular linear transformation of the variables, yielding the diagonal elements

$$\lambda_1 = \frac{T}{c_v}, \qquad \lambda_2 = \frac{1}{v\kappa_T} \tag{2}$$

We note than λ_1 and λ_2 are not eigenvalues,* and inverting the order of the variables $u = u(v, s)$ we arrive at

$$\lambda_1' = \frac{1}{v\kappa_s}, \qquad \lambda_2' = \frac{T}{c_p} \tag{3}$$

Now according to a well-known rule,† the number of positive, negative, and vanishing terms in the diagonalized form must be independent of the particular transformation used.

The determinant of the matrix 1 can be written as

$$D = \lambda_1\lambda_2 = \lambda_1'\lambda_2' = \frac{\partial(T, -P)}{\partial(s, v)} \tag{4}$$

This quantity plays a very important role in MTE. Normal stability depends on $D > 0$. Moreover, the equivalence of the fundamental equations $U = U(S, V)$ and $G = G(T, P)$ requires that D be neither zero nor infinite. (See page 163.) Whereas $D < 0$ corresponds to essential instability and the states in question are presumably never observed, a small subset of the states with $D = 0$ may be stable; if this is the case, the representative points in the P–T plane are observable critical points.‡ This term could not have been better chosen. What was meant originally is that the point is critical with respect to the possibility of liquefaction of a gas. Actually, the term applies in a much deeper sense as well. We are at the point of breakdown of thermodynamic stability and of the macroscopic method of MTE. The critical point is on the

* See pages 187 and 238.

† Sylvester's law of inertia. See I. M. Gel'fand, *Lectures on Linear Algebra* (Interscience Publishers, Inc., New York, 1961).

‡ It is sometimes incorrectly stated that critical points are instable. It is true, however, that critical stability is greatly weakened compared to normal stability.

borderline of the stable and essentially instable regions, and the system escapes instability only by splitting into two modifications.

In order to interpret the breakdown of the Legendre transformation at critical states, let us consider a system in contact with its critical reservoir $R(T_c, P_c)$. Under these conditions the variables s, v of the system are not as sharply determined as usual, we have a weakening of the so-called principle of thermostatic determinism, which is due to the abnormally large fluctuations of the additive invariants (critical fluctuations). To handle the situation we have to resort to statistical theories, of which STE (Paper 7) is the first (see also Paper 8).

Apart from wording and emphasis, the foregoing sketch is within the scope of Gibbs' classical theory and is to a large extent borne out by experiment. The agreement is most remarkable for the complex many-component systems that are not discussed in this volume. It is so much the more significant that the phase rule fails to provide any theoretical niche for the interesting λ-type anomalies, so called because they manifest themselves as λ-shaped peaks in the specific heat and other compliance coefficients (see Figures 2, 3, 6, and 7 in Paper 4). An unbiased inspection of the experimental data should have led at least to a tentative association of λ anomalies with critical states. Instead, for a good number of years, it was traditional to refer to Ehrenfest's ad hoc scheme of higher-order transitions that is postulated to be nonsingular.

This attitude might have been to some extent motivated by the fact that the interpretation of λ anomalies as critical points is contradicted by the phase rule which provides only for an isolated critical *point* in a one-component system. In contrast the λ points form *lines* in the P–T plane; the so-called λ *point* is merely the intersection of the λ line with the line corresponding to $P = 1$ atmosphere, a theoretically irrelevant special case.

The main purpose of Papers 3 and 4 was to resolve this difficulty and to integrate λ phenomena with the phase rule.* The central idea is that in gauging the thermodynamic complexity of the system it is not enough to count the distinct components but also the symmetry of the modifications should be taken into account. Just as the fluid escapes instability by separating into liquid and vapor, the Ising model separates into samples with positive and negative magnetization. Note that the "spin" of this model admits many interpretations in terms of diverse structural elements (see Sections X and XI of (*MTE*)). What is important is the occurrence of two modifications transformed into each other by a symmetry operation in a fashion that ensures both their distinctness and

* Under the impact of precise calculation and experiments (see references 33 and 35 of page 192), the fundamental identity of λ lines and critical points is being accepted now as axiomatic with little awareness of any difficulty that is indeed not apparent from the point of view of statistical mechanics.

their energetic equivalence ("twinning"). In quantum-mechanical language, we have a case of degeneracy.

In the case of crystalline systems these conclusions are evidently true and may be deemed a bit trivial. The symmetry can be interpreted in terms of classical configurations, say, lattice geometry. It is more surprising that the theory ought to apply also to liquid helium calling for the possibility of a twinning (quantum degeneracy) of the superfluid state.

The problem of establishing a microscopic model to account for these fluid-type symmetry properties is still unsolved and is the greatest challenge arising from "Generalized Thermodynamics." It may be of some interest, however, that the newly found experimental c_v singularity also seems to point toward the relevance of hidden symmetries in fluids.

At the critical point $D = 0$, which is equivalent to $\lambda_2 = 0$ and $\lambda_2' = 0$. However, *a priori* it is possible to have (i) $\lambda_1 > 0, \lambda_1' > 0$, or (ii) $\lambda_1 = 0$, $\lambda_1' = 0$. Any other possibility, such as $\lambda_1 = 0, \lambda_1' > 0$, is excluded by Sylvester's law of inertia.*

In the classical theory, and this includes Papers 3 and 4, the option was in favor of case (i). This was entirely reasonable, since one could not see why the additional condition (ii) should be satisfied, unless, of course, this was required by symmetry as in the model of the lattice gas in which particles and holes play a symmetric role.

Now that we have experimental evidence in favor of the alternative (ii) we can invert the argument and infer that the neighborhood of the critical region of the fluid exhibits a structure that is symmetric around the critical point. As indicated already, the model of lattice gas has indeed such a symmetry, which does not mean, of course, that this simple model is altogether adequate.

Instead of speculating on the nature of a more satisfactory model we wish to point out a simple inference concerning the proper parametrization of the critical region even within MTE. It is plausible that one ought to choose "natural variables" in the sense of MTE that are adapted to the symmetry in question. It was recently noticed that all the indications favor an inversion of the traditional roles assigned to the volume V and to the mole number N. Instead of holding N constant and using $v = V/N$ as the independent variable, it is preferable to hold V constant with the density $\rho = N/V$ as variable. Whereas the intensity conjugate to v is the negative pressure $-P$, we have to deal now with the conjugate pair ρ, μ. Although there are very obvious advantages in favor of using the density ρ, the chemical potential is at a disadvantage compared to the pressure on the basis of intuitive appeal. We have seen,

* This fact leads at once to the inference that $c_v \to \infty$ implies that the real part of the sound velocity $(\rho\kappa_s)^{-1/2}$ goes to zero. See C. E. Chase, R. C. Williamson, and L. Tisza, *Phys. Rev. Letters* **13**, 467 (1964).

however, at the end of the last section that from the objective point of view the chemical potential is an acceptable thermodynamic parameter. The advantages of the use of the ρ, μ pair of conjugate variables are discussed in Paper 5.

In the light of these results we have to modify a few statements of Papers 3 and 4. In Figure 5 of the former paper we have indicated some of the isothermals of a fluid. The schematic drawing indicates the required symmetry around the critical point, but in actual fact such a symmetry would be grossly distorted with the indicated labeling of the axes (v, P). The figure would be qualitatively correct, however, if the axes are relabeled as $[\rho, -\mu]$. A semiquantitative comparison of the isothermals in both schemes was plotted by Chase and Williamson.*

A further study of these matters within MTE is hampered by the fact that a prediction of specific functional forms of the fundamental equation is beyond the scope of this discipline. The nonanalytic character of the fundamental equation at the critical point (branch point) can be inferred, however, from quite general arguments. Suppose that the determinant D vanishes at the critical point as $\sim(T - T_c)$, and hence that $C_p \sim (T - T_c)^{-1}$. This implies that $\Delta S = \int C_p/T\, dT$ becomes infinite when integrated across the critical point. This is in conflict with the fact that the fundamental equation is always bounded. It follows that D cannot be expanded into a power series. (This is an improvement over the argument of page 208.) Recently a great deal of progress has been achieved in providing mathematical descriptions of such singularities.† The prevalence of the same type of branch point for widely different physical systems makes one suspect that such model-independent properties should be derivable within generalized thermodynamics. However, this would presumably require new techniques beyond those now available.

* C. E. Chase and R. C. Williamson, *Proceedings of the Conference on Phenomena in the Neighborhood of Critical Points*, National Bureau of Standards, Washington, D. C., 1965 (to be published). We may note that Figure 1*b* of Paper 4 is incorrect: The broken line indicating the two-phase region should be inverted.

† See reference 2 of Paper 5.

Reprinted from ANNALS OF PHYSICS, Volume 13, No. 1, April 1961

ANNALS OF PHYSICS: **13,** 1–92 (1961)

The Thermodynamics of Phase Equilibrium *

LASZLO TISZA

Department of Physics, Research Laboratory of Electronics, Massachusetts Institute of Technology, Cambridge, Massachusetts

Thermodynamics is usually subdivided into a theory dealing with equilibrium and into one concerned with irreversible processes. In the present paper this subdivision is carried further and the Gibbsian thermodynamics of phase equilibrium is distinguished from the thermodynamics of Clausius and Kelvin. The latter was put into an axiomatic form by Carathéodory; the present paper attempts a similar task for the Gibbs theory. The formulation of this theory as an autonomous logical structure reveals characteristic aspects that were not evident until the two logical structures were differentiated. The analysis of the basic assumptions of the Gibbs theory allows the identification and removal of defects that marred the classical formulation. In the new theory thermodynamic systems are defined as conjunctions of spatially disjoint volume elements (subsystems), each of which is characterized by a set of additive conserved quantities (invariants): the internal energy, and the mole numbers of the independent chemical components. For the basic theory, it is convenient to assume the absence of elastic, electric, and magnetic effects. This restriction enables us to define thermodynamic processes as transfers of additive invariants between subsystems. Following Gibbs, we postulate that all thermostatic properties of system are contained in a fundamental equation representing the entropy as a function of the additive invariants. Geometrically, this equation is represented as a surface in a space to which we refer as Gibbs space. In order to make the information contained in the fundamental equation complete, we have to use, in many cases, additional quasi-thermodynamic variables to specify the intrinsic symmetry properties of the system. The walls, or boundaries limiting thermodynamic systems are assumed to be restrictive or nonrestrictive with respect to the transfer of the various invariants. The manipulations of the boundary conditions (imposition and relaxation of constraints) are called thermodynamic operations. In systems with nonrestrictive internal boundaries, the constraints are consistent with infinitely many distributions of the invariants over the subsystems. These virtual states serve as comparison states for the entropy maximum principle. This principle allows us to identify the state of thermodynamic equilibrium, attained asymptotically by real systems. The thermostatic ex-

* This research was supported in part by the U. S. Army Signal Corps, the Air Force Office of Scientific Research, and the Office of Naval Research; and in part by the U. S. Air Force (Office of Scientific Research, Air Research and Development Command) under Contract AF49(638)-95. Reproduction in whole or in part is permitted for any purpose of the United States Government.

tremum principle is the basis of a theory of stability. Stability may be normal or critical. In the latter case, the compliance coefficients (specific heat c_p, expansion coefficient, and isothermal compressibility) tend, in general, to infinity. The phases of thermodynamic systems are each represented by a primitive surface in Gibbs space, the points of these surfaces correspond to modifications of the phase. The actual distribution of the invariants of a system over phases is determined by the entropy maximum principle. Among the fundamental theorems are the two phase rules. The first rule specifies the dimension of the set of points in the space of intensities, in which a given number of modifications can coexist. The second rule specifies the dimensionality of the set of critical points. The phase rules are somewhat more general than those of Gibbs because of our use of symmetry considerations. At a critical point two modifications become identical, and we obtain critical points of two kinds: (i) the modifications differ in densities, as for liquid and vapor; (ii) the modifications differ in symmetry, as the two directions of ordering in the Ising model. The second case relates to the well-known λ-points and λ-lines in the $p - T$ diagram, hence these phenomena fit into the framework of the theory without *ad hoc* assumptions. For the case of liquid helium, this interpretation requires that the superfluid ground stage be degenerate. This conclusion is not inconsistent with the third law, but it requires substantiation by quantum-mechanical methods. Another fundamental theorem is the principle of thermostatic determinism: a reservoir of given intensities determines the densities (energy and components per unit volume) of a system which is in equilibrium with it and, conversely, the densities of the system determine the intensities. The mutual determination is unique, except at critical points and at absolute zero. A more satisfactory description of these singular situations calls for the use of statistical methods. The present approach leads to several statistical theories, the simplest of which is developed in the second paper of this series.

INTRODUCTION

This paper is the first in a series in which thermodynamics is developed beyond its usual scope from a new postulational basis.

The postulational basis of classical thermodynamics is firmly established in tradition and a new departure calls for an explanation of the underlying ideas.

It is widely believed that thermodynamics consists essentially of the implications of the first, second, and third law of thermodynamics. Actually, however, few if any significant results can be derived from these postulates without using a number of additional assumptions concerning the properties of material systems, such as the existence of homogeneous phases, validity of equations of state and the like.

The three laws of thermodynamics are presumably of universal validity, a point that is emphasized by their formulation as impossibility axioms. In contrast, most of the additional assumptions involve approximations of a more or less restrictive character. In the traditional presentations these additional assumptions and their limitations receive but scant attention. Therefore the rigor of the theory that the painstaking establishment of the universal principles was meant to ensure is considerably impaired.

The nature of a theory is determined by its postulational basis and it is to be expected that the reformulation of the basic postulates of thermodynamics will have a bearing on the type of thermodynamic theory that will be obtained.

Within judiciously chosen limitations, a set of basic assumptions can be expressed in terms precise enough to generate a mathematical theory with desirable properties. Some of the limitations of the theory can be overcome by the formulation of sets of successively more refined assumptions generating theories that are adapted to deal with the more exacting requirements. Thus thermodynamics becomes a master scheme consisting of a number of closely knit deductive systems devised for different types of situations. This procedure is not only more rigorous than the classical one, but opens up new areas for investigation where the possibilities of the classical theory seemed already exhausted.

An important step leading toward such a pluralistic conception of thermodynamics was the establishment of irreversible thermodynamics as an autonomous discipline (*1*). The classical formalism is recognized to be valid only for equilibrium situations.

We wish to make the further point that even the classical theory of equilibrium is a blend of two essentially different logical structures: we shall distinguish the theory of Clausius and Kelvin, on the one hand, from that of Gibbs, on the other.

In the first of these theories the thermodynamic system is considered as a "black box"; all relevant information is derived from the amount of energy absorbed or provided by idealized auxiliary devices such as heat and work reservoirs coupled to the system. The main achievement of this theory is the establishment of the concepts of internal energy and entropy from observable quantities.

In the Gibbs theory (*2*) attention is turned toward the system. The concepts of internal energy and entropy are taken for granted, and are used to provide a more detailed description of the system in equilibrium, which includes its chemical and phase structure.

The logical-mathematical structure of the Clausius-Kelvin theory received the most satisfactory expression in the axiomatic investigation of Carathéodory (*3*). We shall speak of the Clausius–Kelvin–Carathéodory (CKC) theory.

No comparable axiomatic investigation of the Gibbs theory has been performed thus far. Accordingly, in the existing texts the CKC and the Gibbsian thermodynamics are intricately interwoven, although the Gibbs formalism is emphasized in texts oriented towards physical-chemical applications.

The first objective of this paper is to formulate a postulational basis from which a theory, almost identical to that of Gibbs, is derived as a self-contained logical structure. We shall call this theory thermostatics, or alternately the macroscopic thermodynamics of equilibrium (MTE).[1]

In order to obtain a preliminary impression of the features in which this modernized Gibbs theory differs from that of CKC, it is convenient to compare the geometrical methods utilized by each.

The application of geometrical methods to thermodynamics is based on a thermodynamic *phase space*, i.e., a space spanned by a number of thermodynamic variables. In the CKC theory one considers a thermodynamic phase space spanned by such variables as pressure, volume, and the mole numbers of the chemical components. The guiding idea is that the variables be directly measurable. Beyond this requirement, however, the theory is quite insensitive to the specific choice of phase space. In particular, there is no geometrical significance attached to the distinction between extensive and intensive variables, as indicated by the use of volume and pressure in an interchangeable manner.[2]

In contrast, in the Gibbs theory a particular phase space spanned by the extensive variables: energy, entropy, volume, and mole numbers plays a quite unique role. In order to underline its importance we shall refer to it as *Gibbs space.*[3]

It was a remarkable discovery of Gibbs that a single surface in this space (the primitive surface) provides a compact representation of all thermostatic properties of a phase. Phases are homogeneous extensions of matter that constitute the building blocks for more complex, heterogeneous systems. Moreover, Gibbs based his theory of thermodynamic stability on the analysis of the curvature of the primitive surface.

From the mathematical point of view, the success of this procedure is somewhat surprising because such elementary geometrical concepts as orthogonality and metric, which are necessary both in the elementary and the Riemannian theory of curvature, are lacking in the thermodynamic phase spaces.

Nevertheless Gibbs, acting apparently on physical intuition alone, developed an *ad hoc* theory of curvature, the exact mathematical foundation for which was discovered nearly half a century later by Pick and Blaschke (*7*). This somewhat esoteric theory operates in terms of affine differential geometry in which parallel projection is substituted for the more common orthogonal projection.

[1] We use capitalized abbreviations to denote theories with explicitly stated postulational bases, whereas "thermodynamics" is used in a more vague, traditional sense.

[2] The geometrization of thermodynamics along these lines has been further developed by Landsberg (*5*).

[3] It has already been recognized by Ehrenfest (*6*) that the precise distinction between extensive and intensive variables calls for an axiomatic investigation going beyond that of Carathéodory.

The Gibbs space alone, of all thermodynamic phase spaces can be considered as an affine space. Thus the idea of attaching a special importance to Gibbs space is supported by the fact that in this space we can establish a far reaching geometrization of the properties of matter.

In this paper we shall confine ourselves to a few remarks concerning the role of affine geometry in thermostatics, and hope to return to this question in the future.[4]

The formulation of the Gibbs theory of phase equilibrium, as an autonomous logical structure, gives justice to aspects of the theory that are obscured when embedded in the CKC theory. At the same time the analysis of the basic assumptions has also brought to light significant defects in the classical theory.

The most important of these defects concerns the question of identity and distinctness of two phases. It is evidently important to establish the criteria by which we decide whether or not a surface can be considered a phase boundary separating two distinct phases. It is the implicit assumption of the classical theory that two volume elements contain identical phases if and only if they have identical chemical compositions and identical densities of the extensive quantities. Actually, however, volume elements of identical composition and density may still be distinct because of their symmetry properties. Situations

[4] The following remarks are for preliminary orientation of the reader who is familiar with differential geometry, we shall not make use of them in the present paper: The elementary theory of curvature of a surface embedded in three-dimensional Euclidian space is based on the two fundamental forms: the first of these (I) determines the line element in terms of the parameters describing the surface; the second form (II) expresses the curvature with respect to the embedding space. Both forms are needed to determine the principal radii of curvature, R_1 and R_2. However, according to a fundamental theorem of Gauss, the product R_1R_2 depends solely on form I, it is invariant with respect to bending of the surface, an operation which preserves the metric I, but changes the embedding specified by II. This theorem is the basis for the development of the Riemannian theory of curvature solely in terms of the intrinsic metric form I of the surface. Neither the elementary, nor the Riemannian theory of curvature can be applied in Gibbs space, in which no physically meaningful metric form is definable. However, in 1917, Pick and Blaschke developed a theory of curvature, within affine differential geometry, based solely on form II. This form is called the affine fundamental form; it is invariant under linear transformations of determinant unity (equiaffine invariance). We shall see, particularly in Section VIII that there is a quadratic form, we call it the fundamental thermostatic form, that plays an important role in the thermostatic theory of stability. This form can be shown to be an affine fundamental form of the primitive surface in Gibbs space. Thus the parallel between the significant thermostatic and geometric concepts is very close. Although the Riemannian and the affine theories of curvature have similar traits, there are also differences. Without entering into geometrical intricacies, Appendix B contains a short discussion of the reduction of a quadratic form in affine space. The treatment and its physical interpretation differ considerably from the eigenvalue method that is customary in spaces in which orthogonal (or unitary) transformations are meaningful.

of this sort arise if there exist two phases that transform into each other by means of inversion or mirroring, as in the case of right and left quartz. Such a symmetry relation ensures the distinctness of the two phases and the exact identity of their densities as well. If the phases are equivalent under a pure rotation, we should not consider them as distinct.

The redefinition of the criterion of distinctness has a significant implication for the phase rule. While the possible types of phase equilibrium predicted by the classical theory remain unchanged, the theory is enriched by a new category: the λ-points and λ-lines arising in order-disorder phenomena. The experimental occurrence of these phenomena was a serious embarrassment to the classical theory. This gap was meant to be filled by the *ad hoc* assumption by Keesom (*8*) and Ehrenfest (*9*) of the concept of higher order transitions.

This scheme has met with wide acceptance in the systematization of experimental facts, although it led to a number of difficulties which could only be removed by making further assumptions of a purely taxonomical nature (*10*).

It is therefore satisfying that the experimental facts in question fit into the present theory without *ad hoc* assumptions.[5] At least this is the case for λ-phenomena in crystals. The λ-anomaly of liquid helium falls into a special class. It fits into the present theory only if the ground-state wave function of the superfluid is assumed to be at least doubly degenerate. The present theory is too phenomenological to provide a substantiation of this conclusion, and the matter requires further study by quantum mechanical methods.

We have mentioned that the key to the continued development of thermodynamics lies in those assumptions that are usually not explicitly mentioned. Among the most significant of these is the assumption of a *constructive principle.*

In the CKC theory systems are considered as undivided units coupled only with auxiliary macroscopic devices, in strong contrast to the constructive principle of statistical mechanics, in which systems are built from constituent invariant particles.

We shall see that thermodynamics has a constructive principle of its own: systems are built up from disjoint volume elements that are described in terms of extensive variables specifying the amount of energy, mass, and chemical components within each volume element.

In MTE these quantities are always assumed to have reached their equilibrium values. However, it is possible to take fluctuations into consideration, and let the extensive variables become random variables. Thus thermodynamics becomes an essentially statistical discipline and the phenomenological formulation appears only as a limiting case.

[5] Some of the ideas concerning the expansion of the classical theory have been reported before (*11–13*). We note also that Landau (Chapter XIV of Ref. *20*) was the first to recognize the connection between symmetry and the λ-lines of crystalline substances. However, we arrive at different conclusions concerning the mathematical character of the singularity.

The simplest instance of statistical thermodynamics will be developed in the second paper of this series (*14*). We shall refer to it as the statistical thermodynamics of equilibrium (STE).

I. DEFINITIONS AND POSTULATES

We now proceed to list the basic definitions and postulates, the *basis* of thermostatics (MTE), amplified by short comments that are required for the clarification of the formal statements. A more extensive discussion of the limits of validity of the postulates, and comparison of the present procedures with traditional ones, follows in the next section.

The postulational basis has a hypothetical character justified by the merits of the theory generated by it. The measure of experimental agreement achieved will be assessed in the final discussion.

A theoretical justification of the postulational basis can be obtained also by reducing the basis to more fundamental theories. Within the context of the present over-all approach this is a program of many stages and is beyond the scope of this paper.

We emphasize that there is a considerable latitude in the formulation of the postulational basis even within thermostatics. The following basis was chosen, partly in view of developing the theory of phase equilibrium, and partly in order to facilitate generalization to the statistical thermodynamics and the quantum mechanics of phase theory.

A. Definitions and Postulates Concerning the Independent Variables of Thermostatics

D a1 *Thermodynamic Simple System:* A finite region in space specified by a set of variables $X_1, X_2, \cdots, X_{r+1}$. The X_i symbolize the physical quantities listed in *P a3*.[6]

D a2 *Composite system:* A conjunction of spatially disjoint simple systems. A composite system is obtained by uniting separate systems or by partitioning a single system. The simple systems that form the composite systems are called its subsystems. The subsystems and their properties will be denoted by primes or superscripts, thus X_i' or $X_i^{(a)}$ denotes the amount of the quantity X_i in the subsystem *a*. The set of quantities $X_i^{(a)}$ specifies the state of the composite system. Geometrically, it is a point in the, say *M*-dimensional vector space spanned by the variables $X_i^{(a)}$, $i = 1, 2, \cdots r + 1$; $a = 1, 2, \cdots$.

D a3 *Transfer quantity* X_i^{ab}*:* The amount of X_i transferred from the system *a* to *b* during an arbitrary, fixed time interval. Such a transfer is called a *thermodynamic process*.

[6] *D a1*, *P a1*, and *C a1* stand for Definition *a1*, Postulate *a1*, and Corollary *a1*, respectively.

P a1 *Conservation postulate:* The quantities represented by the variables X_i are conserved; they obey the continuity equations

$$\Delta X_i^{(a)} + \sum_b X_i^{ab} = 0, \tag{1.1}$$

$i = 1, 2, \cdots, r + 1$, $a = 1, 2, \cdots$, where $\Delta X_i^{(a)}$ is the net increase of the quantity X_i in the system a, and the summation is taken over all systems involved in the exchange of X_i. Both $\Delta X_i^{(a)}$ and X_i^{ab} refer to the same time intervals.

P a2 *Additivity postulate:* In a composite system

$$X_i = \sum_a X_i^{(a)}, \qquad i = 1, 2, \cdots, r + 1, \tag{1.2}$$

where the summation is over the subsystems. The quantity on the left belongs to the composite system.

D a4 *Additive invariant:* a quantity satisfying *P a1* and *P a2*.

D a5 *Wall:* physical system idealized as a surface forming the common boundary of two systems, say a and b, and preventing the transfer of some of the quantities X_k, so that $X_k^{ab} = 0$ regardless of the state or properties of the systems a and b. No such restrictions are valid for the other quantities X_j. The wall is said to be *restrictive of the* X_k and *nonrestrictive of the* X_j.[7] Walls separating two subsystems of a composite system are called *partitions*: those completely enclosing a system are *enclosures*. The set of walls in a system is referred to as the *constraints*. These are said to exert *passive forces* on the distribution of the additive invariants.

D a6 *Thermodynamic Operation:* any imposition or relaxation of a constraint through the uniting or subdividing of systems, or the altering of the type of the enclosures or partitions.

D a7 *Coupled Systems:* systems for which at least one transfer quantity X_i^{ab} can be different from zero in the presence of the existing constraints. The systems are said to be coupled by means of X_i exchange.

C a1 It follows from *P a1* and *P a2* that for a system a in an enclosure restrictive of X_i, the value $X_i^{(a)}$ is constant in time. Hence the conservation laws and the restrictive enclosures impose linear constraints on the variables $X_i^{(a)}$:

$$\sum_a X_i^{(a)} = \xi_i^{(k)} = \text{constant.} \tag{1.3}$$

The summation is taken over the subsystems of an enclosure restrictive of X_i.

[7] These expressions are somewhat elliptic. We should say: "restrictive of the transfer quantity associated with $X_k, \cdots$."

The superscript k identifies the enclosure that may coincide with the entire system or with parts of it. The $\xi_i^{(k)}$ are invariants of the composite system.

The conditions (3) may be supplemented by inequalities, e.g., requiring that some of the X_i be positive.

Let us assume that the composite system has b linearly independent invariants, relabeled with single subscripts $\xi_1, \xi_2 \cdots \xi_b$. Equation (3) can be used to express b of the variables $X_i^{(a)}$ in terms of the ξ_j and the remaining $M - b = f$ variables $X_i^{(a)}$ which we relabel $Y_1, Y_2, \cdots Y_f$. It is convenient to substitute the variables ξ_k, Y_i for the original $X_i^{(a)}$, and we define:

D a8 *Fixed variables:* the variables $\xi_1, \xi_2, \cdots, \xi_b$ appearing in Eq. (3); invariants of the composite system.

D a9 *Free variables:* the variables $Y_1, Y_2, \cdots, Y_f$ obtained by the foregoing construction. Within an indicated range the Y_i specify the *virtual states* of the composite system, i.e., those states that are consistent with the constraints, and with the set of fixed variables ξ_j.

D a10 *Virtual process (displacement):* the increments ΔY_i leading from one virtual state to another. Geometrically, the ratio $\Delta Y_1 : \Delta Y_2 : \cdots$ represents a line in the vector space spanned by the $X_i^{(a)}$.

C a2 The linear combination of two virtual displacements is, again, a virtual displacement. The virtual displacements form a linear manifold specified by conditions of the form

$$\sum_a \Delta X_i^{(a)} = 0. \tag{1.4}$$

The range of summation is the same as in (3).

We consider some important special cases. If a composite system consists of isolated subsystems, all variables $X_i^{(a)}$ are fixed, the space of virtual states reduces to a point.

In the case of two subsystems coupled by the transfer of the positive quantity X_i, we have

$$\xi_1 = X_1^{(a)} + X_1^{(b)}, \tag{1.5}$$

and

$$0 \leqq Y_1 = X_1^{(a)} \leqq \xi_1 . \tag{1.5b}$$

The virtual displacement is

$$\Delta Y_1 = \Delta X_1^{(a)} = -\Delta X_1^{(b)} = X_1^{ba}. \tag{1.5c}$$

Note that the displacement uniquely determines a transfer quantity. This is no longer the case for three coupled systems. In fact, the transfer

$$X^{ab} = X^{bc} = X^{ca} \tag{1.6}$$

produces no displacement: $\Delta X^a = \Delta X^b = \Delta X^c = 0$, hence (6) may be added to any set of transfer quantities without changing the displacement.

Up to this point, the variables X_i have been defined only as additive invariants (*P a1*, *P a2*, *Da 4*). It is indeed possible to develop the formalism of MTE on the basis of such an *implicit* definition by postulation. However, the physical interpretation of the theory calls for an *explicit* interpretation of the variables.

P a3 The variables $X_1, X_2, \cdots, X_{r+1}$ are interpreted as the volume V, the internal energy U and the mole numbers of the independent components $N_1, N_2, \cdots, N_c$; hence

$$r = c + 1. \tag{1.7}$$

The joint use of *P a1*, *P a2* with *P a3* gives rise to new sorts of questions: Are the variables enumerated in *P a3* actually additive invariants? Are there actual walls that are restrictive or nonrestrictive of various combinations of these variables? Neither of these questions is answered by an unqualified "yes," and the theory to be developed is subject to limitations that are discussed in Section II. Here we confine ourselves to a short summary of the conclusions reached.

The conservation of internal energy is postulated in the first law of thermodynamics and is sufficiently well substantiated. The additivity of the energy is, in general, not a valid statement, but it represents a sufficiently good approximation if long-range gravitational, electric or magnetic fields and surface effects are only of negligible importance. The thermodynamic theories of electric, magnetic, elastic, gravitational and capillary systems call for special considerations that are not included in this paper.

In mechanics and electrodynamics the non-additive interaction energies are essential to provide a coupling of systems, but in thermodynamics coupling arises even in the case of strict additivity through the exchange of additive invariants.

The material content of thermodynamic systems appears as a collection of particles (molecules, atoms, ions, nuclei, etc.) of definite identity. The numbers of these particles, or the mole numbers of the corresponding chemical species are additive, but not conserved, since they are subject to change due to chemical reactions. This complication can be easily handled, provided that the relevant reactions can be classified into slow and fast ones, in such a way that the advancement of the former is negligible on the time scale of the actual experiment, while the latter can be assumed to have come to a standstill at chemical equilibrium. Under such conditions it is possible to define additive invariants, the so-called *mole numbers* of the (*independent chemical*) *components* (p. 63 of Ref. *2*). The specification of these numbers will be called the *composition*.[8]

[8] In this paper we do not consider chemical reactions explicitly and therefore we do not go into a detailed definition of independent components. In the absence of reactions, the number of indestructible particles of each species are additive invariants, and can be used

Although we do not enter into an explicit discussion of chemical reactions, we may note that their existence has an implicit bearing on the scope of the thermostatic formalism. The rates of reactions can be speeded up, or slowed down, to an extreme degree by the variation of the conditions. One can vary the temperature and the pressure, and add, or remove catalysts and anticatalysts. On changing the set of relavant reactions, the number of distinct independent components, used according to *P a3* for the description of the system, will change as well. Therefore, we state:

C a3 The choice of thermodynamic variables and the resulting formalism is relative to the set of chemical reactions, that have been used to define the independent components. The formalism is only adequate under conditions at which these, and only these, reactions proceed at a noticeable rate.

The joint use of *P a3* and *D a5* requires special consideration. First, we note that an enclosure may be restrictive of all the *X-s* listed in *P a3* without precluding a change of the shape of the system. Although the present theory is concerned only with shape independent effects, it is desirable to specify whether or not an enclosure is *rigid*, and precludes a change of *shape* of the enclosed system. In fact, we shall find it convenient to assume, for the time being, that all our walls and enclosures are rigid, and that both volume and shape of our systems are constant. This assumption seems drastic, since it excludes macroscopic compressional work from our discussion. (Other types of macroscopic work are excluded in virtue of *P a3*, which does not provide for elastic, electric, and magnetic variables.) We shall consider macroscopic work in a somewhat different context in Section IV. The temporary exclusion of macroscopic work has a number of advantages.

P a4 The following types of rigid (volume and shape preserving) walls and enclosures are actually available:

Impermeable wall: restrictive of matter.

Semi-permeable wall: restrictive of certain chemical species, nonrestrictive of energy.

Permeable wall: nonrestrictive of matter and energy.

Adiabatic wall: restrictive of matter and energy.

Diathermic wall: restrictive of matter, nonrestrictive of energy.

D a11 *Isolated system:* system enclosed in a completely restrictive enclosure.

Closed system: system in an impermeable enclosure.

---

in *P a3*. Part of the usefulness of the concept of components is that it allows us to "save" the simple formalism for inert species, even in the presence of reactions, by establishing a more sophisticated conceptual interpretation of the variables.

Open system: system in which exchange of matter is not precluded by impermeable enclosure.

D a12 *Heat:* transfer quantity, associated with energy, across a rigid diathermic partition (no work):

$$Q^{ba} = U^{ba}. \tag{1.8}$$

B. Definitions and Postulates on Equilibrium and Entropy

P b1 *Entropy maximum principle.* (Composite systems).

Isolated composite systems tend toward a quiescent asymptotic state called thermodynamic equilibrium, in which the free variables assume constant values specified as solutions of an extremum problem. The extremal function is constructed as follows: To each simple system there is assigned an *entropy function*, or *fundamental equation*

$$S^{(a)} = S^{(a)}(X_1^{(a)}, \cdots X_{r+1}^{(a)}), \tag{1.9}$$

a continuous first order-homogeneous function with continuous first, and piecewise continuous higher derivatives. The fundamental equation of the composite system is

$$S(\xi_1, \xi_2, \cdots, \xi_b) = \max \{\sum_a S^{(a)}(X_1^{(a)}, X_2^{(a)}, \cdots, X_{r+1}^{(a)})\}. \tag{1.10}$$

The comparison states of this maximum problem are the virtual states described by the free variables $Y_1, Y_2, \cdots, Y_f$.

We shall see in Section XII that the solution of the maximum problem (10) is unique, under rather general conditions. Moreover, if the conditions of uniqueness are not satisfied, the various solutions are related to each other by rather obvious symmetry relations. We express this by stating that the solution of Eq. (10) is *almost unique.* The precise meaning of this term will be explained later.

C b1 Let $\bar{Y}_1, \bar{Y}_2, \cdots \bar{Y}_f$ be a set of values of the free variables for which expression (10) is a maximum. In virtue of Eq. (10) the $\bar{Y}_j$ are represented as functions of the ξ_k. By using Eq. (3), we obtain the $\bar{X}_i^{(a)}$ corresponding to the $\bar{Y}_i$ and ξ_k. Then

$$S(\xi_1, \xi_2, \cdots \xi_b) = \sum_a S^{(a)}(\bar{X}_1^{(a)}, \bar{X}_2^{(a)}, \cdots, \bar{X}_{r+1}^{(a)}). \tag{1.11}$$

The terms of this sum are the entropy values assigned to each subsystem *only* in the case of equilibrium. The left-hand side of this equation is the entropy of the composite system; thus the entropy is additive.

C b2 Let the entropy of a composite system in equilibrium be S_i.

Consider a thermodynamic operation in which some of the internal constraints of the system are relaxed, and thus the manifold of virtual states is increased. With the increase of the set of comparison states, the maximum (10) either increases or remains unchanged. Denoting the entropy in the relaxed equilibrium state S_f we have

$$S_f > S_i \quad \text{(a)}$$

or (1.12)

$$S_f = S_i \quad \text{(b)}.$$

In Case (a) the relaxation of constraints triggers a process leading to a redistribution of the additive invariants, i.e., to a new equilibrium. This process involves the increase in entropy:

$$DS = S_f - S_i, \tag{1.13}$$

The symbol D indicates an actual change in contrast to virtual changes denoted by Δ.

An imposition of a constraint in a system in equilibrium leaves the distribution of the X_i and hence the entropy, unchanged. In particular, the reimposition of the initial constraint fails to reestablish the initial situation.

In case (b) the relaxation of constraints leads to no process at all.

D b1 *Irreversible operation:* relaxation of constraints producing an entropy increase $DS = S_f - S_i$. This quantity is called the *measure of irreversibility.*

D b2 *Reversible operation:* imposition of constraints in a system in equilibrium, or a relaxation of constraints that does not produce an entropy increase.

D b3 *Path:* Sequence of operations and processes in which a simple system is transferred from an initial to a final state.

An operation is performed only after the equilibrium is attained in the previous step. As a result, the system is transformed from an initial to a final state while the total measure of irreversibility is

$$\sum_{k=1}^{N} DS_k . \tag{1.14}$$

In general, the initial and final states do not uniquely specify the path. In particular, the increase of the number of steps N may lead to a decrease of the measure of irreversibility (14). The path is called reversible if

$$\lim_{N\to\infty} \sum_{k=1}^{N} DS_k = 0. \tag{1.15}$$

We shall show later that reversible paths do exist in this idealized limiting sense.

D b4 *Gibbs space associated with a simple system:* the $(r + 2)$-dimensional space spanned by the variables $X_1, X_2, \cdots, X_{r+1}, S$. The Gibbs space of a composite system is the direct product of the Gibbs space of its subsystems. It is spanned by the variables X_i^a, $i = 1, 2, \cdots; a = 1, 2, \cdots;$ S^a, $a = 1, 2, \cdots$.

D b5 *Extensive variables:* variables proportional to the size of the system. Primarily, the additive invariants and the entropy.

C. Definitions and Postulates on Temperature

D c1 The *temperature* T is defined as

$$T = (\partial S/\partial U)^{-1}. \tag{1.16}$$

P c1 The *range* of the *temperature* is

$$0 < T < \infty. \tag{1.17}$$

Zero and infinite temperatures are admitted for limiting considerations.

P c2 The entropy of every thermodynamic system in equilibrium vanishes in the limit of vanishing temperatures:

$$\lim_{T \to 0} S = 0. \tag{1.18}$$

P c3 The internal energy has a lower bound that is reached at $T = 0$.

D. Definitions and Postulates on Phase Equilibrium

P d1 *Phase postulate.*

The structural elements for building up simple systems in equilibrium are spatially homogeneous extensions of matter called *phases.* A system of given composition exists potentially in a number of phases, each of which is specified by a phase-entropy function or a *primitive fundamental equation*

$$\mathcal{S}^{(\alpha)} = \mathcal{S}^{(\alpha)}(X_1^{(\alpha)}, X_2^{(\alpha)} \cdots X_{r+1}^{(\alpha)}; \eta_1^{(\alpha)}, \eta_2^{(\alpha)}, \cdots), \tag{1.19}$$
$$\alpha = 1, 2, \cdots$$

expressing the entropy of the phase α as a continuous first order homogeneous function of the additive invariants, admitting continuous higher derivatives with respect to these variables. The functions (19) represent $(r + 1)$-dimensional hypersurfaces, called *primitive* or *phase* surfaces in Gibbs space. Each function is defined within a characteristic range of additive invariants. (More precisely: in a characteristic range of the densities defined in *D d2*.) In some cases the functions depend also on additional quasi-thermodynamic parameters $\eta_1, \eta_2, \cdots$. These are introduced for homogeneous states of matter that have identical densities and are nevertheless distinguishable by virtue of their intrinsic symmetry properties (e.g., mirror-image states). The parameters are identified as translational invariants in Appendix B. The set of all primitive surfaces contains all of the thermostatic information concerning a system of given composition.

D d1 *Scale factor:* one additive invariant, say X_{r+1}, singled out to specify the size of the system. Unless otherwise stated, the scale factor is chosen as the volume: $X_{r+1} = V$.

D d2 *Densities:*

$$\begin{aligned} x_i^{(\alpha)} &= X_i^{(\alpha)}/X_{r+1}^{(\alpha)}, \qquad i = 1, 2, \cdots r, \\ s^{(\alpha)} &= \mathcal{S}^{(\alpha)}/X_{r+1}^{(\alpha)}, \end{aligned} \tag{1.20}$$

where $X_{r+1}^{(\alpha)} = V^{(\alpha)}$. The intrinsic, size-independent properties of a phase are represented by the r-dimensional primitive surface

$$s^{(\alpha)} = s^{(\alpha)}(x_1^{(\alpha)}, x_2^{(\alpha)}, \cdots x_r^{(\alpha)}; \eta_1^{(\alpha)}, \eta_2^{(\alpha)}, \cdots). \tag{1.21}$$

We shall refer to the space spanned by the variables (20) also as Gibbs space. If necessary, it may be referred to as the $(r + 1)$-dimensional Gibbs space compared with the $(r + 2)$-dimensional space of *D b4*.

It is possible to choose a *mole number* or the *total mass* as a scale factor. This is particularly convenient in one component systems. The quantities (20) then become *molar* or *specific quantities*.

D d3 *Modifications of a phase:* equilibrium states corresponding to different regions of the surface (1.19); a region may shrink to a single point. Two modifications are *distinct* if they have different densities, or if they differ in the parameters η.

Examples: liquid and vapor are two modifications of the fluid phase. Right and left quartz have identical densities and are distinguished only in terms of the parameters η.

Note that in current usage the terms "phase" and "modification" are used as synonyms.

D d4 *Virtual heterogeneous state:* obtained through the decomposition of an isolated simple system into disjoint volume elements under observation of *P a1* and *P a2*:

$$\sum_\beta X^{(\beta)} = X_i, \quad i = 1, 2, \cdots, r + 1, \tag{1.22}$$

where each term corresponds, in ordinary space, to a volume element $X_{r+1}^{(\beta)} = V^{(\beta)}$, and in Gibbs space to a point on one of the phase surfaces, i.e., to a modification of the corresponding phase. The modifications in different volume elements are distinct in the sense of *D d3*.

P d2 *Entropy maximum principle.* (Heterogeneous systems)

An isolated simple system tends toward a heterogeneous equilibrium state that is singled out among the virtual heterogeneous states (*D d4*) by means of the extremum problem

$$S(X_1, X_2, \cdots, X_{r+1}) \\ = \max \{ \sum \mathcal{S}^{(\beta)}(X_1^{(\beta)}, X_2^{(\beta)}, \cdots, X_{r+1}^{(\beta)}; \eta_1^{(\beta)}, \eta_2^{(\beta)}, \cdots) \}. \tag{1.23}$$

The comparison states for this maximum problem are the virtual heterogeneous states defined in *D d4*.

The remarks following *P b1* apply also in this case, and we state that the solution of the maximum problem (23) is *almost unique*. The precise meaning of this term will be explained later.

C d1 Let the maximum of (23) be attained for the heterogeneous state

$$\bar{X}_1^{(\kappa)}, \bar{X}_2^{(\kappa)}, \cdots, \bar{X}_{r+1}^{(\kappa)}; \qquad \eta_1^{(\kappa)}, \eta_2^{(\kappa)}, \cdots, \qquad \kappa = 1, 2, \cdots, m, \tag{1.24}$$

where m is the number of distinct modifications in the actually realized heterogeneous state. If $m = 1$, the state degenerates into a homogeneous equilibrium; only $m \geqq 2$ corresponds to a heterogeneous state in the strict sense of the word.

The translational invariants η are not conserved, and are unconditionally varied in the maximization process.

The expression (23) for the total entropy of the system can be written also as

$$S(X_1, X_2, \cdots, X_{r+1}) = \sum_{\kappa=1}^{m} \mathcal{S}^{(\kappa)}(X_1^{(\kappa)}, X_2^{(\kappa)}, \cdots, X_{r+1}^{(\kappa)}; \eta_1^{(\kappa)}, \eta_2^{(\kappa)}, \cdots). \tag{1.25}$$

Equations (23) and (25) express the connection between the fundamental equation (9) and the primitive fundamental equation (19).

The heterogeneous equilibrium of a simple system is reminiscent of the composite system in which the distinct properties of the subsystems are maintained by passive forces (*D a5*). In order to emphasize the similarities and dissimilarities of these cases, we introduce the following terminology:

D d5 *Active forces:* phenomenological designation of the sum total of those microscopic factors that bring about and maintain the asymptotically constant values of the free variables in composite and heterogeneous systems.

C d2 It is evident that *the thermostatic formalism does not distinguish between two systems, the variables X_i of which are held constant by active or by passive forces.*

E. Criteria of Stability

According to *P b1* and *P d2*, thermodynamic equilibrium is formally associated with the maximization of entropy. The mathematical characterization of the entropy function around its maximum leads to a classification of the states of equilibrium with respect to stability.

Let $S_0(X_{10}, X_{20}, \cdots, X_{r+10})$ be a "point" in Gibbs space (*D b4*) corresponding to the equilibrium of a simple system, and $\delta X_i = X_i - X_{i0}$, $i = 1, 2, \cdots, r$ a virtual displacement (*D a10*) leading to a constrained equilibrium. Because of the insertion of constraints, we now have a composite system. The corresponding virtual entropy change is

$$\Delta S = \delta S + \delta^2 S + \cdots, \tag{1.26}$$

where the terms on the right correspond to the first-, second-, etc. order expansion terms in the virtual displacement.[9] Equilibrium in the widest sense of the word is characterized by the condition

$$\delta S = 0, \tag{1.27}$$

which is valid for all infinitesimal virtual displacements.

The existence of a virtual displacement with $\delta S < 0$, implies that there is a virtual displacement with the opposite sign for which $\delta S > 0$, hence the entropy can actually increase, and there is no equilibrium.

The equilibrium states can be classified in terms of the following relations:

$$\Delta S \leqq 0, \tag{1.28}$$

$$\delta^2 S < 0, \tag{1.29}$$

$$\delta^2 S = 0, \tag{1.30}$$

$$\delta^2 S > 0. \tag{1.31}$$

We define:

D e1 *Equilibrium:* Equation (27) is valid for all (virtual) displacements.

D e2 *Stable equilibrium:* Equations (27) and (28) are valid for all displacement. The stable equilibrium is *normal* if (29) is valid for all displacements, it is *critical*, if (30) is valid for some displacements. The latter are called *critical displacements.*

D e3 *Metastable equilibrium:* Equations (27) and (29) are valid for all displacements; (28) is invalid for some displacements.

D e4 *Essential instability:* Equation (31) is valid for some displacements.

[9] We denote infinitesimal and finite virtual displacements by δ and Δ, respectively.

II. DISCUSSION OF THE POSTULATIONAL BASIS

Among the most important concepts of thermodynamics is that of equilibrium, an elusive concept, since there are no purely observational means for deciding whether an apparently quiescent system has actually reached equilibrium, or is merely stranded in a non-equilibrium state while imperceptibly drifting toward equilibrium.

The solution of this difficulty, which is developed in the present paper, is different from the one chosen in the fundamental considerations of the classical theory. Yet, our approach is much closer to the point of view that is implicit in practical procedures.

The classical fundamental approach insists that the basic postulates be of universal validity even if this requirement can only be achieved by advancing comparatively weak statements. Thus, in essence, the second law asserts the impossibility of processes in which systems drift away from equilibrium, but it does not claim that equilibrium is actually reached. Although this procedure has proved fruitful of results, it has some shortcomings. The rigorous implications of the second law are only inequalities, and the derivation of the important thermodynamic equalities always involves the assumption that equilibrium is actually reached, whether or not this is explicitly stated as a postulate.

In this paper, we postulate equilibrium in a much stronger form (*P b1* and *P d2*). Starting from these assumptions, we arrive at a theory of normal equilibrium behavior. In general, the theory does not predict the actual behavior of individual systems, but provides primarily a practical criterion for deciding whether or not equilibrium has been reached in the concrete case. Of course, if past experience justifies us in expecting equilibrium in a given situation, the theory has a very precise predictive value.

The difference between the present and the classical points of view comes to sharp focus in the discussion of phenomena near absolute zero. The absolute predictions of the classical theory become particularly vague because the occurrence of frozen-in nonequilibrium states is very common. The standard discussion of these phenomena is further complicated by the fact that the incomplete equilibrium states can be of a great variety. Thus in a solid that is in equilibrium with respect to the phonon distribution, there may, or may not, be equilibrium with respect to the orientation of molecules, radicals, electronic or nuclear spins, or with respect to the isotopic distribution.

The questions concerning the level at which equilibrium is established can be answered without ambiguity by a systematic application of the present point of view. In fact, the phenomena near absolute zero provide a unique opportunity for putting the concept of equilibrium on a firm basis.

By combining a quantum statistical calculation of the entropy in the ideal gas limit with experimental data, as outlined in Section VII, it is possible to check whether or not the strong version of Nernst's heat theorem (*P c2*) is satisfied

An affirmative answer assures us that equilibrium is indeed established. Moreover, if we remember the fact that entropy is relative to a particular choice of independent components (*C a3*), the procedure can be carried out with various choices of components, with or without taking cognizance of, say, nuclear spin and isotopic differences, and therefore reliable answers are obtained to our questions concerning the level on which equilibrium is established.

This procedure, described here very sketchily, has been extensively used in practice and yielded valuable information not otherwise attainable (*15*).

There is at present a virtually general agreement about the practical applications of Nernst's theorem, called also the third law of thermodynamics (*16*). Nevertheless earlier controversies linger on in formulations that involve weaker statements in order to lay claim to universal validity.

Three of the most common statements of this sort are, in substance, as follows:

(i) It is impossible to attain absolute zero.

(ii) Only entropy differences, rather than the absolute entropies, are required to vanish as $T \rightarrow 0$.

(iii) The absolute entropy vanishes in crystalline pure substances.

Statements (i) and (ii) are undoubtedly correct, but too weak to provide the foundation for the practical investigations of low temperature equilibrium mentioned above. Statement (iii) is, in general, not correct, if it is meant as the prediction of the *actual*, rather than of the *normal equilibrium* behavior. In some respects it is not restrictive enough, since there are pure crystalline substances such as ice, which exhibit a finite zero point entropy because of frozen-in orientational disorder (*17*). In other respects it is too restrictive, because ordered alloys and liquid helium are needlessly excluded.

We call attention also to a semantic difficulty, which may have complicated the discussion of the third law. The standard term "absolute entropy" has a double meaning. On the one hand, the entropy of *P c2* is "absolute", in the sense that it involves the *entropy constant* provided by quantum statistics and not only the empirical *entropy differences*. On the other hand, the entropy is *relative* to the choice of independent components (*C a3*). Far from causing difficulties, this circumstance renders the use of *P c2* particularly effective in providing subtle structural information about the level of equilibrium.

We turn now to a methodological problem connected with the fact that the thermostatic formalism deals only with states of equilibrium while nontrivial results can be obtained only if processes are considered at least to a limited extent. The nature of this problem is illustrated by the following paradox: How are we to give a precise meaning to the statement that entropy tends toward a maximum, whereas entropy is defined only for systems in equilibrium? Thus in an isolated simple system (*D a1*, *D a11*) the entropy is constant, if it is defined at all.

This difficulty is resolved in a natural way by the artifice of composite systems

(*D a2*) that enables us to deal with more or less constrained equilibria. Other methods involving phenomena of phase and chemical equilibrium to define constrained equilibria will be considered below.

Gibbs made no explicit use of the concept of composite systems, and his formulation of the increase of entropy is not without a certain obscurity.

The concept of composite systems plays an essential part in Carathéodory's theory (*3*) and is widely used in more recent works (*18*). However, the systematic incorporation of this concept into the foundations of the theory necessitates a considerable revision of the classical conceptual framework. This is evident from the set of definitions grouped under Ia. In particular, we have to consider *thermodynamic operations* as well as *thermodynamic processes*. The latter are always irreversible, and quasi-static processes, basic in the traditional theory, appear now as rather complicated limiting constructs (*Db 3*).[10]

The device of composite systems allows us to formulate precise theoretical statements that parallel closely the actual experimental operations. However the quantum-mechanical interpretation of the distinction between thermodynamic processes and operations raises some new problems to be discussed elsewhere.

From the phenomenological point of view, the entropy maximum principle (*P b1*, *P d2*) is so thoroughly corroborated by experiment that we are confident in interpreting any deviation in an actual case as an indication of incomplete thermodynamic equilibrium. The situation is different if we envisage observations on a finer scale and take cognizance of fluctuations. The entropy maximum principle states, in essence, that the continuum of virtual states (*D a9*, *D d4*) is narrowed down by the active forces (*D d5*) to an "almost" uniquely defined actual macroscopic state. (This statement is explained in a remark following *P b1* and in greater detail in Section IX.)

The actual occurrence of fluctuations indicates that this picture is oversimplified[11]: the virtual states have more reality, even in equilibrium, than has been asserted by MTE. In STE (*14*) the phenomenological *P b1* is replaced by statistical postulates. Equilibrium is no longer a state in the thermostatic sense, but rather a distribution over all virtual states. In spite of its statistical character, this theory is still closely related to MTE.

[10] The fact that the thermodynamic processes triggered by thermodynamic operations are irreversible, suggests the possibility of a natural generalization of MTE to irreversible thermodynamics. In the latter theory, just as in MTE, we consider the transition from a more constrained to a less constrained equilibrium. However, in MTE, we deal only with the initial and final equilibrium states, ignoring both the intermediate nonequilibrium situations and also the time delay involved in the transition.

[11] We shall see below that critical phenomena manifest themselves as singularities in the mathematical formalism: a danger signal that thermostatics is not entirely correct, appearing already within the macroscopic theory.

According to the postulates of Section Ia, the basic independent variables of the theory are additive invariants (*D a4*), and thermodynamic processes are defined as exchanges of these scalar invariants among coupled systems (*D a3*, *D a7*). The fact that homogeneous phases are the structural elements for building thermodynamic systems (*P d1*) is another aspect of additivity.

The requirement of additive invariance is a characteristic feature of the present theory, in which thermodynamic systems are constructed from spatially disjoint subsystems coupled by the exchange of conserved quantities. Additive invariance is of as much importance to the present formalism as the assumption of nondissipative forces in analytical mechanics. We develop an "analytical thermodynamics" formulated in terms of additive invariants; effects not fitting into this framework are at first excluded, although they can be handled by specialized considerations and corrections.

In *P a3* we have listed the variables that we wish to use in MTE. We have to consider, now, the conditions under which these variables are actually additive invariants. Moreover we have to investigate the status of those elastic, electric, magnetic variables which are not listed in *P a3* although they are generally considered as extensive thermodynamic variables.

The only geometric variable listed in *P a3* is the volume. This means that in the basic theory we neglect surface phenomena and shape-dependent effects. It goes without saying that capillary effects have to be included in a realistic theory of phase equilibrium, and elastic strains are important for the phase equilibrium of crystals. The basic theory of this paper will have to be corrected to include these additional effects.

The volume is obviously an additive variable. For the time being we consider it only as a scale factor (*D d1*), serving merely to define the size of the system. The volume will be introduced as a full-fledged independent variable in Section IV, and we shall consider its "conservation" in Section V.

We turn to the discussion of the internal energy U; this is a variable of unique importance that is relevant for every thermodynamic system. The internal energy is obtained if the total energy is evaluated in the barycentric coordinate system, and if the potential energy arising from external fields is subtracted. For the purposes of MTE, macroscopic kinetic energies will be ignored. We shall assume also the absence of external fields, say, due to gravitation. For small systems, in which the potential energy differences due to internal displacements of matter are negligible compared with the internal energy, the potential energy represents only an irrelevant additive constant. In large systems, however, a gravitational field impairs the spatial homogeneity of the system. It is therefore appropriate in a theory dealing primarily with homogeneous phases to assume the absence of external fields. The generalization to inhomogeneous systems built up from homogeneous volume elements presents no difficulties.

Within the approximations stated there is no distinction between "energy" and "internal energy" and the principle of energy conservation can be invoked to support the validity of the conservation postulate *P a1* for the internal energy. Henceforth, we shall refer to the latter briefly as "energy".

The situation is different for the additivity postulate *P a2*, the validity of which cannot be inferred from general principles. We have to require that the interaction energy between thermodynamic systems be negligible. This assumption is closely related to the homogeneity postulate *P d1*. From the molecular point of view, additivity and homogeneity can be expected to be reasonable approximations for systems containing many particles, provided that the intermolecular forces have a short range character. This conjecture is confirmed by the following phenomenological considerations: As we shall see, thermostatics implies that the stability of phases is an intrinsic property, independent of the size and shape of the system. This result, which follows essentially from additivity (*P a2*) and homogeneity (*P d1*), will be called the "scale invariance" of homogeneous phases. It provides a convenient criterion for gauging the limits of validity of *P a2* and *P d1*.

Within the range of common physical-chemical experimentations scale invariance is satisfied to such an extent that it is usually taken for granted: e.g., equations of state are given for unit amounts of each substance and scaled to any size of interest. Yet, scale invariance breaks down and stability becomes size-dependent in small droplets because of surface effects, in stars because of long-range forces, and in heavy atomic nuclei because of the combination of both factors. In all these cases the basic theory is to be modified.

We turn now to the discussion of the variables describing the distribution of matter. In the absence of chemical reactions, the mole numbers of the chemical species are additive invariants. In the presence of chemical reactions, these numbers do not satisfy the conservation law *P a1*. However, it is possible to define chemical components, the mole numbers of which are invariants of the chemical reactions, and are appropriate thermostatic variables.

The existing presentations of thermodynamics are not as restrictive in the choice of extensive variables as our *P a3*. Instead of requiring additive invariance, the guiding principle is to account for the complete energy balance, and all variables that are needed for describing the performance of work on or by the system are considered. Thus the electric and magnetic polarization vectors, and the components of the eastic strain tensor are joined to the set of independent variables.

We shall refer to these additional extensive variables of the conventional theory as *pseudo-thermodynamic variables*. The variations of these variables describe the performance of work on the system, but these processes cannot be formalized as exchanges of conserved quantities. The pseudo-thermodynamic variables, e.g., magnetic and electric vectors, are subject to boundary conditions that do not conform to the simple type of constraint (1.3). Consequently, the translation of the theorems of MTE to magnetic systems does not always yield correct theorems. The thermodynamics of magnetic systems will not be systematically considered in this paper. In fact, it is advisable to develop the thermodynamics of magnetic systems within a theory that has a certain amount of autonomy. In this theory one has to identify the thermostatic results that can be translated automatically to magnetic systems. Moreover, entirely new problems have to be considered such as the equilibrium of magnetic phases, and, in particular, ferromagnetic domains. These problems have complicated shape dependent aspects, quite foreign to basic MTE (*19*).

The same remarks hold for electric, elastic and capillary systems. None of these will be considered in the present paper.

One of the central ideas of MTE is that spatially extended homogeneous portions of matter are the structural elements for the construction of thermodynamic systems (*P d1*). The spatial homogeneity of phases is taken in thermostatics as an exact mathematical statement. In reality, of course, homogeneity is limited for more than one reason. Because of atomic structure, homogeneity is to be understood in the "coarse-grained" sense. (see Appendix B.) A limitation of homogeneity arises also because of the long range interactions discussed above in connection with the additivity of the energy. All of the limitations mentioned leave a wide margin of validity for the homogeneity assumption.

An important application of the phase concept is that it leads to an alternative formulation of the entropy maximum principle (*P d2*). The requirement that a system be in a definite phase is comparable to a constraint imposed in composite systems by restrictive partitions.

In comparing the maximum principles *P b1* with *P d2*, and the entropy function (fundamental equation) with the phase-entropy function (primitive fundamental equation), we are struck by the similarity of the corresponding concepts. It is indeed a characteristic feature of the thermostatic formalism that the same formulas can be used with different sets of variables appropriate to a more or less detailed description of the system. Thus, if the entropy function is expressed as a function of a number of free variables, some of these can be eliminated by maximizing the entropy function with respect to the variables in question, and inserting for these their constant extremal values. Consequently the entropy appears as a function of a reduced number of variables. For example, it would be possible to make our formalism even more detailed and express the primitive fundamental equation in terms of the mole numbers of the chemical species. By maximizing this expression for constant values of the mole numbers of the components, one obtains the equilibrium values of the chemical species. For molecular reactions, these results are standard in chemical thermodynamics, and we do not consider them in this paper.

Granted the similarity of *P b1* and *P d2*, we note the difference that the phase-entropy function (the primitive fundamental equation) is entirely regular with continuous higher derivatives, whereas the second and higher derivatives are only piecewise continuous for the entropy function. These singularities of the entropy function follow from *P d2* and, in fact, the main content of thermostatic phase theory is to derive the topological properties of the sets of singular points in Gibbs space.

The distinction between the primitive fundamental equation and the fundamental equation was recognized by Gibbs. His terminology was "primitive surface" and "surface of dissipated energy". The rationale for the last expression will be explained in Section IV. This distinction between the two surfaces is essential, to account for the phenomenon of metastable phases.

We shall see in Section VII that the phase-entropy function can be established from experimental data,[12] and theoretically, it can be obtained from quantum mechanics, a problem to be discussed elsewhere.

We have seen that MTE is more restrictive in its choice of independent variables than the traditional theory, and we have excluded, at least from the basic MTE, the discussion of pseudo-thermodynamic variables.

In another respect, however, the present theory operates with more variables: namely, with the quasi-thermodynamic parameters of *P d1*, to account for the distinctness of modifications that have identical densities. While the pseudo-thermodynamic variables are bound by more complex constraints than those provided by Eq. (1.3), the quasi-thermodynamic parameters are not bound by any constraints at all. They are discussed in more detail in Appendix A.

The definitions and postulates listed in Ic will be discussed in Section III.

III. THERMAL EQUILIBRIUM

The principal method for drawing inferences from our postulational basis consists of the application of the equilibrium and stability criteria Ie to concrete situations. We consider, at first, the special case of a composite system in which the volume and composition of each subsystem is fixed, and the internal energies $U^{(a)}$ are the only free variables (*D a9*). The problem of establishing the equilibrium distribution of the internal energy is commonly known as that of thermal equilibrium (see *D a12*). The discussion of this case provides an opportunity to clarify the meaning and the limits of validity of the postulates listed in I, C.

Let us consider an isolated composite system divided by diathermic rigid partitions into subsystems of constant volume and composition. Thus the extremum problem (1.10) reduces to

$$S(U) = \max \sum_a S^{(a)}(U^{(a)}), \tag{3.1}$$

[12] Provided the phase is stable or at least metastable. The absolute value of the phase entropy is obtained from a quantum statistical calculation.

under the constraint

$$\sum_a U^{(a)} = U. \tag{3.2}$$

For the sake of simplicity, we have suppressed the fixed variables $V^{(a)}$ and $N_k^{(a)}$.

Let $\delta U^{(a)}$ be a virtual displacement leading from the equilibrium values $U^{(a)}$ to the constrained equilibrium $U^{(a)} + \delta U^{(a)}$ with

$$\sum \delta U^{(a)} = 0. \tag{3.3}$$

The condition of equilibrium (*D e1*) is

$$\delta S - \lambda \sum \delta U^{(a)} = \sum_a \left(\frac{\partial S^{(a)}}{\partial U^{(a)}} - \lambda \right) \delta U^{(a)} = 0, \tag{3.4}$$

where λ is the Lagrange multiplier associated with the constraint (3). Hence the condition (4) reduces to

$$\frac{\partial S^{(a)}}{\partial U^{(a)}} = \lambda, \qquad a = 1, 2, \cdots. \tag{3.5}$$

The parameter λ is common to all systems that are in thermal equilibrium with each other. It is advantageous from the conceptual point of view to define the temperature as λ^{-1}, i.e., as an intensity that stipulates thermal equilibrium without reference to the properties of any particular system. According to our original definition (*D c1*) the temperature is a zero order homogeneous function of the additive invariants of individual systems. Of course, Eq. (5) ensures the consistency of the two points of view.

Let us now consider a system consisting of two isolated subsystems of slightly different temperatures

$$T' \gtrsim T''. \tag{3.6}$$

After removing the constraint, an actual process associated with an increase of entropy takes place:

$$dS = dS' + dS'' = \frac{\partial S'}{dU'} dU' + \frac{\partial S''}{dU''} dU'' = \left(\frac{1}{T'} - \frac{1}{T''} \right) dU' > 0. \tag{3.7}$$

In view of (6) and *P c1*, we have

$$dU' < 0. \tag{3.8}$$

Hence energy is displaced from the high to the low temperature system. The energy transferred is a heat quantity $đQ$ (*D a12*). We have

$$\frac{1}{T} (dU)_{V,N} = \frac{1}{T} đQ = dS, \tag{3.9}$$

the well-known relation indicating that T^{-1} is the integrating factor of the incomplete differential $đQ$.

The inequality (8) is valid for

$$T' > T'', \tag{3.10}$$

regardless of the magnitude of $T' - T''$, although the expansion (7), broken off after the first term, fails in this case. The flow of energy between the systems produces changes in their temperatures. We claim that the temperature increases, (to be more precise, it does not decrease) for the system that receives the energy:

$$(\partial T/\partial U)_{V,N} \geqq 0. \tag{3.11}$$

This relation can be proved formally from the stability criterion (1.28), as we shall see in Section VIII where the stability problem will be discussed for the general case of many variables. In order to avoid repetition, we invoke at this point only an intuitive argument in favor of (11). This relation assures us that the actual process (8) leads to an equalization of the temperatures T' and T'', and hence to equilibrium. If (11) were violated, the process (8) would produce increasingly different temperatures and violate the condition (5), and *P b1*.

Combining (11) and *D c1*, we have

$$\frac{\partial^2 S}{\partial U^2} = -\frac{1}{T^2}\frac{\partial T}{\partial U} \leqq 0, \tag{3.12}$$

and thus the entropy versus energy curve is concave to the U axis.

We define the *heat capacity at constant volume*

$$C_V = (\partial U/\partial T)_V = T(\partial S/\partial T)_V\,. \tag{3.13}$$

The heat capacity per mole or per unit mass is called *molar heat* and *specific heat*, respectively.

Condition (12) is now written as

$$0 \leqq C_V \leqq \infty. \tag{3.14}$$

In Eq. (13) the temperature is considered to be the independent variable. Thus it is implied that (1.16) which provides the function $T = T(U)$ can be inverted to give $U = U(T)$. This depends on the condition

$$-\frac{1}{T^2}\frac{\partial T}{\partial U} = \frac{\partial^2 S}{\partial U^2} \neq 0 \tag{3.15}$$

or

$$C_V \neq \infty. \tag{3.16}$$

Thus complications may be expected for $C_V = \infty$, that is if (11) is to be taken with the equality sign. We shall consider such singular situations later.

We define the limiting concept of a heat reservoir as a system, the size of which tends to infinity, as compared with the finite systems that are being studied. With the size of the system its extensive variables, in particular U and S, tend to infinity and reservoirs are specified only in terms of their temperatures. If a finite system is coupled to a reservoir, the temperature of the latter stays constant and the same temperature is reached asymptotically by the finite system.

We turn now to the discussion of the postulates listed in Ic. Postulates *P c2* and *P c3*, although phrased in phenomenological terms, are expressions of the quantum mechanical nature of thermodynamic systems.[13] We pointed out already in Section II that *P c2* is of fundamental importance for the concept of equilibrium. We shall return to this question later.

Until recently most texts of thermodynamics have implicitly assumed that the thermodynamic temperature is positive (*P c1*). This statement appears as a theorem in the presentation of Landau-Lifshitz (*20*). It is proved for systems in which the conversion of the internal energy into translational kinetic energy is a virtual process. If such systems had a negative temperature: $\partial S/\partial U < 0$, an increase of the entropy would be brought about by converting the internal energy into the kinetic energy of internal macroscopic motion. This is a violation of *P b1*.

We should note in this context that Ramsay (*21*) recently demonstrated that negative absolute temperatures are useful for the description of the properties of certain rather unusual systems. The thermodynamic system is a collection of nuclear spins, thermally isolated from the lattice occupying the same region of space. In Ramsay's system the proof mentioned above fails, since there is no conversion of internal energy into molecular kinetic energy (*22*). The concept of two isolated systems (spin and lattice) occupying the same region of space is in conflict with our *D a1*. As in a number of other instances, our basic theory is restricted to a narrower class of systems, in order to yield a larger number of theorems. However, our procedure is consistent with the use of different definitions to suit specialized situations.

After relating the various types of thermostatics definitions of the temperature to each other, the question of measurement and the establishment of an absolute scale is still left open. We shall briefly indicate in Section VII that this

[13] The quantum-mechanical character of the third law of thermodynamics is well known. On the other hand, it is often stated that the lowest state of classical systems is the crystalline state. Actually, from the classical point of view, a collection of nuclei and electrons has no stable state, and the energy would tend to negative infinity due to the collapse of opposite charges.

question is soluble in a practical fashion. Meanwhile, we anticipate the result that temperature is measurable on an absolute scale.

Finally, we consider the procedure of heating a thermodynamic system from an initial temperature T_i to a final temperature T_f, and illustrate the existence of different paths characterized by their measure of irreversibility (*D b1* to *3*). Let us first perform the heating in the simplest manner by coupling the system with a heat reservoir of temperature T_f.

The measure of irreversibility defined in (1.13), in this case, is

$$DS = \int_{T_i}^{T_f} \frac{C_V}{T}\, dT - \frac{1}{T_f} \int_{T_i}^{T_f} C_V\, dT , \tag{3.17}$$

where according to (13) the first term represents the entropy increase of the system, the second is the entropy decrease of the reservoir. The net entropy change is obviously positive.[14]

Let us consider now the same initial and final states of the system and perform the transition over a path consisting of n steps by coupling the system consecutively with n reservoirs of temperatures $T_1, T_2, \cdots, T_n$, where

$$T_i < T_1 < T_2 < \cdots < T_n = T_f .$$

The contact with the reservoir T_k is to be brought about after the system has reached its equilibrium at T_{k-1}. The total measure of irreversibility is now

$$DS^{(n)} = \int_{T_i}^{T_f} \frac{C_V}{T}\, dT - \sum_{k=1}^{n} \frac{1}{T_k} \int_{T_{k-1}}^{T_k} C_V\, dT, \tag{3.18}$$

hence,

$$\lim_{n\to\infty} DS^{(n)} = 0. \tag{3.19}$$

Thus the path is reversible in the ideal limit in which it is performed in infinitely many infinitesimal steps (*D b3*).

IV. THE ENERGY SCHEME

It is an essential feature of MTE that the fundamental equation and the associated stability considerations can be cast in different forms that depend on the choice of independent variables. Although these forms are all consistent with each other, their suitability for solving various problems differs greatly. Thus before extending the discussion of the last section to the general stability problem, we have to expand the basic formalism of the theory.

As a first step toward developing a systematic transformation theory we solve

[14] In the special case in which C_v is constant in the interval $T_i \leqq T \leqq T_f$, the entropy increase is represented geometrically as an area in the diagram in which C_v/T is plotted against T.

the fundamental equation

$$S = S(U, X_2, \cdots, X_{r+1}) \tag{4.1}$$

for the energy:

$$U = U(S, X_2, \cdots, X_{r+1}). \tag{4.2}$$

This procedure is applicable also to the primitive fundamental equation and Eq. (1) may represent either (1.9), (1.19), or (1.21). However, we omit the superscripts if the identification of subsystems or phases is not essential.

We shall refer to the formalism based on (1) and (2) as the *entropy* and *energy scheme*, respectively. Although, formally, the transition from (1) to (2) consists merely of the rotation of the Gibbs space resulting in the interchange of the roles of S and U as dependent and independent variables, there are essential differences in the physical interpretation of the two schemes.

First, we note that Eqs. (1) and (2) are mathematically equivalent in the following sense: The power series expansion of Eq. (1) around a "point" U_0, S_0, can be resolved to give the power series expansion of Eq. (2), provided that the condition

$$(\partial S/\partial U)_0 \neq 0 \tag{4.3}$$

is satisfied. Conversely, Eq. (1) can be regained from Eq. (2) if

$$(\partial U/\partial S)_0 \neq 0. \tag{4.4}$$

According to (3) the energy scheme is not unique for $T \to \infty$, a limitation of little consequence. The neighborhood of absolute zero is of greater practical importance. At first sight, the breakdown of condition (4) would seem to indicate here a less than complete equivalence of the two schemes. Given the energy fundamental equation, the entropy function can be established only with a large uncertainty. However this difficulty is removed by *P c2* that assigns a vanishing entropy to all systems in equilibrium for $T \sim 0$.

We proceed now to compare the thermodynamic processes in the two schemes. Considering the exchange of additive invariants between two subsystems of a composite system, we obtain the virtual processes:

$$\Delta S' + \Delta S'' = 0, \qquad \Delta X_i' + \Delta X_i'' = 0, \qquad i = 2, 3, \cdots, r + 1. \tag{4.5}$$

If we compare these processes with those of Eq. (1.5c), we see that the conservation of energy has been replaced with that of the entropy. The latter condition means that the process is reversible. The nonconservation of the energy is not objectionable for virtual processes. Nevertheless, it is intuitively satisfying to introduce an auxiliary system to provide the excess energy $\Delta U' + \Delta U''$ that may be positive or negative. In order to avoid any irreversible entropy increase, this system is chosen as a "work reservoir," an idealized device that is not the seat of thermal effects. Of course, in *actual* processes the presence of the work reservoir is necessary to maintain the energy balance.

It is interesting to compare the virtual processes (5) with those of the entropy scheme (1.5c). In the second scheme the energy is conserved along with the other additive invariants, while the entropy is allowed to vary. Whereas in the entropy scheme composite systems were essential for the discussion of processes, in the energy scheme one may also consider simple systems coupled with a work reservoir.

The difference between the processes of the two schemes is illustrated also by the following example: in the energy scheme we have adiabatic volume changes $(\Delta V)_S$, e.g., in sound waves; in the entropy scheme, we have free expansions, $(\Delta V)_U$. Evidently, the two schemes deal with different types of processes. We note also the analogy to the Clausius and to the Kelvin–Planck formulations of the traditional theory.

The stability of a thermodynamic system depends on whether or not any of the virtual processes can actually occur. We have seen that this question is decided by the entropy maximum principle, the counterpart of which in the transformed formalism is the *energy minimum principle*. Stating it loosely; in equilibrium the entropy is a maximum at constant energy, while the energy is a minimum at constant entropy.

In Section I we expressed the entropy maximum principle in three different forms (*P b1*, *P d2*, and *Ie*). We shall confine ourselves to the reformulation of the stability criterion *Ie* and we shall prove the equivalence of the two formulations.

Criterion of Stability. Consider a simple system mentally divided into two subsystems by a nonrestrictive mathematical wall. Transfer this state by a virtual process (5) into a constrained equilibrium that is "trapped" by the insertion of a restrictive wall. *The variation of the volume by a movable piston* is now included among the virtual processes. Let the process be associated with the absorption of the virtual work

$$\Delta W = \Delta U' + \Delta U'' \tag{4.6}$$

from the work reservoir.

We consider the relations that are analogous to (1.26) to (1.31):

$$\Delta W = \delta W + \delta^2 W + \cdots, \tag{4.7}$$

$$\delta W = 0, \tag{4.8}$$

$$\Delta W \geqq 0, \tag{4.9}$$

$$\delta^2 W > 0, \tag{4.10}$$

$$\delta^2 W = 0, \tag{4.11}$$

$$\delta^2 W < 0. \tag{4.12}$$

D f1. *Equilibrium:* Equation (8) is valid for all (virtual) displacements.

D f2. *Stable equilibrium:* Equations (8) and (9) are valid for all displacements. The stable equilibrium is *normal* if (10) is valid for all displacements; it is *critical* if (11) holds for some displacements. The latter are called *critical displacements*.

D f3. *Metastable equilibrium:* Equations (8) and (10) are valid for all displacements, (9) is invalid for some displacement.

D f4. *Essential instability:* Equation (12) is valid for some displacement.

Expression (6) represents the minimum work required to produce the displacement (5) leading from the equilibrium to the constrained equilibrium state.

If the roles of initial and finite states are interchanged, $-(\Delta U' + \Delta U'') = |\Delta U' + \Delta U''|$ is the maximum work (available work) which may be extracted from the system on transferring it from the constrained to the unconstrained equilibrium.

The foregoing criterion is obviously identical in wording to the energy minimum principle of macroscopic physics (mechanical statics and electrostatics) where thermal effects are ignored. Hence the thermostatic energy-minimum principle can be expressed concisely as follows:

The energy-minimum principle of macroscopic physics maintains its validity in thermostatics if the comparison states are provided by reversible virtual processes in which the total entropy is held constant.

We shall prove the equivalence of the criteria. Let us at first assume that a state is instable according to the energy criterion: thus there is a varied state specified by (5) for which $W = \Delta U' + \Delta U'' < 0$ and $\Delta S' + \Delta S'' = 0$. After dissipating the energy $|W|$ in the varied state, the energy is brought back to its original value and, by virtue of *P c1*, the entropy has increased. Hence the state is instable according to the entropy criterion. Conversely, we assume that there is a state with $\Delta S = \Delta S' + \Delta S'' > 0$ and $\Delta U' + \Delta U'' = 0$. Removing the entropy ΔS through contact with a heat reservoir leads to a varied state with $\Delta S = 0$, $\Delta U < 0$, and hence to instability with respect to the energy criterion.

The above proof relies essentially on the assumption that the temperature is positive (*P c1*). Although the discussion of negative temperatures is beyond the scope of this paper, it is instructive to consider such systems for comparison. The positive and negative temperature cases are represented in Figs. 1(a) and 1(b), respectively.

Let us consider a system represented by point A. We interpret this situation as a composite system in a constrained equilibrium, where the states of the separate parts are represented by B and C. The relative amounts of these parts are given by the ratio of the line segments AC/BA (lever rule). If the internal constraint inhibiting the passage of energy is relaxed, the system performs an irreversible process into the state E, and $\Delta S = AE$ is the measure of irreversi-

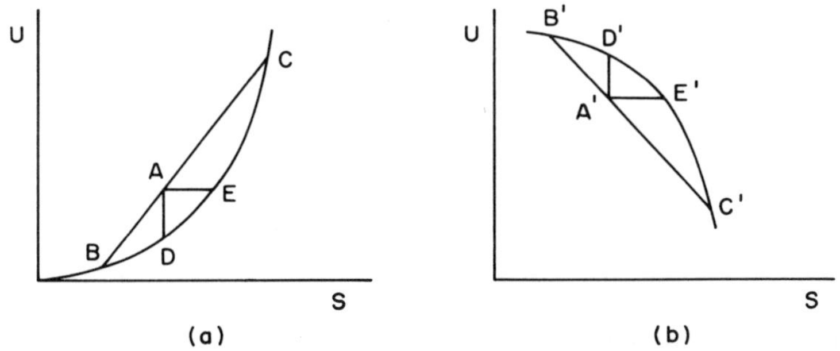

FIG. 1. The fundamental equation, $U = U(S)$, with X_i = constant ($i = 2, 3, \cdots$); (a) positive temperatures (b) negative temperatures.

bility. If the system is coupled with a work reservoir and the process is led reversibly, the maximum available work is represented by $-\Delta U = AD$.

The case for negative temperatures is presented in Fig. 1(b). The irreversible attainment of equilibrium is essentially the same as in the positive temperature case, but the reversible transition to equilibrium requires the work input $\Delta U' = A'D'$.

On the other hand, in the negative-temperature case, the system is capable of evolving spontaneously from B' towards E' by simultaneously increasing its entropy and performing work. However, from the kinetic point of view, this process must proceed slowly (long relaxation time) because otherwise the system would not be found at all in a negative-temperature state.

Finally, we note that Fig. 1(a) has already been used by Gibbs (p. 51 of Ref. 2). However, point A is interpreted there as a system the parts of which are in state of macroscopic motion, and the process $A \to E$ is the dissipation of the kinetic energy. Accordingly Gibbs calls the surface $BDEC$ the "surface of dissipated energy." This approach has the disadvantage that the initial state is not an equilibrium, and strictly speaking, cannot be represented in the diagram.

V. THE INTENSIVE VARIABLES

We write the fundamental equation (4.2) in the differential form

$$dU = \sum_{i=1}^{r+1} P_i \, dX_i \,, \tag{5.1}$$

where the

$$P_i = \frac{\partial U}{\partial X_i}, \qquad i = 1, 2, \cdots, r+1 \tag{5.2}$$

are the intensities conjugate to the extensive variables X_i. The entropy is included here among the extensive variables: $X_1 = S$.

Taking into account the physical interpretation of the variables (*P a3*), we obtain corresponding interpretations for the intensities:

$$dU = TdS - pdV + \sum_{j=1}^{c} \mu_j \, dN_j \, . \tag{5.3}$$

The temperature T is the conjugate of the entropy. This relation is equivalent to *D c1*, a definition that can be expressed by saying that in the entropy scheme $1/T$ is conjugate to U. The definition of the intensities within the entropy scheme will be briefly discussed later.

The chemical potentials μ_j are defined as the intensities conjugate to the mole numbers of the chemical components. The intensity conjugate to the volume is the negative pressure $-p$. This follows from the requirement that in the absence of caloric and chemical changes Eq. (3) should reduce to the well-known energy relation of fluid mechanics.

We shall derive a number of formal properties of the intensive variables, and follow it, at the end of this section, by a few more specific results that are related to their physical interpretation.

We consider a composite system in which one of the variables, say X_k, is free to vary on account of nonrestrictive partitions. In other words, we have the virtual processes:

$$\sum_a \delta X_k^{(a)} = 0$$

and (5.4)

$$\delta X_i^{(a)} = 0, \quad \text{for all } i \neq k \text{ and all } a.$$

Applying the equilibrium condition formulated in the last section, we have

$$\delta W = \sum_a \delta U^{(a)} = 0 \tag{5.5}$$

under the constraints (4). By using the method of Lagrange multipliers, we obtain

$$\delta U - \lambda_k \sum_a \delta X_k^{(a)} = \sum_a (P_k^{(a)} - \lambda_k) \delta X_k^{(a)} = 0. \tag{5.6}$$

Hence for all subsystems coupled with each other through the exchange of X_k we arrive at the equilibrium condition

$$P_k^{(a)} = \lambda_k = \text{constant for all } a. \tag{5.7}$$

In other words the free extensive variables will attain those equilibrium values which ensure that their conjugate intensities are constant. This result is immediately generalized to the case of virtual processes involving the simultaneous variation of two or more extensive variables X_i, provided that the different variations $\delta X_i^{(a)}$ and $\delta X_k^{(a)}$ are independent of each other. The conjugate intensities are always constant over the entire composite system.

In order to appreciate the significance of the intensive variables, it is convenient to single out one subsystem for detailed consideration, and join all the others to provide the "surroundings" of this system. We require also that the over-all size of the surroundings be very large, in the limit infinite, compared with the system of interest. The surroundings is specified by the set of its intensities $P_1, P_2, \cdots$. This description is quite schematic; it ignores the densities of the subsystems constituting the surroundings. The specification is adequate, nevertheless, for deciding whether or not a system is in equilibrium when coupled to this surroundings. Moreover, because of the large size of the latter, the intensities are not influenced by any exchange of quantities between system and surroundings. (This plausible statement is formally proved in Section VIII.) We shall refer to "surroundings" also as (*generalized*) *reservoirs*.

According to Eq. (2) the intensive variables are functions of the extensive variables of the system. Since the intensities are zero order homogeneous functions of the extensive variables, and do not depend on the size of the system, the intensities depend actually only on the densities (1.20). Hence

$$P_i = P_i(x_1, x_2, \cdots, x_r) \qquad i = 1, 2, \cdots, r + 1. \tag{5.8}$$

Relations of this sort that arise through differentiation of fundamental equations we shall call *generalized equations of state*. We shall see that the relations commonly called equations of state arise from (8) through transformations of the variables. However, the common usage is more restricted and does not include the complete set of Eqs. (8). The information contained in this complete set is equivalent to that of the fundamental equation. In fact, applying Euler's relation to the first-order homogeneous equation (4.2), we obtain

$$U = \sum_i^{r+1} X_i \frac{\partial U}{\partial X_i} = \sum_i^{r+1} X_i P_i. \tag{5.9}$$

However, the different equations of state are not independent from each other, but satisfy the so-called Maxwell relations

$$\frac{\partial P_i}{\partial X_k} = \frac{\partial P_k}{\partial X_i}, \qquad i, k = 1, 2, \cdots, \tag{5.10}$$

since these quantities are the mixed second derivatives of the fundamental equation.

Differentiating (9) and subtracting (1), we obtain the Gibbs-Duhem (GD) relation

$$\sum_{i=1}^{r+1} X_i \, dP_i = 0. \tag{5.11}$$

Dividing by the scale factor X_{r+1}, we obtain

$$dP_{r+1} = -\sum_{i=1}^{r} x_i\, dP_i\,. \tag{5.12}$$

We assume that Eqs. (8) can be solved for the densities:

$$x_i = x_i(P_1, P_2, \cdots, P_r), \qquad i = 1, 2, \cdots, r. \tag{5.13}$$

The condition of solubility is the nonvanishing of the Jacobian

$$D_r = \frac{\partial(P_1, P_2, \cdots, P_r)}{\partial(x_1, x_2, \cdots, x_r)} \neq 0. \tag{5.14}$$

Inserting (13) in (9), we obtain P_{r+1} as a function of the other intensities:

$$-P_{r+1} = \psi(P_1, P_2, \cdots, P_r). \tag{5.15}$$

where the minus sign is for convenience.

This function is an integral of the differential equation (12), which indicates that the integrability condition

$$\frac{\partial x_i}{\partial P_k} = \frac{\partial x_k}{\partial P_i}, \qquad i, k = 1, 2, \cdots \tag{5.16}$$

is satisfied. However, a direct integration of (12) does not yield (15) but involves undetermined integration constants.

Conversely we can regain Eq. (8) from (13), provided that

$$D_r^{-1} = \frac{\partial(x_1, x_2, \cdots, x_r)}{\partial(P_1, P_2, \cdots, P_r)} \neq 0. \tag{5.17}$$

Conditions (14) and (17) are significant. We shall see that (14) breaks down for heterogeneous and critical equilibria, and (17) near absolute zero. We shall refer to states in which these conditions are satisfied as *regular*.

We have from (8) and (5)

$$\psi(P) = \sum_{i=1}^{r} x_i P_i - u(x), \tag{5.18}$$

where x and P each stand for the entire set $x_1, x_2, \cdots, x_r$ and $P_1, P_2, \cdots, P_r$, respectively. Thus (15) is the Legendre transform of the fundamental equation $u = u(x)$ expressed in density variables. Obviously, the size of the system cannot be expressed in terms of intensities alone. In regular states the transformation (18) allows one to obtain $\psi = \psi(P)$ from $u = u(x)$ and vice versa, the Legendre transformation preserves the information contained in the fundamental equation.

The basic extremum principle of thermostatics is easily translated into the present context. We define the function

$$w(x'; P) \equiv u(x') + \psi(P) - \sum_1^r x_i'P_i, \tag{5.19}$$

where x' represents the set of densities x_1', x_2', $\cdots$, x_r'. In general the x' are distinct from the x that represent the densities of the system when in equilibrium with the reservoir; we say also that the x and P are "matched." It is easy to see that

$$w(x', P) \geqq 0 \tag{5.20}$$

In fact, according to (18) we have $w(x, P) = 0$ for matching variables. Otherwise

$$w(x'; P) = w(x'; P) - w(x; P) = u(x') - u(x) - \sum_i (x_i' - x)P_i. \tag{5.21}$$

This is the minimum work required to transfer the system from the state x to x' while it is in contact with a reservoir in equilibrium with the initial state. This quantity, of course, must be non-negative, hence (20) is proved. This relation can be considered as a minimum principle in which the x' are varied at constant P, or vice versa.

One often uses fundamental equations in terms of a mixed set of variables, say P_1, P_2, $\cdots$, P_k, X_{k+1}, $\cdots$, X_{r+1}. We shall use occasionally for the corresponding Legendre transform the short notation

$$-U[P_1, P_2, \cdots, P_k] = \sum_{i=1}^k X_iP_i - U(X_1, X_2, \cdots, X_{r+1}) \tag{5.22}$$

More specifically by using the variables made explicit in (3), we have

$$U = TS - pV + \sum_j \mu_jN_j. \tag{5.23}$$

Thus the Gibbs function is

$$G(T, p, N_1, N_2, \cdots, N_c) = U[T, p] = U - TS + pV. \tag{5.24}$$

We shall consider now some of the properties of the intensive variables that are connected with their specific physical interpretation.

In Section III temperature, or rather $1/T$, was seen to govern the level of thermal equilibrium. In the energy scheme the equality of temperature throughout a composite system is the condition of equilibrium with respect to the performance of work, while entropy is reversibly transferred between subsystems of definite temperatures. Such processes are called Carnot cycles. The temperature concept emerging from the analysis of this process is identical to the one established in Section III and we shall not pursue this line of thought any further.

We turn to the discussion of pressure which is the intensity governing mechanical equilibrium. The formal analogy between temperature and pressure is incomplete. Although the mechanical coupling of two fluid systems can often be formalized as an "exchange of volume" between subsystems, this is not necessarily the case for all types of coupling. Thus the volume of such systems could be varied by a separate piston of cross section A' and A'', while the pistons are rigidly connected with each other. The virtual process is now

$$\delta V'/A' + \delta V''/A'' = 0. \tag{5.25}$$

The corresponding equilibrium condition is easily shown to be

$$A'p' = A''p''. \tag{5.26}$$

Only for $A' = A''$ do we get the standard thermostatic situation.

The possibility of a volume coupling of the type (25) was first pointed out by Ehrenfest–Afanassjewa (*23*) in order to emphasize the contrast with the case of temperature, a parameter governing thermal equilibrium independently of the mechanism providing the energy coupling.

In case of a heterogeneous system in which one phase can expand at the expense of the other, we have $A' = A''$ and the general case $A' \neq A''$ is of no importance in the context of this paper. It is nevertheless interesting that the formalization of mechanical interaction, as an exchange of an additive invariant, has its limitations. We return to this point at the end of this section.

We note that the use of adiabatic pistons enables one to consider volume exchange as a one-variable process, provided that one is within the energy scheme. Thus we have

$$\delta V' + \delta V'' = 0, \qquad \delta S' = 0, \qquad \delta S'' = 0. \tag{5.27}$$

Within the entropy scheme one has to consider the two-variable process

$$\begin{aligned} \delta U' + \delta U'' &= 0, \\ \delta V' + \delta V'' &= 0. \end{aligned} \qquad \text{(b)} \quad (5.28)$$

since adiabatic walls do not ensure the constancy of the energies of the subsystems when work is performed through the exchange of volume. In order to ensure $\delta U' = \delta U'' = 0$, the energy lost or gained through work has to be compensated by the right amount of heat in each subsystem. This, of course, cannot be automatically achieved through a constraint.

Finally we consider equilibrium processes, in which there is a displacement of mass, i.e., of mole numbers. The conjugate intensity is the chemical potential. The analogy between chemical potential and temperature governing mass and energy flow, respectively, is quite far reaching. However the chemical potential involves a number of additional features which we shall briefly survey.

For a one-component system the chemical potential $\mu = P_{r+1}$ can be reduced to quantities introduced earlier. Choosing the mole number N as a scale factor, we obtain from (9) and (12)

$$\mu = u - Ts + pv \tag{5.29}$$

and

$$d\mu = -s\,dT + v\,dp. \tag{5.30}$$

Accordingly, the chemical potential, besides being an intensity, is also the Legendre transform of the molar energy, or, in standard terminology, it is identical to the molar Gibbs function. For multicomponent systems, the chemical potential can be identified with the partial molar Gibbs function.

An interesting distinction between energy and mass flow is connected with the availability of restrictive walls. While there exist walls that are restrictive of mass and nonrestrictive of energy and entropy, the converse is generally not true. If a wall is nonrestrictive of at least some chemical species (semipermeable membrane), it is nonrestrictive of entropy and energy also. Consequently, the virtual processes involve simultaneous independent variations of the mole numbers and of the entropy:

$$\begin{aligned} \delta N_i' + \delta N_i'' &= 0, & \text{(a)} \\ \delta S' + \delta S'' &= 0. & \text{(b)} \end{aligned} \tag{5.31}$$

The criterion of equilibrium consists of the simultaneous conditions

$$\begin{aligned} \mu_i' &= \mu_i'', & \text{(a)} \\ T' &= T''. & \text{(b)} \end{aligned} \tag{5.32}$$

Although these considerations are all but universally valid, there is at least one exception demonstrating the contingent character of our assumption on the availability of walls.

A narrow capillary acts in liquid helium II as an entropy filter, it inhibits the flow of the normal fluid that contains the entropy and transmits the superfluid, the entropy content of which is presumably zero (*24*). Thus the capillary is nonrestrictive of mass and restrictive of entropy flow. Accordingly, for two containers of helium II thus connected, $\Delta\mu = 0$, but we may have $\Delta T \neq 0$. Hence from (26) we obtain

$$\frac{\Delta p}{\Delta T} = \frac{s}{v} = s\rho. \tag{5.33}$$

This is the well known H. London relation experimentally verified in very narrow capillaries from the λ-point to 0.1°K (*25*).

Capillaries wide enough to allow the flow of entropy carried by the normal liquid no longer act as entropy filters, and (29) has to be replaced by a relation based on irreversible thermodynamics (*26*).

At this point, we are able to explain more fully the restrictive character of our definition of the independent variables as contained in *P a3*. Following the usual practice, we could supplement Eq. (11) by additional terms $P_i\, dX_i$, where the X_i are, say, the components of the electric and magnetic polarization, or of the elastic strain tensor. The corresponding P_i are the components of the electric and magnetic field intensity and of the elastic stress tensor. These additional variables satisfy the Maxwell relations (10) and (16). However the processes involving electric, magnetic, and elastic work cannot be formalized as exchanges of additive invariants, and the equilibrium condition (7) is not valid for the conjugate P_i. Thus in a heterogeneous system the magnetic intensity is not constant throughout the system. Moreover, the homogeneity of the system is impaired by the application of elastic stresses. For this reason, we denote these P_i as pseudo-intensities, the conjugate X_i as pseudo-thermodynamic variables, and we exclude them from present consideration.

VI. THE FIRST PHASE RULE

The basic problem of the thermostatic theory of phase equilibrium can be formulated as follows: We consider a thermodynamic system of given composition. We assume that the *possible* phases of this system are known and are specified in terms of its primitive fundamental equations (*P d*1). The problem is to predict the *actual* heterogeneous structure of a simple system specified in terms of its extensive variables.

This question is answered in principle in *P d*2. However, this answer is unwieldy, it involves, on equal terms, all possible phases of a system.

In order to arrive at more useful general results, we simplify the problem by using the concept of a generalized reservoir, which was introduced in the last section. We single out one modification in a heterogeneous system for special consideration, and study its equilibrium with the rest of the system which is assumed to be large enough to be taken as a reservoir.

The intrinsic properties of the modification are specified by the set of densities $x_1, x_2, \cdots, x_r$ that represent a point in x-space. The size of the system is given by the scale factor X_{r+1}.

The modification can be kept in equilibrium either by completely restrictive walls or by contact with a generalized reservoir of an appropriate set of intensities $P_1, P_2, \cdots, P_{r+1}$. These intensities are related to each other by the differential GD relation (5.12). The integral relation (5.15) associates an r-dimensional surface in P-space with the modifications of a single phase. This means that only r intensities are capable of independent variations while still providing the environment for the same phase in one of its modifications.

These results can be generalized at once to the case of heterogeneous system consisting of, say, m modifications. If each of these modifications is in equilibrium with the same reservoir, the modifications are also in equilibrium with each other. In usual parlance, the modifications are said to coexist.[15]

The equilibrium of the m modifications with the same reservoir is formally expressed in terms of the GD relations

$$dP_{r+1} = -\sum_{i=1}^{r} x_i^{(\kappa)}\, dP_i\,, \qquad \kappa = 1, 2, \cdots, m, \tag{6.1}$$

where the intensities have the same value in all coexisting modifications but the densities $x_i^{(\kappa)}(P_1, P_2, \cdots, P_r)$ are characteristic of each modification.

Each of these equations can be used to eliminate one of the intensities, and therefore, we are left with $\delta_I = \text{r} + 1 - m$ intensities capable of independent variations. The number δ_I is equal to the dimensions of the region of P-space that represents the heterogeneous equilibrium of the m distinct modifications. This conclusion rests on the tacit assumption that Eqs. (1) are linearly independent. Assuming more generally that only $\tilde{m}$ of the m equations (1) are linearly independent, we obtain

$$\delta_I = r + 1 - \tilde{m} = c + 2 - \tilde{m}, \tag{6.2}$$

in which we have made use of (1.7).

The non-negative number δ_I is called the variability of the heterogeneous system, or the number of its thermostatic degrees of freedom. We shall refer to (6.2) as the "first phase rule."

In deriving this rule Gibbs had assumed (p. 96 of Ref. *2*) that $\tilde{m} = m$, i.e., that the GD equation associated with different phases are linearly independent. This assumption is indeed plausible if it is taken in conjunction with the tacit assumption of the classical theory that two modifications are distinct if and only if they have different densities.

The situation has to be reconsidered from the point of view of the present definition of distinctness (*D d3*). We agree with the classical point of view in precluding an accidental linear dependence of GD relations of modifications having different densities. However, modifications related to each other by inversion or mirroring are energetically equivalent and have the same GD relation. Moreover, they are distinghishable by virtue of their symmetry. The last condition is not satisfied in the case of equivalence under a pure rotation.

We shall refer to such modifications or phases as modification doublets, or phase doublets. The distinction between these two concepts depends on the

[15] Note that we consider the phase surfaces "in the small." Thus coexisting liquid and vapor modifications are represented by separate GD relations that yield two surface elements upon integration. When they are investigated by methods "in the large" (Section IX), these surface elements turn out to be parts of the same surface that represent the fluid phase. However, this connection "in the large" is irrelevant in the present context.

analysis of the properties of the phase surface in the large and will be explained in Section IX.

The present version of the first phase rule differs from the classical formulation only to the extent that, when the system contains doublet modifications, or phase doublets only one of each pair is to be counted when the effective number of modifications m is being established.

We consider somewhat closer the properties of a system consisting of two coexisting modifications that differ in their densities. The complete description of the system is provided by the $r - 1$ independent intensities $P_1, P_2, \cdots, P_{r-1}$ and the scale factors X'_{r+1} and X''_{r+1}. Let us consider a phase transition, i.e., a process in which the boundary between the modifications is displaced, and

$$\Delta X'_{r+1} + \Delta X''_{r+1} = 0 \tag{6.3}$$

while the system is in contact with the same reservoir. In the course of this process the reservoir has to supply the quantities $\Delta X_1, \Delta X_2, \cdots, \Delta X_r$ which are in definite proportion to each other and define a direction in Gibbs space:

$$\Delta X_1 : \Delta X_2 : \cdots : \Delta X_r = (x_1'' - x_1') : (x_2'' - x_2') : \cdots : (x_r'' - x_r'), \tag{6.4}$$

where x_i', x_i'' are the densities of the coexisting modifications.

The differences $x_i'' - x_i'$ are characteristic properties of the phase transition: increments of volume, concentration and entropy. The latter is related to the latent heat l: $s'' - s' = l/T$.

Finally, we note that after subtracting the GD relation of a phase from that of a coexisting phase, we obtain the generalized Clausius–Clapeyron equation

$$\sum_{i=1}^{r} (x_i'' - x_i')\, dP_i = 0, \tag{6.5}$$

where the dP_i are intensity variations in the region of coexistence in P-space.

The first phase rule expresses an upper limit for the number of coexisting modifications. We turn now to the study of the intrinsic stability of homogeneous systems, in order to find out the conditions under which the heterogeneous equilibrium is realized rather than the homogeneous one.

VII. THE THERMOSTATIC FUNDAMENTAL FORMS. CONNECTION WITH EXPERIMENT.

The study of the intrinsic stability of a homogeneous phase involves the discussion of the quadratic expansion terms of the primitive fundamental equation. This quadratic form is also important in establishing the connection of the theory with experiment. Therefore we investigate its properties before turning to our central problem.

We consider a single homogeneous phase, and expand its primitive fundamental equation about a point $X_{10}, X_{20}, \cdots, X_{r+1,0}$ in Gibbs space. We vary only the first r variables, keep the scale factor X_{r+1} constant, and write the aforementioned expansion in terms of densities as

$$u(x_1, x_2, \cdots, x_r) = u_0 + \sum_1^r P_{i0}\xi_i + \frac{1}{2}\sum u_{ik}\xi_i\xi_k + \cdots, \tag{7.1}$$

where

$$\xi_i = \delta x_i = x_i - x_{i0}, \tag{7.2}$$

$$P_{i0} = (\partial u/\partial x_i)_0, \tag{7.3}$$

and

$$u_{ik} = (\partial^2 u/\partial x_i \partial x_k)_0 = (\partial P_i/\partial x_k)_0 = (\partial P_k/\partial x_i)_0. \tag{7.4}$$

We denote the quadratic form in (7.1) as the thermostatic *fundamental form*, or *stiffness form*. The corresponding matrix $\| u_{ik} \|$ will be called the *stiffness matrix*. Its physical meaning is evident from

$$\pi_i = \sum_1^r u_{ik}\xi_k, \tag{7.5}$$

where

$$\pi_i = \delta P_i = P_i - P_{i0}. \tag{7.6}$$

The determinant of the stiffness matrix is

$$\det | u_{ik} | = D_r, \tag{7.7}$$

where D_r is the Jacobian (5.14). We shall consider at first the regular case, for which $D_r \neq 0, \infty$. The cases $D_r = 0$ and $D_r = \infty$ are singular and will be considered below.[16]

In the regular case the fundamental equation has a Legendre transform (5.18) that can be expanded as follows:

$$\psi(P_1, P_2, \cdots, P_r) = \psi_0 + \sum_1^r x_{i0}\pi_i + \frac{1}{2}\sum_1^r \psi_{ik}\pi_i\pi_k + \cdots, \tag{7.8}$$

where

$$\psi_{ik} = (\partial^2\psi/\partial P_i \partial P_k)_0 = (\partial x_i/\partial P_k)_0 = (\partial x_k/\partial P_i)_0, \tag{7.9}$$

and

$$\xi_i = \sum_1^r \psi_{ik}\pi_k. \tag{7.10}$$

[16] In case the full set of $r + 1$ extensive quantities is subject to variation, the determinant of the expanded set of Eqs. (5) vanishes: $D_{r+1} = 0$. This follows from the fact that the GD equation (5.11) constitutes a linear relation among the $r + 1$ equations. This result is intuitively obvious: the scale factor X_{r+1} cannot be expressed solely in terms of size independent intensities.

The quadratic form in (8) we call the *compliance form*. We also refer to the stiffness and compliance forms jointly as the "fundamental forms." The matrix $\| \psi_{ik} \|$ is the *compliance matrix*. It is the inverse of the stiffness matrix and will be denoted also as

$$\| \psi_{ik} \| = \| u_{ik} \|^{-1} = \| u^{ik} \|. \tag{7.11}$$

In the next section we shall develop the general properties of the fundamental forms. Meanwhile, we consider their role in establishing contact between theory and experiment.

For the sake of simplicity, we confine the argument to one-component systems and interpret the variables as $x_1 = s$, $x_2 = v$, $P_1 = T$, $P_2 = -p$. We choose the scale factor as the mole number $X_3 = N$. Thus u, s, v are molar quantities. The fundamental equation is

$$u = u(s, v). \tag{7.12}$$

Its Legendre transform is

$$\psi(T, p) = -g(T, p) = Ts - pv - u, \tag{7.13}$$

where $g(T, p)$ is the molar Gibbs function. We shall refer to the formalism developed from the fundamental equation (14) as the T, p scheme. The equations of state in this scheme are

$$\frac{\partial \psi}{\partial(-p)} = \frac{\partial g}{\partial p} = v(p, T), \tag{7.14}$$

$$\frac{\partial \psi}{\partial T} = -\frac{\partial g}{\partial T} = s(p, T). \tag{7.15}$$

Equation (14) is the conventional thermal equation of state, and (15) is sometimes referred to as the caloric equation of state. The former is directly measurable,[17] and is tabulated for reference, but the latter is not, since caloric experiments yield only entropy differences.

The stiffness matrix is

$$\left\| \begin{matrix} \left(\dfrac{\partial T}{\partial s}\right)_v & \left(\dfrac{\partial T}{\partial v}\right)_s \\ -\left(\dfrac{\partial p}{\partial s}\right)_v & -\left(\dfrac{\partial p}{\partial v}\right)_s \end{matrix} \right\| = \left\| \begin{matrix} \dfrac{T}{c_v} & \left(\dfrac{\partial T}{\partial v}\right)_s \\ -\left(\dfrac{\partial p}{\partial s}\right)_v & \dfrac{1}{v} B_s \end{matrix} \right\|, \tag{7.16}$$

where B_s is the adiabatic bulk modulus.

[17] We assume here that the thermodynamic temperature is measurable. Actually, the measurement of T poses theoretical problems, which we shall briefly discuss later in this section.

The compliance matrix is

$$\left\| \begin{matrix} \left(\dfrac{\partial s}{\partial T}\right)_p & \left(\dfrac{\partial s}{\partial(-p)}\right)_T \\ \left(\dfrac{\partial v}{\partial T}\right)_p & \left(\dfrac{\partial v}{\partial(-p)}\right)_T \end{matrix} \right\| = \left\| \begin{matrix} \dfrac{c_p}{T} & v\alpha \\ v\alpha & v\kappa_T \end{matrix} \right\|, \tag{7.17}$$

where c_p, α, and κ_T are molar heat at constant pressure, expansion coefficient and isothermal compressibility, respectively.

The elements of the compliance matrix are directly measurable, and are also tabulated for reference. The caloric equation of state (15) and the fundamental equation (13) follow from the empirical quantities through direct integration. This procedure introduces two arbitrary integration constants into the fundamental equation. These are most conveniently taken as the entropy and enthalpy constants s_0 and $h_0 = g_0 + T_0 s_0$, where the subscript refers to a reference point T_0, p_0.

We recall that our discussion refers to the primitive fundamental equation of a single phase. The arbitrariness in the constants s_0, h_0 is reduced by the fact that the constants of different phases are related to each other by means of the latent heat:

$$l = T_0(s_0'' - s_0') = h_0'' - h_0'. \tag{7.18}$$

It is assumed that the two phases coexist at T_0. Similar relations hold for chemical species undergoing chemical reactions.

Within the context of classical molecular chemistry the foregoing procedure leaves the entropy and enthalpy constants of the chemical elements arbitrary. It would seem plausible to eliminate the arbitrariness of the entropy constant by using the strong formulation of the third law (*P c2*) and set $s_0 = 0$ for $T = 0$

Although quantum mechanics supports this procedure, there are some valid objections against it if one takes a purely empirical attitude. In the first place, absolute zero cannot be reached, and while the entropy of a system may practically vanish at attainable temperatures, there is no purely empirical method to verify this fact.

A second point is that there is no purely empirical method to establish whether or not a given system has actually reached equilibrium. This difficulty is particularly great at low temperatures, and would seem to render a fruitful application of *P c2* virtually impossible.

However, the problem of the entropy constant takes on a completely new aspect if one makes use of some simple results of quantum statistics.[18] We confine

[18] The injection of quantum statistical calculations seems to be against the traditional ground rules of the phenomenological theory. We believe that this objection is not serious. We shall return to this question in the final discussion.

ourselves to outlining the essential ideas involved. It is an important fact that in states of sufficiently low density, and not too low temperatures, practically every system breaks up into molecules that are, on the average, so far from each other that the system behaves as an ideal gas. Its equation of state is

$$\lim_{\rho \to 0} \frac{p}{T} \propto \rho. \tag{7.19}$$

Moreover, its entropy is the sum of the molecular entropies. Under these ideal conditions, the entropy can be computed from quantum statistics. With the exception of monatomic gases, this calculation is not based on first principles, but utilizes spectroscopic data, and its results are very reliable (*27*).

The calculated entropy contains no arbitrary constant, but it depends on the choice of independent components in the sense of *C a3*. With the entropy value pegged in the low-density limit, say at room temperature, the empirical data can be used to integrate to the lowest available temperatures. If the resulting entropy value is practically zero, we conclude that: (i) the system is in equilibrium and (ii) the temperature is low enough for the system to behave almost as it would at absolute zero. A finite positive entropy indicates that at least one of these conditions is not satisfied. It usually takes a combined experimental-theoretical study to separate the role of the two conditions (*15*).

To arrive at a negative entropy value is unusual; nevertheless, it is conceivable that the variables of the problem are being chosen in an inadequate fashion.[19]

The most interesting aspect of this result is that only through the detour over quantum statistics is the concept of thermodynamic equilibrium placed on a solid empirical basis.

Summing up we can say that the fundamental equation can be established from actual measurements supplemented by quantum statistical calculations carried out in the ideal gas limit.

The fundamental equations (12) and (13) are not the only ones available. Other equations are the Helmholtz free energy as a function of T, V, and the enthalpy as a function of S, p. Still more possibilities arise if the volume is chosen as a scale factor. Since all of these schemes are equivalent, we arrive necessarily at a huge number of thermodynamic identities. These identities have important uses. One of them is to express quantities of interest in terms of the compliance coefficients or other measurable quantities of the p, T scheme.

[19] Thus it is usual to ignore the contribution of nucear spins to the entropy, since in most cases this contribution does not diminish in the available temperature range. If further cooling leads to a freezing out of the spin entropy, one might arrive at negative entropies. This result is, of course, spurious and is eliminated by including the relevant type of entropy in the statistical calculation.

We do not enter into the derivation of these identities in this paper, but merely note that they allow us to short-circuit the explicit construction of the fundamental equations (14) and (13).

A second important use of the thermodynamic identities is the establishment of the thermodynamic temperature scale. From the empirical point of view, a temperature scale is based on the properties of a particular substance. Let the temperature defined in such a fashion be T^*, and the thermodynamic temperature calibrated on this empirical scale be given in terms of a function $T(T^*)$. Let us consider, now, a thermodynamic identity that involves only measurable quantities. Since thermodynamic identities are valid only in terms of the thermodynamic temperature, any identity of this sort constitutes a differential equation for $T(T^*)$. This equation can be integrated if a numerical value is assigned, by convention, to a fixed point.

According to the first phase rule, the triple point of a one component system is an isolated point in P-space ($c = 1$, $\tilde{m} = 3$, $\delta_I = 0$) and may serve as a fixed point. Recent international agreement has fixed the triple point of water as exactly 273.16°. The fact that different identities and different substances lead to the same temperature scales is a further verification of the theory.

VIII. LOCAL STABILITY

We proceed now to investigate the stability of homogeneous modifications by using the criteria of stability formulated in Section IV.

We mentally subdivide the system into two parts specified by the fixed scale factors X'_{r+1} and X''_{r+1}, and consider the virtual process

$$\delta X_i' + \delta X_i'' = 0, \qquad i = 1, 2, \cdots r. \tag{8.1}$$

The virtual work associated with this displacement is

$$\begin{aligned}\Delta W = \Delta U' + \Delta U'' = \sum_1^r P_i'\delta X_i' + \frac{1}{2}\sum_1^r U'_{ik}\delta X_i'\delta X_k' \\ + \cdots + \sum_1^r P_i''\delta X_i'' + \frac{1}{2}\sum_1^r U''_{ik}\delta X_i''\delta X_k'' + \cdots .\end{aligned} \tag{8.2}$$

We have made use of the expansion (7.1), but rewritten it in terms of the extensive variables instead of the densities. The stiffness matrix is now

$$U'_{ik} = (\partial^2 U'/\partial X_i'\partial X_k')_0 \tag{8.3}$$

that is a homogeneous function of order (-1).

We let size of one of the systems tend to infinity:

$$X''_{r+1}/X'_{r+1} \to \infty . \tag{8.4}$$

Since $u'_{ik} = u''_{ik}$, we see from (1)–(4) and (7.4) that the quadratic term associated with the large system is negligible. This system can be considered as a reservoir that is adequately described solely in terms of its intensities $P_i'' = P_{i0}$.

We obtain from (2), after dividing by X'_{r+1}, omitting the primes and using the notation (7.2):

$$\Delta w = \sum_1^r (P_i - P_{i0})\xi_i + \frac{1}{2}\sum_1^r u_{ik}\xi_i\xi_k + \cdots. \tag{8.5}$$

The condition of equilibrium provides $P_i = P_{i0}$, a result already obtained in Section V. According to *D f2* (Section IV), stability requires $\Delta w \geqq 0$.

The implications of this relation have to be discussed both for infinitesimal and finite displacements. The second question concerning *stability in the large* is taken up in the next section. Meanwhile, we turn to the discussion of infinitesimal displacements (local stability), determined primarily by the quadratic form (5).

It is shown in Appendix B (See also Ref. *11*), that this quadratic form can be brought to a diagonal form

$$\Delta w = \sum_{k=1}^r \lambda_k \eta_k^2, \tag{8.6}$$

where the η_k are linear combination of the ξ_k explicitly defined in Appendix C and

$$\lambda_k = \left(\frac{\partial P_k}{\partial x_k}\right)_{P_1, P_2, \cdots, P_{k-1}, x_{k+1}, \cdots, x_r} \equiv \left(\frac{\partial P_k}{\partial x_k}\right)_{P_1, P_2 \cdots, P_{k-1}}. \tag{8.7}$$

The second expression is used as a short notation, in which only the intensities held constant are made explicit, while the constant densities are suppressed.

The diagonalization of the form (5) provided by (6) and (7) replaces the coupled processes $\delta X_1, \delta X_2, \cdots, \delta X_r$ by independent processes

$$\delta X_1, (\delta X_2)_{P_1}, \ldots (\delta X_r)_{P_1, P_2, \cdots P_{r-1}}. \tag{8.8}$$

This interpretation of the diagonalization procedure is indeed suggested by (7); it is confirmed by a formal proof, since it follows from (B5), (B11), and (7.5) by a simple algebra, that η_i with $\eta_k = 0$ for $k \neq i$ is equivalent to $(\delta x_i)_{P_1 P_2 \cdots P_{i-1}}$.

Of course, the processes (8) involve implicitly a coupling of the δx_i. Thus for $(\delta x_2)_{P_1}$ we have

$$u_{11}\delta x_1 + u_{12}\delta x_2 = \delta P_1 = 0. \tag{8.9}$$

In each subsequent process one more extensive variable is involved. The substitution of independent basic processes for the coupled processes of the original problem is, of course, reminiscent of the diagonalization of quadratic forms by the eigenvalue method. However, it is evident from Appendix C that the two methods are substantially different.

Since the expansion (5) can be centered around any point of the primitive surface, we can classify the points of this surface as follows:

(a) Elliptic points: all $\lambda_k > 0$,

(b) Parabolic points: all $\lambda_k \geqq 0$, at least one $\lambda_k = 0$, (8.10)

(c) Hyperbolic points: at least one $\lambda_k < 0$.

The application of the definitions *D f1-4* (Section IV) allows us to draw the following conclusions. Elliptic points are stable with respect to small displacements (local stability), but may be either stable or instable with respect to finite processes (absolute stability and metastability, respectively).

Hyperbolic points represent states of essential instability. They are, in general, inaccessible both from the experimental and the theoretical points of view, and we shall always exclude them from consideration.[20]

The case of parabolic points is the most complex and also the most interesting. Whether or not parabolic points are stable depends on the large displacements which will be discussed in the next section.

While most parabolic points turn out to be instable, there are stable limiting situations. Yet, the stability of parabolic points is of a lower order than the one found in stable elliptic points. We refer to the latter as *normal*, to the former as *critical equilibrium* (*D f2*).

We shall compare now the properties of normal and critical equilibrium to the extent that this can be established from the properties of the fundamental quadratic form.

For the sake of simplicity, we consider the case of two independent variables, and elaborate on the foregoing conclusions. An elliptic point is characterized by the conditions

$$\lambda_1 = (\partial P_1/\partial x_1)_{x_2} > 0, \tag{8.11a}$$

and

$$\lambda_2 = (\partial P_2/\partial x_2)_{P_1} > 0. \tag{8.11b}$$

According to (*B*-15) we have

$$\lambda_2 = \frac{D_2}{D_1} = \left(\frac{\partial P_2}{\partial x_2}\right)_{x_1} - \frac{(\partial P_1/\partial x_2)^2_{x_1}}{(\partial P_1/\partial x_1)_{x_2}} > 0. \tag{8.12}$$

Hence

$$(\partial P_2/\partial x_2)_{x_1} \geqq (\partial P_2/\partial x_2)_{P_1} > 0, \tag{8.13}$$

[20] A widely used procedure involving the essentially instable states of a Van der Waals gas will be critically discussed in the next section. From the mathematical point of view, points with all of the $\lambda_k < 0$ have an elliptic character. In the present context these are essentially instable, as are hyperbolic points proper with positive and negative λ_k.

where the equality sign is relevant only if

$$(\partial P_1/\partial x_2)_{x_1} = 0. \tag{8.14}$$

Relations (13) are equivalent to (11a) and (11b). They are usually referred to as the *principle of Le Chatelier*. For a further discussion we refer to p. 63 of Ref. *20*.

For a system with the fundamental equation (7.12) we may choose $x_1 = v$, $x_2 = s$, or $x_1 = s$, $x_2 = v$. The relations (13) yield, in this case,

$$\frac{1}{c_v} \geqq \frac{1}{c_p} > 0, \tag{8.15}$$

and

$$B_s \geqq B_T > 0, \tag{8.16}$$

where B is the bulk modulus. The equivalence of (15) and (16) follows at once from the thermodynamic identities

$$\frac{B_s}{B_T} = \frac{c_p}{c_v}, \; c_p - c_v = Tv \left(\frac{\partial p}{\partial T}\right)_v^2 \Big/ B_T. \tag{8.17}$$

The equality sign in (15) and (16) becomes relevant only if the expansion coefficient vanishes.

The foregoing results are easily generalized to several variables. In elliptic points we have the generalized Le Châtelier principle:

$$\frac{\partial P_k}{\partial x_k} \geqq \left(\frac{\partial P_k}{\partial x_k}\right)_{P_1} \geqq \left(\frac{\partial P_k}{\partial x_k}\right)_{P_1 P_2} \geqq \cdots \geqq \left(\frac{\partial P_k}{\partial x_k}\right)_{P_1, P_2, \cdots P_{k-1}} > 0. \tag{8.18}$$

We express this relation in words as follows: A locally stable system is displaced from its equilibrium by the displacement dx_k. The system responds by changing its conjugate intensity by dP_k. This response is the largest if all the other x_i are fixed, and it decreases upon relaxation of each constraint that frees a variable x_i by coupling the system to a reservoir of intensity P_i.

We turn now to the discussion of critical equilibrium, when the stiffness matrix is singular and is of rank $(r-1)$[21]:

$$D_r = 0, \qquad \lambda_r = (\partial P_r/\partial x_r)_{P_1, P_1, \cdots P_{r-1}} = 0, \qquad \lambda_k \neq 0 \;\text{ for }\; k \neq r. \tag{8.19}$$

The development of the thermostatic formalism made it repeatedly necessary to require that the condition $D_r \neq 0$ be satisfied. The breakdown of this condition in critical equilibrium indicates that in such states the thermostatic formalism reaches its limits of validity. An instance of this breakdown becomes evident from the discussion of Eqs. (7.5) which, in general are no longer soluble,

[21] This is the only case that has been found to be of physical interest, thus far.

while the associated homogeneous equation

$$\sum_{k=1}^{r} u_{ik}\delta x_k = 0 \tag{8.20}$$

has a nontrivial solution

$$\delta x_1 : \delta x_2 : \cdots \delta x_r = u^{1k} : u^{2k} : \cdots u^{rk}, \tag{8.21}$$

where the u^{ik} are proportional to the $(r - 1)$ dimensional minors of $\| u_{ik} \|$, and may be taken as the elements of the compliance matrix. Since the rank of this matrix is $(r - 1)$, the u^{ik} cannot all vanish and (21) defines a direction, the *critical displacement*, in Gibbs space (*D f2*).

The inhomogenous equations are soluble only if the condition

$$\sum_{i=1}^{r} u^{ik}\delta P_i = 0 \tag{8.22}$$

is satisfied. Hence, the transformation (7.5) maps an r-dimensional domain of x-space to an $(r - 1)$-dimensional domain of P-space. Consequently, the Legendre transformation becomes singular here. The implications of this fact will be discussed in the next section.

The most conspicuous symptom of critical equilibrium is that the compliance matrix becomes singular, its elements in general tend to infinity. In the system (7.12) this means that

$$c_p \to \infty, \quad \kappa_T \to \infty, \quad \alpha \to \infty. \tag{8.23}$$

We note that c_v and κ_s are finite. The assumption $c_v \to \infty$ would imply that the rank of the stiffness matrix is $(r - 2)$, that is $\lambda_r = \lambda_{r-1} = 0$.[21, 22] However, the parabolic character of critical equilibrium does not necessarily imply (23). If the minor $u^{ik} \approx 0$, the corresponding compliance coefficients will not tend to infinity. The displacive transitions discussed in the next section belong in this category (see Ref. *11*).

IX. STABILITY IN THE LARGE. SECOND PHASE RULE

We turn to the second part of our study of stability and consider processes in which the densities of a homogeneous system change by finite amounts.

One method of dealing with this problem is to consider the higher order terms of the expansion (8.5) of the fundamental equation.

A second approach, first developed by Gibbs, deals with the properties of the primitive surfaces in the large by geometrical methods.

[22] The conclusion that c_v is finite seems to be at variance with the exact theoretical findings concerning the critical point of the two-dimensional Ising model. This contradiction is only apparent; in the Ising model there is no distinction between c_v and c_p.

The first method alone is insufficient for our present problem. Usually, the two methods are used in combination (*2*, *4*, *28*). We shall show, however, that the purely geometrical method is by itself sufficient to establish the second phase rule, our main objective in this section.

While the expansion method has its use, it is unreliable in the most interesting case: in critical equilibrium.

If a simple system is prepared with such values of the extensive variables that correspond to an instable region of a phase surface, the system escapes unstability by breaking up into a heterogeneous system consisting of two or more modifications. These modifications may belong to the same phase surface or be on different phase surfaces. The vaporization and the crystallization of a liquid, furnish examples for the two situations.

We shall consider in detail the instability that arises within a single primitive surface. Of the two cases this one is richer in variants, and the results are easily extended to the instability involving several surfaces.

Heterogeneous equilibria of two modifications of the same phase occur under two kinds of conditions: (i) the coexisting modifications differ in their densities and, (ii) the modifications have equal densities, as ensured by symmetry relations, and are distinguished in terms of quasithermodynamic parameters η. We shall refer to the first case as a *condensation type equilibrium*, to the second as an *equilibrium of modification doublets* (or multiplets).[23]

We shall, at first, give a brief account of the Gibbs construction for condensation type equilibria (p. 44 of Ref. *2*) partly because this construction is not easily accessible (see, however, Chapters 2 and 3 of Ref. *4* and Chapter 12 of Ref. *28*), and partly because we wish to develop the analogy between the two cases.

The requirements of geometrical representation suggest that the simplest case with the smallest number of variables, be discussed first. The generalization to several variables offers no difficulties.

The simplest condensation type transition is the condensation of a one-component fluid with $c = 1$, $r = 2$. The primitive fundamental equation in molar quantities is

$$u = u(x_1, x_2) \tag{9.1}$$

where $x_1 = s$ and $x_2 = v$. We shall assume that the primitive surface contains a connected sheet of elliptic points bounded by a one-dimensional locus of para-

[23] Conforming to the time-independent character of thermostatics, we usually speak of phase equilibrium. The widely used terms condensation and phase transition have a temporal connotation: The extensive or intensive variables of a system are varied as a function of time and the system is assumed to run through a corresponding sequence of equilibrium states. The distinction between these concepts is of some importance, since the same heterogeneous equilibrium may be crossed in the course of different transitions produced by the variation of different variables.

bolic points.[24] Empirically, the phase surface of every fluid turns out to be of this sort.

In the elliptic domain the u surface is convex toward the x_1, x_2, plane. Outside, it is concave. According to the Gibbs construction the concave part of the surface is replaced by a ruled surface consisting of infinitesimal strips of tangent planes each of which has two common points with the u surface.

A geometrical representation is given in Fig. 2.

We single out one tangent plane with the tangent points A', A''. These so-called conjugate points represent distinct modifications of the same phase which have the same intensities, and hence can coexist. The line $A'A''$ is an isothermal, isobaric line. The projections of A', A'' on the x_1-x_2 plane are $\tilde{A}'$, $\tilde{A}''$. Figure 2(a) represents the intersection with the plane x', u. The parabolic points contained in this plane are B', B''. Their projections are again marked by a tilde. The loci of the points A', A'' and B', B'', respectively, are called the *binodal* and the *spinodal* curve (Ref. *4*). The former marks the limit of absolute stability, the latter the limit of essential instability. The region between the two curves is that of metastability.

In constructing the derived fundamental equation, the portion $A'B'B''A''$ of the primitive surface is replaced by the double tangent $A'A''$. As A' and A'' move along the binodal curve, the tangent lines produce a ruled surface. The point E of this surface represents a heterogeneous state consisting of the modifications A' and A''. The relative amount of these modifications is $EA'':A'E$ (lever rule).

The heterogeneous equilibrium has a discontinuous character that manifests itself in discontinuities of the molar quantities $\Delta x = x_i'' - x_i'$. In Ehrenfest's terminology (*9*) we have a first order transition.

The discontinuous aspect of the equilibrium is particularly evident if it is considered as a transition in the sense explained in footnote 23, provided that the transition is performed by varying the intensities. In the P-space representation (i.e., in the p-T diagram for the system (1)), the entire segment $A'A''$, and hence the modifications A' and A'' are represented by a single point. The conjugate modifications are locally stable, and no intrinsic property foreshadows the fact that an infinitesimal variation of the intensities produces an abrupt transition into a modification, the densities of which differ by the finite amounts Δx_i from those of the original modification.

However, the basically discontinuous character of the heterogeneous equilibrium is masked by a spurious continuity if the transition is performed by a variation of the extensive variables at constant intensities. In such a process

[24] Phases consisting exclusively of elliptic points do not exhibit the intrinsic instabilities that will be discussed. Surfaces containing disjoint sheets of elliptic points can be considered as consisting of two primitive surfaces. This is the alternative mentioned, and set aside for later discussion, at the beginning of this section.

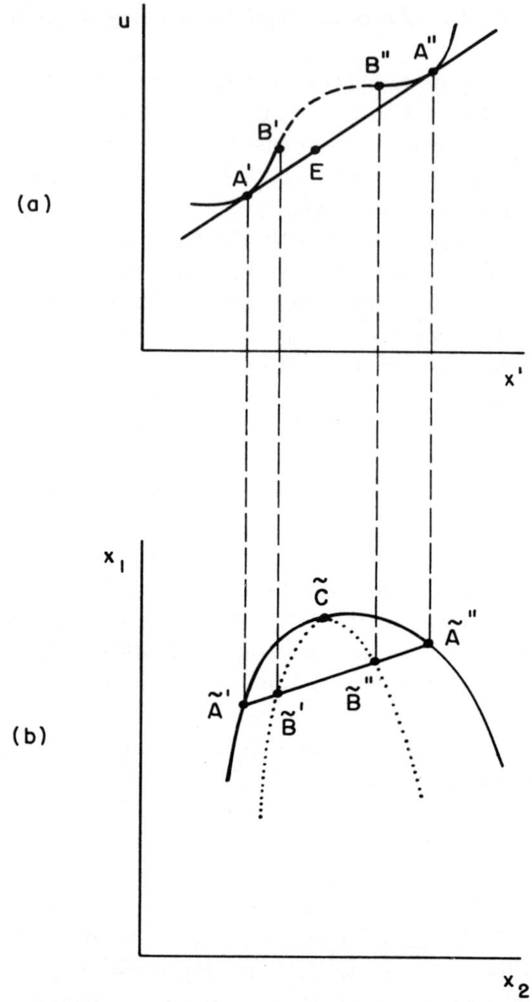

FIG. 2. Critical point of the first kind in terms of the fundamental equation $u = u(x_1, x_2)$; (a) intersection of the primitive surface with the $u - x'$ plane; the axis x' connects points A' and A''. (b) projection to the $x_1 - x_2$ plane; $\tilde{A}'\tilde{C}\tilde{A}''$ and $\tilde{B}'\tilde{C}\tilde{B}''$ are projections of the binodal and spinodal curves; $\tilde{C}$ is the projection of the critical point. The scales along x' in (a) and (b) are not the same.

the point E representing the system moves from A' to A'', while the extensive variables $X_i = N'x_i' + (N - N')x_i''$ change continuously as N' decreases from N to zero.[25]

[25] Thus the condensation of an ideal Bose–Einstein gas is a first-order transition. If it is cooled at a constant pressure $(dT)_p$ the system collapses and its volume vanishes at the

The region between the binodal and spinodal curves is metastable. Part of this region is observable: supersaturated vapor, superheated and supercooled liquid are well-known instances of this sort. However, in general, observation does not extend to the parabolic points that form the boundary between the metastable and the instable domains. We have seen in Section VIII that in parabolic points critical displacements are, in first approximation, not opposed by thermodynamic restoring forces. Such a displacement would move a point B' or B'' into the essentially instable region.[26]

This argument does not hold for parabolic points that are on the border between the absolutely stable and the instable domain, for example, point $\tilde{C}$ in Fig. 2(b). Such states are stable, and are indeed observed. Designated as critical points in the last section, they are seen to have two properties: (i) Critical points are stable parabolic points in which the determinant of the stiffness matrix vanishes and the elements of the compliance matrix are infinite. They are at the limit of the region of stability. (ii) Critical points are at the end of the coexistence region in P-space (coexistence line in the p-T diagram), where two modifications of different densities become identical.

We claim that, property (i) implies property (ii), since, passing through a stable parabolic point, a system can escape instability only by breaking up into coexisting modifications.[27]

We propose to establish an upper bound for the dimension of the set of stable parabolic points in P-space. According to the result obtained, we can substitute in this problem property (ii) for property (i). The problem transformed in such a way can be readily solved.

Since the coexistence region of a one-component, two-modification, system is one-dimensional and the end of this line an isolated point, we see that critical points can form only a zero-dimensional set (isolated points) in the p-T diagram.[28]

This result can be immediately generalized to systems with the fundamental

condensation temperature (see p. 52 of Ref. *24*). The same result is obtained for $(dp)_T$. The fact that the transition is continuous for $(dV)_T$ and $(dT)_v$ does not warrant the designation of the BE condensation as a third-order transition. By the same token we ought to consider the van der Waals type condensation as a continuous rather than as a first order transition.

[26] This conclusion concerning the instability of parabolic points is confirmed by the discussion of the higher terms in (8.5) (see p. 259 of Ref. *20*). A brief discussion is also given at the end of this section.

[27] It is usually assumed or deduced that the converse statement also holds, and property (ii) implies property (i). (see pp. 114 and 129 of Ref. *2* and pp. 79–81 of Ref. *20*). Although certainly correct for a wide variety of cases, we prefer to leave the limits of validity of this statement in abeyance. This question is irrelevant for the argument of this paper.

[28] No statement of such generality can be made about the points in x-space, since the contact between the binodal and the spinodal curves may extend over a one dimensional region in the space of densities (*29*).

equation $u = u(x_1, x_2, \cdots, x_r)$, where the variables are all densities. Homogeneous modifications of such a system can exist in an r-dimensional manifold. Along this surface the heterogeneous equilibrium has the same abrupt, discontinuous character as described above for the one-component system.[29]

The edge of the region of coexistence is $(r - 2)$-dimensional and it is to this edge that critical equilibrium is restricted. Thus, for the dimension of the manifold of critical points of the condensation type, we obtain

$$0 \leqq \delta_c \leqq r - 2 = c - 1. \tag{9.2}$$

This result was first obtained by Gibbs. We shall refer to (2) as an instance of the *second phase rule*.

We arrive at a more general form of this rule if we make use of the new definition of phase ($P\ d1$) and repeat the foregoing construction for the case of modification doublets, that are distinguished from each other only in terms of the quasi-thermodynamic variable η. The fundamental equation of a single-component system is now

$$u = u(x_1, x_2, \eta), \tag{9.3}$$

where $x_1 = s$, $x_2 = v$ and the number of particles is selected as a scale factor.

For the purposes of presentation it is convenient to start the discussion in terms of the well known Ising model of ferromagnetism (*30*). The external magnetic field is assumed to vanish. We shall see that the generality of our final results will not be affected by the special features of the model. One of these features is that the properties of the system are volume independent. Thus the fundamental equation reduces to

$$u = u(x_1, \eta). \tag{9.4}$$

The quasi-thermodynamic parameter η is discussed in some detail in Section X and Appendix *A*. Meanwhile, it is satisfactory to think of it as the parameter of long-range order, or as the magnetization of the system. The interactions arise only between nearest neighbors, and, among other things, the long-range dipole-dipole interactions are neglected. We assume, also, the absence of an external magnetic field. In contrast to a true ferromagnet, the Ising model satisfies our postulates with $r = 1$. According to (1.7) we should call this a "zero-component" system, an expression of the idealized character of the model. There is no objection against its use in thermodynamics; another zero-component system is, e.g., a volume filled with blackbody radiation.

[29] A fluid mixture of two components is an example of this sort. As explained above, the discontinuity of the transition becomes apparent in P-space, or in p, T, μ space in the special example considered. It is more usual to plot the states of the system in the p, T, c space, where c is the concentration. Thus a spurious continuity is introduced into the transition and the system boils, and freezes in extended ranges of temperatures.

We assume that the function (4) is defined also for nonequilibrium values of η. The equilibrium values of this parameter are determined from

$$\frac{\partial u}{\partial \eta} = 0, \qquad \frac{\partial^2 u}{\partial \eta^2} > 0. \tag{9.5}$$

In the Ising model, u is a function of η^2, hence the nonvanishing solutions of (5) appear in pairs: η and $-\eta$. This situation is indeed realized for sufficiently low entropy, or temperature.

Equation (4) is represented Fig. 3, the functions $\pm\eta(x_1)$ are the two solid lines $\tilde{C}\tilde{A}'$ and $\tilde{C}\tilde{A}''$ in Fig. 3. These, again, are binodal curves, the points A', A'' represent *doublet modifications* (see Section VI), referred to also as domains.

The $\tilde{B}'\tilde{C}\tilde{B}''$ curve is the locus of the parabolic points for which $\partial^2 u/\partial\eta^2 = 0$. The analogy between Figs. 2 and 3 is apparent. In general, the parabolic points are again instable, and, at best, the point C, and possibly another intersection with the $\eta = 0$ axis, will correspond to a stable order-disorder type of critical point.[30] This point is the limit between states which do and which do not exhibit a domain structure of doublet modifications.

It is evident that the construction of the critical point point depends only on the condition that the minimum problem (5) have at least two (or possibly more) branches of energetically equivalent solutions. Otherwise, it is entirely independent of the physical interpretation of the parameter η.

These results can be easily generalized to thermodynamic system of r degrees of freedom with the fundamental equation

$$u = u(x_1, x_2, \cdots, x_r, \eta). \tag{9.6}$$

This surface represents a modification that is stable in an r-dimensional domain of x-space, or alternatively, in an r-dimensional domain of P-space. We assume that this modification is characterized by a non-vanishing value of η. The modification coexists over its entire r-dimensional range of stability with the modification specified by $-\eta$, or with any other modification that is related to (6) by symmetry. We refer to the coexisting modifications as domains. The critical equilibria associated with the disappearance of the domain structure are at the $(r - 1)$-dimensional edge of the r-dimensional domain in P-space. Thus the dimensionality of order-disorder type critical point is

$$0 \leqq \delta_{0\text{-}d} \leqq r - 1 = c. \tag{9.7}$$

Equations (2) and (7) can be expressed in terms of a single equation as

$$0 \leqq \delta_{II} \leqq r - 2 + \sigma = c - 1 + \sigma, \tag{9.8}$$

[30] Note the logic of the argument. We do not attempt to predict that C has to be a critical point, although this may well be true. What turns out to be of importance is to set an upper limit to the set of stable parabolic points.

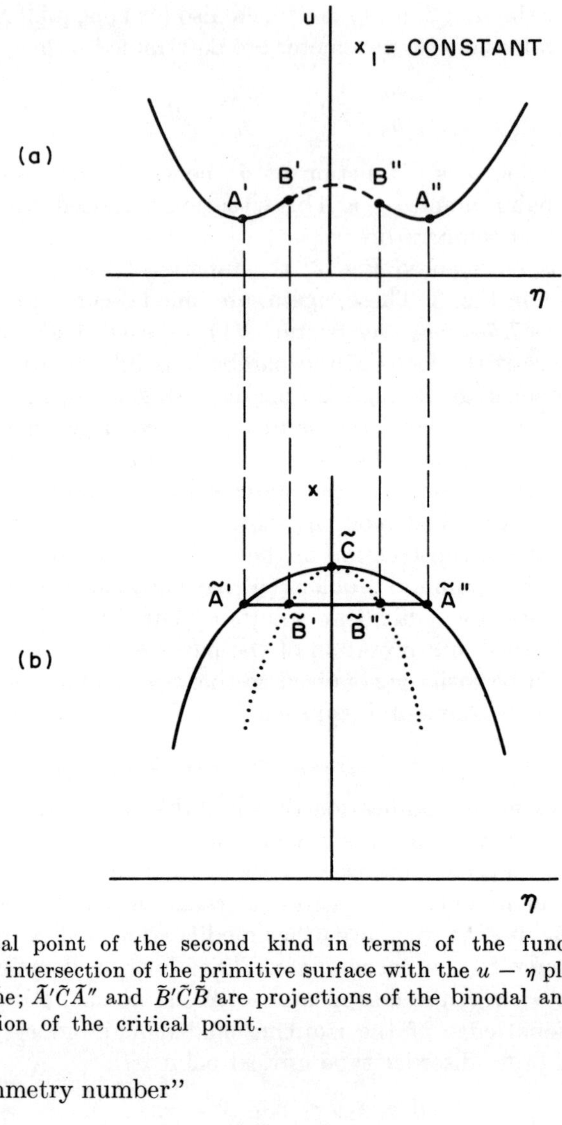

FIG. 3. Critical point of the second kind in terms of the fundamental equation $u = u(x_1, \eta)$; (a) intersection of the primitive surface with the $u - \eta$ plane. (b) projection to the $x - \eta$ plane; $\tilde{A}'\tilde{C}\tilde{A}''$ and $\tilde{B}'\tilde{C}\tilde{B}$ are projections of the binodal and spinodal curves, $\tilde{C}$ is the projection of the critical point.

where the "symmetry number"

$$\sigma = \begin{matrix} 0 & \text{if} & x' \neq x'', \\ 1 & \text{if} & \eta' = -\eta'' \end{matrix} \tag{9.9}$$

Or, in words, σ is zero for the condensation type, and unity for the order-disorder type of transitions. We shall refer to these cases also as *critical points of the first and second kind*, respectively.

The meaning of the second phase rule (9) is not to predict critical behavior in a given instance, but to provide an upper bound for the dimension of the P-space manifold in which such singular behavior can arise.

In practice, critical equilibria are readily recognizable experimentally as λ-points. Thus the number δ_{II} will be observed experimentally, and (9) indicates the system of minimum complexity measured in terms of r and σ, in which the observed singularities can, indeed, occur.

It is evident from Figs. 2 and 3 that the geometrical representation of an isolated critical point ($\delta_{II} = 0$) calls for a three dimensional Gibbs space: the critical phenomenon arises out of the interplay of processes described in terms of two independent variables such as s and v, where one variable may be replaced by a quasi-thermodynamic parameter. The dimensionality of the space increases and the representation becomes impractical, for systems with more than one chemical component. This situation cannot be simplified by keeping certain variables x_i constant. Thus along the abscissa x' in Fig. 3a, in general, all the $x_1, x_2, \cdots, x_r$ vary.

Nevertheless, the diagonalization (8.6) of the fundamental form suggests a representation of critical equilibrium, in terms of a well-chosen one-variable process. This is indeed possible.

Consider the Legendre transform

$$-\varphi(P_1, P_2, \cdots, P_{r-1}, x_r) = \sum_{i=1}^{r-1} P_i x_i - u. \tag{9.10}$$

We have

$$\lambda_r = \frac{\partial^2 \varphi}{\partial x_r^2} = \left(\frac{\partial P_r}{\partial x_r}\right)_{P_1 P_2 \cdots P_{r-1}}. \tag{9.11}$$

Assuming that $\varphi(x_r)$ is sufficiently regular, to admit an expansion into a power series at the critical point, we obtain the following conditions for critical (stable) equilibrium:

$$\text{(a)} \qquad \frac{\partial^2 \varphi}{\partial x_r^2} = 0,$$

$$\text{(b)} \qquad \frac{\partial^3 \varphi}{\partial x_r^3} = 0, \tag{9.12}$$

and

$$\text{(c)} \qquad \frac{\partial^4 \varphi}{\partial x_r^4} > 0.$$

The situation is geometrically represented in Fig. 4, in terms of a one parameter family of curves $\varphi_{P_1}(x_r)$ where each curve corresponds to a different

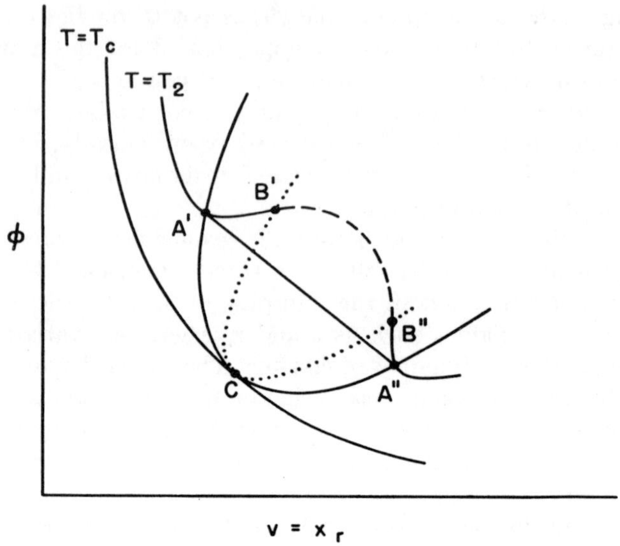

FIG. 4. Critical region in terms of the fundamental equation $\varphi = u[P_1, P_2, \cdots, P_{r-1}]$; curves of constant $P_1, P_2, \cdots, P_{r-1}$ plotted against x_r. $P_1 = T$, $P_2 = p$.

constant value of P_1, conveniently taken to be the temperature or pressure. We have chosen the slope of the curves negative, in order to apply it to the case of ordinary condensation with $x_r = v$, $P_r = -p$.

The presence of condition (12b) supports our earlier conclusion that the dimension of the set of critical points is by one less than that of the parabolic points defined by (12a) only.

For order-disorder type critical points the variable x_r in (10)–(12) replaced by η. In this case condition (12b) is satisfied by symmetry and the dimension of the set of critical points is increased by one, in accordance with (9).

However, it is worth noting that the regularity assumption inherent in the formulation of conditions (12) is not necessarily justified. Thus it has been suggested by Zimm (*29*) that at the ordinary condensation type critical point

$$(\partial^n p/\partial v^n)_T = 0 \tag{9.13}$$

for all values of n.

If this interpretation of the experimental facts were justified, conditions (12) would be inapplicable. However, there is no objection to the use of Fig. 4 in terms of the correct empirical functions, and the second phase rule is unaffected.

The geometrical representation of the condensation phenomenon that is most commonly used in thermodynamics texts arises through the differentiation of the curves of Fig. 4. The result is represented in Fig. 5. This method operates

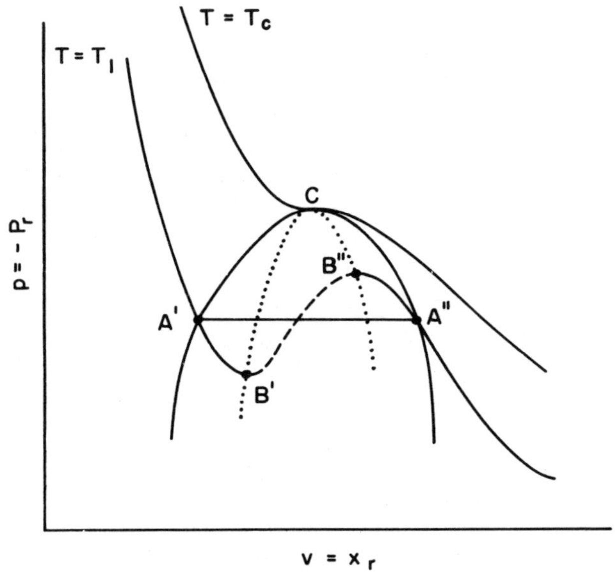

FIG. 5. Critical region in terms of the curves $-P_r = -P_r(x_r)$ at constant $P_1, P_2, \cdots, P_{r-1}$; $P_1 = T$.

in terms of the equation of state, and is ubiquitous in its application to the van der Waals gas.

The comparison of the two geometrical representations provides a graphic illustration for the superiority of the method of fundamental equations.

In Fig. 5, the conjugate points A' and A'' are located by means of the well known Maxwell rule of equal areas, the application of which involves an integration along the essentially instable branch $B'B''$ of the isothermal section of the equation of state. This procedure is objectionable because the instable region is not accessible to observation. Moreover, the instable branch of the isothermal cannot be considered as a legitimate theoretical construct, since a rigorous evaluation of the partition sum leads automatically to the absolutely stable heterogeneous states.

As we have seen, the Gibbs construction of the ruled surface, "boarding up" instable regions, relies only on the knowledge of the absolutely stable regions of the phase surface. The difference between the two cases stems from the fact that the equation of state contains less information than the fundamental equation. Thus a "van der Waals gas" does not constitute a fully defined thermodynamic system. A complete definition of the system would include, the specific heat as a function of, say, temperature and volume:

$$c_v = c_v(T, v). \tag{9.14}$$

With the aid of this information, the conjugate points A' and A'' can be identified by integrating around the critical point and always staying in the absolutely stable domain. In the concept of a "van der Waals gas" a spurious interpolation through the instable range is substituted for the missing information (14).

We turn now to the discussion of the equilibrium of modifications belonging to different phases. It is evident that the Gibbs construction of the ruled surface can also be carried out in this case to identify the conjugate points. The corresponding densities are, in general, different from each other. The modifications never become identical and we have no critical equilibrium.

We mention also the case of phase doublets. Their primitive surfaces are identical to each other, insofar as the dependence on densities is concerned, and they are distinguished only in terms of quasi-thermodynamic parameters. Phase doublets are related to each other by inversion or mirroring, since two systems equivalent under a pure rotation are considered to be identical to each other. A modification of such a phase automatically coexists with its mirrored modification. In contrast to the case of modification doublets, phase doublets cannot gradually become identical to each other, and do not give rise to critical points. An example will be discussed in the next section.

Up to this point we have considered critical equilibrium as a singularity of the behavior of a thermodynamic system. Actually, however, the critical phenomenon constitutes a singularity in the structure of the theory itself.

This is evident from the fact that the condition of validity $D_r \neq 0$ (see (5.14)) of the Legendre transformation leading from the variables $x_1, x_2, \cdots, x_r$ to $P_1, P_2, \cdots, P_r$ breaks down at the critical point. We may add that the condition of validity $D_r^{-1} \neq 0$ (see 5.17) of the inverse transformation fails in the vicinity of absolute zero.

While under normal conditions the x-space and the P-space representations of the fundamental equation provide an equivalent description of the thermodynamic system, the P-space representation contains less information at the critical point and the x-space representation provides less information near absolute zero.

The significance of this fact is best discussed in connection with the minimum principle (5.20). The conditions for an extremum are

$$\frac{\partial w(x'; P)}{\partial x_i'} \equiv \frac{\partial u(x')}{\partial x_i'} - P_i \equiv P_i' - P_i = 0, \tag{9.15}$$

$$\frac{\partial w(x'; P)}{\partial P_i} \equiv \frac{\partial \psi(P)}{\partial P_i} - x_i' \equiv x_i - x_i' = 0 \tag{9.16}$$

In such stable regular states in which the system is in a homogeneous modification these equations have a unique solution. The assumption that (15) is satisfied in distinct states, say x_i' and x_i'', implies that P_i' vanishes along the connecting line in x-space. Hence $D_r = 0$, a contradiction against the assumption of regularity.

Thus we have arrived at a result that, in a way, can be considered as the fundamental theorem of thermostatics. *If a system in a regular state is in equilibrium with its surroundings, the intensities of the surroundings uniquely determine the densities of the system and vice versa.* We propose to call this theorem the *principle of thermostatic determinism.* To motivate this terminology, we point out that the aforementioned extremum problems relate to two main types of thermodynamic investigations.

In the first type of investigation we predict and measure the properties (densities) of a system as a function of the intensities of the surroundings.

In the second kind of investigation we measure the system densities and infer from them the intensities of the surroundings. If the intensity is the temperature, the system plays the role of a thermometer.

The principle of thermostatic determinism assures us that in regular states both procedures lead to a unique answer.

We turn now to the discussion of the singular cases and consider in particular $D_r = 0$. According to Appendix B this condition implies that

$$(\partial P_r/\partial x_r)_{P_1,P_2,\cdots,P_{r-1}} = 0. \tag{9.17}$$

The Legendre transformation maps an r-dimensional domain of x-space into an $(r - 1)$-dimensional domain of P-space and Eq. (15) cannot be uniquely solved for the x_i'.

In order to clarify the physical meaning of condition (17) we have to keep in mind that this equation applies to two essentially different situations: (i) A heterogeneous equilibrium of two or more modifications, each of which is described in terms of elliptic points of the primitive fundamental equation. (ii) A critical equilibrium corresponding to a stable parabolic point.

In case (i) the failure of thermostatic determinism is only apparent, and the principle can be "saved" by a minor improvement of its formulations. We consider, for simplicity, a one-component system that is, under regular conditions, adequately described in terms of the fundamental equation $G = G(T, p, N) = U[T, p]$. Given T, p and the scale factor N, the extensive variables X_i (i.e., S and V) and hence also the molar quantities s, v are uniquely determined. This is no longer true in the heterogeneous region of, say, two modifications. In this case it is sufficient to specify one of the intensities T or p, since the other is determined from the equation $p = p(T)$ of the coexistence region. However, for the molar quantities we obtain two sets of solutions: x_i' and x_i'' instead of one. Accordingly the extensive quantities range over

$$X_i = N[x_i' + (1 - \alpha)x_i''], \tag{9.18}$$

where α varies from 0 to 1.

However, this indeterminacy in the description of the system can be removed. According to Section VI, a complete description of the system is provided in terms of a single intensity, say T, and *two* scale factors, say N' and N''. In other words, given T, N', and N'', the variables p, V', V'', S', S'' are uniquely determined.

An appropriate fundamental equation in terms of the Helmholtz function $A = U[T]$ is

$$A(T, V, N) = N'a'(T, v') + N''a''(T, v''). \tag{9.19}$$

The generalization to several components and to more than two coexisting modifications is straightforward.

Case (ii), that of critical equilibrium, is essentially different from the one just considered. The uncertainty of the inference leading from the intensities P_i' to the densities x_i' is not just a shortcoming of the formalism, but has a basis in experiment. The molar quantities or densities x_i' exhibit anomalous large fluctuations in critical equilibrium. This fact is not accounted for in our postulational basis, and hence the latter needs revision.

The qualitative aspects of this revision can be easily explained. In *P b1* and *P d2* it was assumed that the virtual states of the system have no reality and serve only as comparison states for the entropy maximum principle, that assigns a single set of values of the extensive variables to the equilibrium state.[31] Fluctuation phenomena and, in particular, critical fluctuations, force us to assign physical reality to the virtual states. The state of equilibrium of thermostatics is actually not a well-defined state, but rather a statistical distribution over the virtual states. In elliptic points this distribution is sharply peaked and thermostatics, operating with the average values of these distributions, leads to quantitatively correct answers. Nevertheless the elimination of the statistical elements renders the theory from the qualitative, conceptual point of view inadequate. As we shall briefly outline in the final discussion, this difficulty is the point of departure for the development of statistical thermodynamics.

It is interesting to note that statistical mechanics predicts infinite fluctuations whenever (17) holds, that is in critical *and* in heterogeneous equilibrium. This property is sometimes used to provide a criterion for the heterogeneous state. While critical fluctuations in finite systems are finite (see e.g. Ref. *12*), they do assume anomalously large values. However, this is not the case in heterogeneous systems where the fluctuations are normal. The theoretical fluctuation is cal-

[31] Some qualifications of this uniqueness, irrelevant for the present purposes, will be given below.

culated from an ensemble that contains systems with arbitrary amounts of the two modifications. This corresponds to the incomplete description of the system in terms of the Gibbs function. Thus the "fluctuation" in question has a subjective character and is "transformed away" on transition to the Helmholtz function as explained above. Such an elimination of the objective critical fluctuations is of course impossible.

In the language of the statistical theory we may say that the ensemble of virtual states in homogeneous systems has the ergodic property, that is, the time average over the states of an individually given system is equal to the average over all virtual states; the ensemble corresponding to heterogeneous states is not ergodic and the results of statistical calculations are to be interpreted with care.

The distinction between the two situations is very clear in terms of the derived and primitive fundamental equations. At the derived surface

$$\mathrm{Det} \mid U_{ik} \mid = 0 \tag{9.20}$$

in the heterogeneous *and* in the critical equilibrium. In contrast, at the primitive surface

$$\mathrm{Det} \mid \mathfrak{U}_{ik} \mid = 0 \tag{9.21}$$

for the critical point and

$$\mathrm{Det} \mid \mathfrak{U}_{ik} \mid \neq 0 \tag{9.22}$$

for each of the coexisting modifications. The situation is obscured in the statistical mechanical procedure which deals only with absolutely stable states and fails to distinguish the primitive and the derived fundamental equations.

We turn now to the discussion of the case that the inverse Legendre transformation leading from the P_i to the x_i becomes singular because of $D_r^{-1} = 0$. In analogy with (17) we have now

$$(\partial x_r/\partial P_r)_{x_1, x_2, \cdots, x_{r+1}} = 0. \tag{9.23}$$

If x_r is interpreted as the specific entropy, this relation means that the specific heat vanishes. It follows indeed from *P c2* that this requirement is asymptotically satisfied near absolute zero. More precisely, we can associate with various substances characteristic temperatures θ in such a way that the specific heat tends to zero at least linearly in T/θ. As T falls significantly below θ, the inference of the temperature of the surroundings, obtained from the density of the system, becomes more and more uncertain. This is a well-known difficulty of low-temperature thermometry.

In conclusion, we indicate briefly that the principle of thermostatic determinism can be adapted to other situations. Thus in *P b1* and *P d2* we have tentatively stated that the maximization of the entropy function leads to an *almost unique* determination of the free variables of the problem. We are now in a position to make these statements more definite. However, we shall not fill in all the details of the simple proof.

It is easy to show that the heterogeneous phase structure of a simple system is uniquely determined, except for the unavoidable ambiguity connected with symmetry. Thus the relative amount of each modification doublet (or multiplet) remains undetermined.

Moreover, a homogeneous modification is distributed in a unique fashion over a composite system, the partitions of which are fixed, that is, if the volume of each subsystem is a fixed variable. On the other hand, ambiguities of the solution arise in systems partitioned by adiabatic movable pistons. If such a piston is released from an arbitrary initial position, generally a damped oscillation will result. The conservation laws are not sufficient to determine a unique asymptotic equilibrium position. The latter will depend on the mechanical details of the process that determine how the dissipated macroscopic kinetic energy is shared by the subsystems separated by the piston. Also, the heterogeneous equilibrium in a composite system exhibits considerable ambiguity, which, however, is of a rather trivial nature.

X. LAMBDA LINES AS LOCI OF CRITICAL EQUILIBRIUM

The replacement of the classical version of the second phase rule (9.2) has significant experimental implications. The simplest types of critical points predicted by the theory are summarized in Table I. The column with $\sigma = 0$ corresponds to the classical phase rule, the one with $\sigma = 1$ contains the categories predicted by the present theory. We shall survey some representative nonclassical cases and compare the theoretical predictions with experimental evidence.

TABLE I

VALUES OF δ_{II} WITH THE DESIGNATION OF THE CORRESPONDING TYPES OF PARABOLIC POINTS

r \ σ	0	1
1	<0 none	0 critical point in Ising model
2	0 critical point in fluids	1 λ-line
3	1 critical line of mixing in two component fluids	2 λ-surface in two component systems

From the practical point of view by far the most important class is the one with $r = 2$, $\delta_{II} = 1$, $\sigma = 1$; more explicitly, the theory predicts the possibility of lines of parabolic points in the p - T diagram of one-component systems. We propose to consider this theoretical category as the interpretation of the well known λ-phenomenon[32] observed in many crystals (*31*, *32*) and in liquid helium (*24*, *25*). Plotting the compliance coefficients $\psi_{ik}(T)$ of certain one-component systems as functions of the temperature at constant pressure, it is found that these functions exhibit singularities of a shape reminiscent of the letter λ. The sharp maximum of this curve is called the λ-point: $T_\lambda(p)$. As the pressure varies, the λ-point traces in the $p - T$ plane the λ-line. We recall that ψ_{ik} stands for c_p, α, and κ_T.

In evaluating the merits of this interpretation we have to examine first, whether there is an empirical justification for considering λ-points as parabolic singularities, and second, whether the role assigned to symmetry is borne out by experiment.

Considering at first the matter of the singularity, we have to remember that the establishment of a mathematical singularity from experimental curves is always difficult. We cannot expect to observe actual infinities of the $\psi_{ik}(T)$ functions since the effects that limit the perfect homogeneity of a phase (see Section III) will tend to round off the ideally sharp singularity. Moreover, a smoothing is brought about by the experiment itself, which proceeds by discontinuous steps: a finite DU is measured against a finite DT.

In spite of these reservations, the appearance of most λ-points is sufficiently sharp to warrant the interpretation as a parabolic singularity. This conclusion has been confirmed most convincingly for helium. According to the precision measurements of Fairbank *et al.* (*33*) the specific heat of this substance has a logarithmic singularity at the λ-point.

Notwithstanding the fact that the correspondence between theory and experiment is most satisfactory, we have to dwell somewhat more on this question, because a point of view, that is substantially different from the one here advanced, has been widely accepted in the literature.

It is evident from the foregoing discussion that the classical phase rule (9.2) does not provide the possibility of critical lines in single-component systems.

The need for extending the classical phase theory was recognized by Keesom (*8*) and Ehrenfest (*9*). However, since the comparison of λ-*lines* with critical *points* must have appeared incongruous, λ-*lines* were compared with *lines* of ordinary heterogeneous equilibrium. These are characterized by the differences of the densities

[32] These are commonly called "second-order transitions." This terminology is based on an inadequate theoretical conception. At any rate, we prefer to designate the experimental findings by a theoretically noncommittal descriptive term.

$$\Delta x_i = x_i'' - x_i' = \Delta \frac{\partial \psi}{\partial P_i} = -\Delta \frac{\partial g}{\partial P_i}. \tag{10.1}$$

The analogy was conceived by postulating along the λ-line second order discontinuities:

$$\Delta \psi_{ik} = \lim_{t \to 0} [\psi_{ik}(T_0 + t) - \psi_{ik}(T_0 - t)] = \Delta \frac{\partial^2 \psi}{\partial P_i \, \partial P_k}. \tag{10.2}$$

Generally speaking, a transition of the nth order was supposed to have discontinuities in the nth derivative of the Gibbs function.

Ehrenfest also derived his well known analog to the Clausius–Clapeyron relations relating the slope of the λ-line to the $\Delta\psi_{ik}$.

Keesom and Ehrenfest advanced the concept of second-order transitions in an *ad hoc* manner to describe experimental facts. Subsequently this concept received theoretical support. The λ-phenomenon was given a microscopic interpretation in terms of order-disorder transitions and the approximate statistical calculations invariably led to discontinuities of the specific heat (*30*). Moreover Landau derived the Ehrenfest scheme from thermodynamic stability considerations (*34*). This theory seemed particularly significant since it did not appear to involve any specialized assumptions.

A radically new element entered the picture with Onsager's rigorous solutions of the two-dimensional Ising model (*35*), resulting in a logarithmic singularity of the specific heat. Although it was not immediately obvious to what an extent this result was applicable to actual systems, it was hard to avoid the inference that Landau's theory contained implicit assumptions that impaired its apparent generality.[33] It became necessary to reconsider the fundamental assumptions of the theory,[34] and also the method of analysis of the experimental results.

[33] The present theory agrees in some respects with that of Landau. Both theories derive the theory of phase equilibrium from principles of thermodynamic stability. Moreover, Landau was the first to recognize the importance of symmetry for the λ-phenomenon in crystals, a point to which we return shortly. In other respects there are deviations. In particular Landau's conclusions concerning the nature of the λ-singularity differ from those reached in the present paper. The origin of this discrepancy is that Landau expresses the fundamental equation of his system in terms of the Gibbs function. This presentation is equivalent with that in the energy scheme if (5.14) and (5.17) are satisfied. However, at critical points the Jacobian in (5.14) vanishes, and the compliance coefficients, that is, the second derivatives of the Gibbs function become infinite. Hence, the power series expansion of the Gibbs function does not have the regularity properties assumed by Landau (p. 434 of Ref. *20*).

[34] An attempt of this sort was advanced a few years ago by the author (Ref. *11*). Some of the ideas of the present paper can be found already in this earlier publication, in particular the identification of the λ-lines as singular, critical states. However, in that paper, the logical structure of the theory was left in a rudimentary state. Thus no distinction was made between the thermodynamic extensive variables listed in *P a3*, and the pseudo-

The numerous attempts to test the Ehrenfest relations were always inconclusive since it proved to be difficult, if not impossible to deduce definite $\Delta\psi_{ik}$ from the experimental curves. In particular, the $\Delta\psi_{ik}$ tend to increase as dT, the variation of the temperature between two consecutive measurements is decreased. Since the ideal thermostatic results can be expected to obtain only in the limit $dT \rightarrow 0$, the sharpening of the λ-anomaly in the course of this limiting procedure can be taken as a strong argument in favor of a parabolic singularity.[35] In view of this situation it is common usage at present to classify phase transitions in terms of categories that are more inclusive than those of Ehrenfest. Such categories have been advanced in particular by Pippard who developed also the analogs of Ehrenfest's relations adapted to λ-singularities rather than to second order discontinuities (*36*). In Appendix C relations practically identical to those of Pippard are derived within the present theory.

In spite of the usefulness of such *ad hoc* taxonomical considerations, we wish to emphasize that the most important results of thermodynamics are those in which taxonomical categories are derived from thermodynamic principles. The analysis of experimental results in such terms allows one to extract significant structural information concerning the systems involved. A case in point is the discussion of the role of symmetry in one component systems exhibiting singular λ-lines in the p, T plane.

The discussion of the symmetry of specific thermodynamic systems can be given only in a microscopic, structural sense. Accordingly the rest of this section as well as Section XI go beyond the limitations we have observed thus far. However, the discussion will be qualitative and conceptual throughout.

The connection between the macroscopic and the microscopic points of view is provided by the ensemble concept. In particular, an isolated thermodynamic system in equilibrium is described by a microcanonical ensemble Γ that has the ergodic property. By this we mean that there are incessant transitions among the microstates of constant energy that are associated with Γ, and that the

---

thermodynamic variables, such as the electric and magnetic moment and the elastic strain tensor. Accordingly, the number r defined in (2.7) differs from the definition given earlier. The present alternative is essential for the precise statement of the phase rules.

[35] The differences $\Delta\psi_{ik}$ are obviously not invariant with respect to the variation of the experimental conditions. In earlier attempts of testing the Ehrenfest relations, one often proceeded by prescribing special experimental conditions in order to arrive at definite $\Delta\psi_{ik}$. However, this procedure misses the point: the issue to be decided is not the validity of the Ehrenfest relations, since these are exact thermodynamic relations whenever the $\Delta\psi_{ik}$ are meaningful. The question is rather, whether or not these quantities are physically meaningful and whether the "order" of transitions is a relevant concept for their classification. The above mentioned difficulties indicate that, generally speaking, the answer to this question is in the negative.

average properties of the latter provide the macroscopic properties of the system.[36]

Let us consider now a symmetry operation R, such as the translation, rotation or inversion of the spatial coordinates, or the inversion of time. The ensemble Γ is said to be invariant under R, that is $R\Gamma = \Gamma$, if each microstate of the ensemble is transformed into a microstate that belongs to the same ensemble. In particular, some microstates may be themselves invariant under R.

The present theory predicts that the λ-lines of one-component systems are associated with at least one symmetry operation R in such a way that on one side of the λ-line in the p, T plane the ensemble Γ of the system is invariant under R, whereas on the other side we have modification doublets of the same energy, the ensembles of which satisfy the relations.[37]

$$R\Gamma^+ = \Gamma^-, \qquad R\Gamma^- = \Gamma^+. \tag{10.3}$$

The ensembles Γ^+ and Γ^- constitute a decomposition of the energy surface into two subsets, each of which has the ergodic property, while the entire surface is not ergodic. The sign of the quasi-thermodynamic parameter η serves to identify the subset relevant in an actual situation. The need for using such nonclassical parameters is connected with the limitation of validity of the classical ergodic hypothesis.

Assuming that the surrounding favors neither of the doublets over the other, from the intrinsic point of view the chances for their formation are equal, and in strict equilibrium one is realized with the exclusion of the other. In practice, of course, the expenditure of a small amount of surface energy produces a course grained mixture of both doublet forms. In crystallography this common phenomenon is called twinning. Assuming that the operation R is the inversion, the two forms Γ^+ and Γ^- are called *enantiomorphic*.

In ferromagnetic and ferroelectric crystals the long range depolarizing forces

[36] We note that our postulation of the ergodic property serves merely to define the ensemble Γ: we confine ourselves to such subsets of the energy surface in phase space for which the ergodic property is valid. The mathematical ergodic theorem asserts the ergodic property for metrically transitive (indecomposable) parts of the energy surface. (See Chapter III of Ref. 37.) The ergodic hypothesis of classical statistical mechanics goes beyond this statement and claims in effect, that the entire energy surface is metrically transitive. This conjecture may be plausible for chemically inert gases. However the occurrence of doublet modifications represents a physically significant counter example against its general validity.

[37] Strictly speaking, the thermodynamic argument requires only that there should be modification multiplets. Thus we may have a polar group pointing in any of the six cubic directions (1, 0, 0). This case could be considered as an accidental degeneracy of three doublet modifications. Empirically the modification always appears in pairs, but we have not been able to prove this result in sufficient generality.

compensate for the surface energy and one usually has a so-called domain structure of modification doublets even in strict equilibrium (*19*).

We shall occasionally refer to the modification Γ as the *high form* and to Γ^+, Γ^- or to their twinned mixture as the *low form*. These terms have a convenient double meaning. Strictly speaking they refer to forms of higher or lower symmetry, but, apart from a single known exception, they can be interpreted also as high and low temperature forms.[38] In order to understand the scope and limitation of this connection between symmetry and temperature, we have to discuss somewhat more specifically the structural properties of systems undergoing the high-low symmetry change.

We note that in case of crystals the specification of Γ ensembles that is adequate for the present purpose consists of three steps.

(i) The first step is the purely geometrical characterization of an ideal crystal in terms of a Bravais lattice and of atomic parameters determining the equilibrium positions of atoms in the primitive unit cell, the basis.

(ii) The sites of the ideal lattice are not actually occupied by atoms. Hence, as a second step one has to specify for the ensemble Γ the probabilities of finding an atom at a definite location with respect to the ideal lattice.

There are several mechanisms responsible for the deviation of the actual from the ideal configuration, such as small vibrations and crystal defects. The mechanism of particular interest in the present context is the ordering-disordering process arising in crystals in which the number of sites for atoms of a given kind is in excess over the number of atoms.

(iii) Certain crystals are not adequately characterized in terms of their configuration, but require also the specification of the distribution of momentum. This term is used here and in the following discussion in a generic sense to denote the linear momentum, or the orbital and spin angular momentum of the elec-

[38] The role of symmetry in the λ-phenomenon of crystals was first recognized by Landau (*34*), who showed that the continuous variation of the densities x of a system on crossing the λ-line is consistent with the discontinuous variation of symmetry if the symmetry of the low form is a subgroup of that of the high form. Although the abrupt change of the symmetry group defines a sharp λ-temperature $T_\lambda(p)$ at a constant pressure for any experimental run, the value of this temperature is subject to thermal hysteresis and, in contrast to temperatures of ordinary heterogeneous equilibrium, $T_\lambda(p)$ cannot be chosen as a reliable temperature fixed point. The reason for this situation was pointed out by Justi and Laue (*38*). A homogeneous phase can be supercooled below T_λ, just as it can be supercooled below the temperature of heterogeneous equilibrium. In the latter case the appearance of the low temperature phase releases enough latent heat to bring the system back to the equilibrium temperature, that is maintained as long as there are two phases present in equilibrium with each other. In the case of λ-points there is no latent heat, nor are two phases coexisting with each other in equilibrium. Hence there is no mechanism for assuring that the $T_\lambda(p)$ should be identical in different experimental runs, particularly for heating and cooling experiments.

trons, or the angular momentum arising from the degeneracy of the wave functions of the molecules or ionic groups. To describe such states we have to use quantum mechanical wave functions, that is probability amplitudes of states, rather than probabilities of geometrical configurations. The ensembles Γ are represented now in terms of density matrices.

Corresponding to the three levels in the description of the crystal we distinguish three classes of λ-transition.

(i) *Displacive transitions.* The change of the crystal symmetry is adequately described in terms of a slight distortion of the ideal lattice.

(ii) *Transitions involving configurational ordering.* The change of symmetry occurs because of the variation of the probability of occupancy of sites that are in excess over the number of atoms that occupy those sites. The lattice may become distorted in the course of the ordering process, that is in certain cases the ordering process is superposed over a displacive transition.

(iii) *Transitions involving momentum ordering.* The transition is brought about by a change in the degree of order of the distribution of momentum, understood in the generic sense explained above. This ordering of momentum may be in addition to a certain configurational ordering.

The majority of λ-lines studied so far belong into class ii. These cases of configurational ordering are also the best known and we confine ourselves to a sketchy discussion aiming mainly to establish the contrast and analogy with classes i and iii.

The prototype of a system with configurational ordering is the Ising model. We assume that each unit cell is capable of two geometrical configurations symbolically denoted by "+" and "−". For the purposes of statistical calculations it is usually assumed that the energy of the system depends only on the relative configuration of nearest neighbors and favors like neighbors. For the present qualitative discussion the law of interaction may be more general, so long as the interaction energy falls off with distance strongly enough to be consistent with the additivity of the macroscopic energy.

In each microcanonical ensemble there are definite probabilities p^+ and $p^- = 1 - p^+$ for finding the two cell configurations, respectively. Thus we can define the well-known long range order parameter

$$\eta = p^+ - p^- = 1 - 2p^-. \tag{10.4}$$

In the high form of the system $\eta = 0$, whereas in the low form $0 < |\eta| \leqq 1$ and the two signs of η correspond to the two doublet modifications.

We consider the operation R that interchanges the two configurations in each unit cell. Ensembles with $\eta = 0$ are invariant under R, while (3) holds for ensembles with $\eta \neq 0$. Thus the connection with the foregoing general discussion is established.

The Ising model admits a variety of concrete interpretations. We survey some of the most important ones in order to indicate the extent of experimental corroboration of the general theory.

First, we have *substitutional transitions* occurring in certain 50 atomic percent binary alloys, *AB*. The lattice is subdivided into two sublattices and the two configurations are obtained as an *A* atom occupies a site of the first or second lattice, respectively.

We denote another important subclass as *orientational transitions*. These arise in crystals containing molecules or ionic groups that admit two equilibrium orientations in each unit cell. Examples are the hydrogen halides and a number of ammonium salts (*32*).

An alternate theory of the transitions just mentioned was put forward by Pauling (*39*) and Fowler (*40*). This theory associates the transition temperature with a change from rotational oscillation of the molecules or ionic groups below the λ-point to a modification in which most of the molecules are freely rotating. This conception of "rotational" rather than "orientational" transitions found wide acceptance. Thus the review papers of Ref. *32* analyze the experiments from this point of view.

While the idea of orientational transitions is consistent with the present theory, that of rotational transitions is not, unless it is supplemented by additional assumptions guaranteeing the occurrence of modification doublets.

From the experimental point of view a decision between the two theories was until recently very difficult. The particles partaking in the reorientation or rotation are usually protons or deuterons, which are notoriously hard to locate experimentally.

This gap was filled to a large extent by the powerful method of nuclear magnetic resonances (*41*), that provides information on the dynamic behavior of protons in solids and liquids. The application of this method to the hydrogen halides, to NH_4Cl and to a number of other crystals (*42*) decided the question in favor of the orientational theory. It was found in particular that the λ-points are not associated with a marked change in the dynamic properties of the protons, and, conversely, whenever there is a change from a more to a less hindered rotation in a relatively narrow interval of temperature, this dynamic change does not manifest itself in caloric effects and certainly gives rise to no λ-transition.

While not every system can be analyzed in such an unambiguous way, there is no instance of a verified Pauling type transition.

Among the crystals investigated by the nuclear magnetic resonance techniques the case of methane requires special attention. The experimental results (*43*) were interpreted by Nagamiya (*44*) and Tomita (*45*) as follows. There is a marked transition from a state of coupled rotations to a relatively free rotation at about 65°K. There is no anomaly of the specific heat in this region.

The λ-transition at 20°K is associated with a change in ordering, but this is not of the purely orientational type, since the molecules are in a state of coupled rotation or resonance among equivalent states even at the lowest temperatures. These states are constructed by taking into account the degeneracy connected with the proton spin.

The author believes that methane is an example for a class iii λ-point involving momentum ordering, a case that will be discussed below. It is very likely that the Nagamiya model is consistent with our requirement that the low form of CH_4 should appear in doublet modifications. However, the problem needs more experimental and theoretical study.

We turn now to the class i λ-points denoted above as displacive transitions. These are conceptually simpler than order-disorder transitions, but they are less common and from the experimental point of view more elusive. The transitions can be described in geometrical terms. Thus in barium titanate the cubic crystal suddenly becomes tetragonal. The deviation from cubic symmetry is measured by the parameter

$$\eta = \frac{c}{a} - 1, \tag{10.4}$$

where a, c are the tetragonal lattice constants. The tetragonal axis is polar and we have $R\Gamma^+ = \Gamma^-$ for the low form and $R\Gamma = \Gamma$ in the high form. Thus the symmetry properties of the transition are the same as those of order-disorder transitions. In particular, η decreases continuously to zero as the λ-line is approached from below.

The displacive transitions known at present are all ferroelectric: Rochelle salt, barium titanate, and the ferroelectrics discovered by Mathias and his collaborators (*46*). The dielectric constant exhibits the same λ-singularity as, say, KH_2PO_3 in which the spontaneous polarization is due to the ordering of hydrogen bonds (*47*). The transitions are recognized to be displacive by the absence of λ-singularity in the specific heat. This is due to the fact that, in contrast to class ii transitions, the high and low forms do not differ substantially in their entropy.

A striking manifestation of this circumstance is the lower critical point T_l of Rochelle salt. Below T_l the crystal reverts to the nonpolarized high symmetry form that is stable above the upper Curie point T_u. Although there is no satisfactory theory for this lower critical point, the geometrical character of the symmetry change at least represents no conflict with thermodynamics. This is in contrast with order-disorder transitions, for which the high symmetry form is more disordered, has the higher entropy and always appears at the high-temperature side of the λ-point.

It is likely that there exist also nonferroelectric displacive transitions. However, because of the absence of anomalous physical properties these might easily escape detection. The change in symmetry should be observable by crystallographic means.

We conclude the discussion of displacive transitions by noting that this terminology was advanced originally by Buerger (*48*) to designate the high-low transition of quartz (*49*).

In its low form quartz is trigonal (D_3) and transforms at $t = 574°C$ into the hexagonal high form (D_6). The trigonal form exists in the equivalent "obverse" and "reverse" forms which are modification doublets in the present terminology. These modifications are transformed into each other by atomic displacements that are small compared to the lattice constant. This is in contrast to the orientational and substitutional transitions in which the distance between equivalent positions of atoms partaking in the ordering process is comparable to the lattice constant. This is of course the basis for designating this transition as displacive. At the same time, however, the transition has a pronounced λ-singularity of the specific heat. In fact, this is presumably the first instance of a λ-point for which an infinite specific heat was reported on purely empirical grounds (*50*). Thus we have a clear indication of an order-disorder process among the slightly displaced equivalent positions. We find it advantageous to reserve the descriptive term "displacive" to the above mentioned class of purely geometrical transitions. The modification doublets of low quartz often occur simultaneously in the same crystal, they are called electrical, or Dauphiné twins. The twinning disappears at the λ-point.

Another property of quartz is that both the D_3 and the D_6 forms are optically active, and appear in right and left forms because of a helical arrangement of the constituent atoms. The optical twinning persists over the entire range of stability of the crystal without ending in a λ-point. In our terminology, the situation is to be described as a *phase doublet.*

The most interesting, but also the most problematic λ-transitions are those of class iii involving momentum ordering. Ferromagnetic and antiferromagnetic substances exhibiting definite patterns in the distribution of electron spin (*51*) belong in this class. The symmetry requirements of the present theory are evidently satisfied. In fact, we may interpret the symmetry operation R as the operation of time reversal $t \rightarrow -t$, that transforms every spin distribution into an energetically equivalent, yet distinct distribution.

The interpretation of R as time inversion may not be the only possibility. The same result could in some cases be achieved in terms of spatial operations. However, since time inversion is independent of the special geometrical properties of particular systems, it is presumably the appropriate operation for all class iii transitions.

The importance of including time inversion in the discussion of the symmetry of thermodynamic systems has been pointed out before (*52*). Also "magnetic space groups" and "magnetic symmetry classes" have been derived by joining the time inversion to the spatial rotation and inversion operations (p. 428 of Ref. *20*).

We have pointed out above that the λ-point of solid methane is likely to represent an instance of momentum ordering. The occurrence of this phenomenon is certainly not common, but depends on a delicate balance of the interaction forces tending to bring about configurational ordering with the effect of quantum mechanical zero point energy opposing such a configurational order. Thus the replacement of one fourth of the protons with deuterons (CH_3D), or the application of hydrostatic pressure is sufficient to upset this balance and produce another transition (see Clusius et al. *15*; *53*), below which the system is presumably in a state of configurational order (*54*).

We wind up the discussion of crystalline λ-lines with the conclusion that in all cases where the structural properties of the system are sufficiently understood, the connection between λ-singularity and symmetry is borne out by experiment. This conclusion forms the basis for the more significant but at the same time more problematic, inferences concerning liquid helium.

XI. SUPERFLUIDITY

The p, T diagram of helium exhibits a λ-line $T_\lambda(p)$ marked by an anomaly in the specific heat $C_p(T)$ and in the other compliance coefficients. This anomaly is similar to the one found at the critical gas-liquid temperature T_c, provided the heating is performed at the exact critical pressure p_c. Moreover, the entire λ-line is thermodynamically identical to the λ-lines that are otherwise found only in solid and liquid crystals.[39] In fact, it was the anomaly of the specific heat of helium that was first designated as a λ-point (*55*) and was used as the experimental basis for the concept of second order transitions (*8*, *9*). Finally, it is for helium, that the singularity of the specific heat was experimentally established in a most convincing fashion by the precision measurements of Fairbank *et al.* (*33*).

[39] The discussion of this section is based on the assumption that the λ singularities in helium form a one-dimensional manifold. Actually, the singularity of the function $C_p(p, T)$ has been experimentally established only at the saturation vapor pressure. In the absence of direct measurements of $C_p(T)$ at higher pressures, we can support the above assumptions by plausibility arguments. First, we conjecture that the saturation vapor pressure has no special significance for the intrinsic properties of the liquid, and these are unlikely to undergo a qualitative change upon application of a slight excess pressure. Second, the Pippard relations (see Appendix C) are derived from the assumption that the λ-singularity is only weakly influenced by pressure. According to personal communication of Chase and Maxwell, the relation (C9) is satisfied if their experimental values of the coefficient of expansion $\alpha(T)$ are plotted against the $C_p(T)$ of Ref. *33*.

Under such conditions, it is plausible to assume that the λ-line of helium belongs thermodynamically to the same category as the λ-lines of crystals and that the second phase rule of the present theory applies to it. This assumption leads to the conclusion that the λ-line consists of critical points of the second kind, the symmetry number is $\sigma = 1$, and the superfluid state exists in at least two energetically equivalent modifications that are related to each other by symmetry. Since this situation prevails even at absolute zero, we have to conclude that the ground state wave function of the superfluid state is at least doubly degenerate.

This conclusion is remarkable for more than one reason. We state in essence that the existence of a singular λ-line cannot be inferred from the density of low-lying excitations, but is closely connected with the structure and degeneracy of the ground-state wave function of the many body system. The prediction of a degenerate ground state for helium is at variance with the existing microscopic theories of this system and conflicts also with the widely accepted statement of Nernst's law, according to which the ground state wave function of a quantum mechanical system is nondegenerate. (See p. 66 of Ref. *20*.) However, the argument raised against the present ideas from the point of view of Nernst's law is easily dismissed, since it stems from an inadequate formulation of this law.

The degree of degeneracy of a quantum state with the angular momentum quantum number J is $2J + 1$. Since J cannot be zero for systems containing an odd number of electrons, we have obvious counter examples to the statement that the ground-state wave function is nondegenerate.

In order to arrive at a proper formulation of Nernst's law we consider for the sake of definiteness a crystal of N unit cells with a free spin per cell. The degree of degeneracy g_0 of the lowest ferromagnetic state is between the limits:

$$2 \leqq g_0 \leqq N + 1, \tag{11.1}$$

where the upper limit refers to the embedding of the angular momentum $N\hbar/2$ into an isotropic medium. The double degeneracy of the lower limit is a consequence of time reversal and persists in any anisotropic medium. The third law requires that the entropy per unit cell should tend to zero in the limit of an infinite crystal. This is indeed the case:

$$s_0/k = \lim_{N\to\infty} (\log g_0)/N = 0. \tag{11.2}$$

This result is to be compared with the entropy per unit cell in the high-temperature paramagnetic limit

$$s/k \approx \lim_{N\to\infty} (N \log 2)/N = \log 2. \tag{11.3}$$

In the high-temperature region the wave functions associated with the cells are joined with random phases, while in the region of validity of Nernst's law there are strict phase relations to be observed, and the degeneracy of microscopic units does not manifest itself in a zero point entropy.

Our thermodynamic conclusion that the ground state wave function of helium is degenerate and presumably a doublet, could be substantiated only by an explicit construction of such a wave function. This task is beyond the scope of the thermostatics developed in this paper. Nevertheless we are in a position to discuss a number of heuristic ideas which shed some light on the intuitive meaning of our general results, and which provide us also with clues as to the direction in which a quantum mechanical theory is to be sought.

Our starting point is the Ising model, the only system for which a parabolic singularity has been rigorously demonstrated (*35*). This model can be adapted to a variety of physical situations. In particular, Lee and Yang (*56*) have shown that the Ising model in a magnetic field is mathematically identical with that of a "lattice gas," in which each lattice point is either vacant or occupied by an atom. To each configuration of the lattice of spins there corresponds a configuration of the lattice gas in which a lattice point is vacant or occupied according to whether the corresponding spin is "+" or "−".

The density of the gas in appropriate units is formally identified with

$$\rho = (1 - \eta)/2, \tag{11.4}$$

where η is the long-range order parameter, or the intensity of magnetization. The pressure of the gas is

$$p = -F - H, \tag{11.5}$$

where F is the free energy per spin and H is the magnetic field.

Lee and Yang have shown that the lattice gas is, on the one hand, formally identical to the Ising model, and has, on the other hand, properties that are remarkably similar to those of a real fluid. In particular, the critical temperature of magnetization (Curie temperature) corresponds to the fluid critical point.

It is apparent that the correspondence established by Lee and Yang is consistent with the second phase rule as summarized in Table I. Both systems have isolated critical points ($\delta_{II} = 0$), but in the Ising model $r = 1$, $\sigma = 1$ while in the real fluid $r = 2$, $\sigma = 0$. The lattice gas provides the link between the systems. The pressure (11.5) is on the one hand analogous to the fluid pressure, but formally it has symmetry properties around the critical pressure that allow one to connect it with the magnetic field. The symmetry of the lattice gas, which is spurious, so far as the real fluid is concerned, manifests itself in the "law of constant diameters" according to which the sum of the conjugate densities of the coexisting gas and liquid is independent of the temperature. This law is a fairly good approximation to the "law of rectilinear diameters" of the real fluid, stating that the sum of the conjugate densities is a slowly varying linear function of the temperature (*57*). The residual discrepancy between the law of constant diameters and the experimental situation is inherent in the model. In contrast to the symmetry of opposite spins in the Ising model, the symmetry between particles and holes in the real fluid is only approximate. Therefore, the author cannot accept the contention of Lee and Yang that a real continuum gas can be considered as the limit of a lattice gas as the lattice constant becomes infinitesimally small.*

The representation of the properties of real fluids in terms of the lattice gas poses another problem. Strictly speaking a lattice corresponds to a crystal. If the lattice gas is to represent a fluid, the term "lattice" is not to be taken too literally. In ambiguous situations it is useful to qualify that term and to distinguish "metric" and "topological" lattices. The term "metric lattice" should indicate a strict interpretation and imply the existence of long range correlations of the distances of lattice points which produce sharp x-ray diffraction lines. The topological lattice is merely a graph, in which the number of lines branching out of lattice points represent coordination numbers. The lattice of the Ising model can be given such a topological interpretation and coupling energies assigned to bonds without any reference to the distance of the spins.

The representation of fluids in terms of lattices is understood in the same topological sense. According to (11.4) the density is defined in combinatoric rather than metric terms. In addition, the lattice may have also approximate metric properties ranging over a few lattice constants.

Of course, this handling of the situation by means of a semantic refinement is only the formulation of the problem, rather than its solution. In a satisfactory theory of fluids the topological lattice should be given a rigorous mathematical expression.

Lee and Yang have shown under rather general assumptions concerning the coupling forces, that the lattice gas cannot have more than one critical point. In order to find a model exhibiting a one dimensional continuum of critical points which is to represent the helium λ-line, we have to utilize the Ising model in a different fashion.

We start from an idea suggested by Fröhlich (*58*), whose point of departure was a crystalline model of liquid helium advanced earlier by London. For a critical discussion of this theory we refer to Section B of Ref. *24*. The idea is briefly this: liquid helium is considered a binary alloy consisting of N atoms and N holes constituting congruent lattices. Above the λ-point the atoms and holes would be distributed at random over the 2 N sites, whereas below the λ temperature the atoms would statistically prefer one of the sublattices. We should have a class ii substitutional λ-transition.

In contrast to the isolated critical point T_c of the lattice gas, the Fröhlich model leads to a λ-line $T_\lambda(p)$. The crux of the difference is that: in the lattice

* Note added: The difficulty is removed for long-range forces rather than nearest-neighbor interactions.

gas, the magnetization, a *pseudo-thermodynamic* variable, is replaced by the density, a *thermodynamic* variable in the sense of *P a3*. In the Fröhlich model, the magnetization is turned into a long range order parameter, a *quasi-thermodynamic* variable. The existence of a λ-temperature is assured even at a constant value of the density, considered as an arbitrary parameter of the problem. A variation of the density, or pressure leads to a one dimensional continuum of λ-points $T_\lambda(p)$.

Either one of the two sublattices may be selected for occupancy and thus the model leads to the doublet modifications required by thermostatics. In spite of this success in explaining the λ-line, the Fröhlich model has serious shortcomings and was almost at once abandoned.

In the first place, no evidence of lattice structure was revealed by x-ray studies. Even more important, the energetic conditions in helium do not favor the postulated configurational ordering. If the sites of lattice A are all occupied, those of lattice B are all vacant, the transitions of atoms from lattice A to B lowers the energy of the system. This is, of course, in conflict with the idea of configurational ordering.

In view of this situation London rejected the idea of configurational order and suggested that helium might be the seat of an order in momentum space. Moreover, in rediscovering the condensation phenomenon of the Bose–Einstein gas, he provided a model exhibiting such an order. The theoretical work following up London's initiative is so extensive and so well known that a documentation is neither practical, nor necessary in the present context.

In spite of the success of the boson gas theories, we have to conclude from our thermostatic analysis that the structureless nondegenerate ground state of the ideal Bose–Einstein gas is inconsistent with the experimental λ-singularity. The strength of the thermodynamic argument leads us to formulate the conjecture, that under the coupling conditions prevailing in liquid helium, the correct solution of the many boson problem has a degenerate ground state. This degeneracy might be due to the time inversion symmetry similar to the well known case of the Kramers degeneracy (*59*).

The fact that the theories of the low density imperfect boson gas have not yet confirmed the above conclusions should not be taken as a serious argument against them. We believe that the ideal crystal and the ideal gas theories represent complementary limits of the correct theory. It should be extremely difficult to demonstrate subtle structural properties from power series expansions centering around the ideal gas limit, and it is an attractive possibility to bracket the correct theory from both ends. The exploration of the crystalline limit as a point of departure for improvements is so much the more promising, as the neglect of this approach is in curious contrast with the high degree of sophistication reached by the gas theories.

As an introduction to such a program, we reconsider the above mentioned two arguments levelled against the Fröhlich model. First, we note that the lattice should be interpreted in the weaker, topological sense and thus the conflict between fluidity and lattice structure is neither more, nor less serious than that arising in connection with the lattice gas. Second, we hold that London's argument for replacing configurational order by an order in momentum space is convincing. However, the Bose–Einstein condensation in momentum space provides only an incomplete picture of the class iii λ-transition involving momentun ordering, discussed in the last section.

The relation between momentum ordering and Bose–Einstein condensation is conveniently explained in terms of ferromagnetism. The low energy excitations of a ferromagnet can be represented as superposition of spin waves (*60*). Starting from a crystal exhibiting complete spin alignment, a spin wave corresponds to a single inverted spin traveling by means of resonance across the crystal. These waves have the character of bosons. Their condensation temperature corresponds to the Curie temperature.

The spin waves lack any spatial localization and condense in momentum space. Hence a theory that would conceive of ferromagnetism merely as an instance of Bose–Einstein condensation would not provide any information about the structure, symmetry and degeneracy of the ground state. Yet there is compelling experimental evidence for a spatial pattern of spin distribution both for ferromagnetic and antiferromagnetic substances (*51*).

We conjecture that the situation in liquid helium is analogous, although the structure, symmetry and degeneracy of the ground state are extremely elusive from the experimental point of view. Therefore, at the present juncture, the clarification of these ideas is primarily a task for theory. The central issue is establishing the nature of the localized "momentum" that in helium plays the role of that of the electron spin in ferromagnets. Closely connected with this problem is the nature of the order that is emerging in the λ-transition.

We wind up our discussion by formulating with due reserve an intuitive qualitative picture of liquid helium that is suggested by the foregoing considerations and is consistent with the experimental situation.

According to this tentative picture liquid helium is considered as a topological lattice in which each elementary cell is the seat of a quantized angular momentum. In the lowest state the moments are lined up parallel to each other ensuring that the large zero-point kinetic energy does not contribute to randomness and entropy. Time reversal would transform this state into another of opposite momentum. Thus we arrive at an interpretation of the doublet modifications of thermostatics. In the bulk liquid in rest there are equal amounts of both modifications forming a domain structure. An application of a torque would bring about an unbalance of the domains of opposite angular moments in analogy to the action of a magnetic field on a ferromagnet. The critical velocity (p. 198 of Ref.

25) might represent the transition from an orderly domain structure to a turbulent regime in which the domains are tangled in a complex fashion.

Finally the system will exhibit elementary excitations analogous to spin waves, in which a single cell has an elementary momentum with a sign opposite to that of the domain. Such excitations might be identified with Landau's rotons.

While the above picture is in qualitative agreement with experiment, we were not able so far to deduce quantitative experimental predictions. A satisfactory theory could be developed only in quantum mechanical terms.

Finally we digress to make a short remark on superconductivity. The well-known analogy with superfluidity would suggest the conjecture that here again we deal (a) with a λ-anomaly (b) with degenerate ground-state wave functions transformed into each other by time reversal.

However, this analogy is incomplete and superconductors seem to be the only examples for an Ehrenfest type discontinuity of the specific heat. Under such conditions the second phase rule cannot be invoked either to assert or to deny conclusion (b), although we conjecture that the conclusion is correct.

XII. DISCUSSION

Technically we should at this point evaluate our postulational basis by surveying the success of the theory in describing experimental facts. However, to the extent that the present theory covers traditional ground, we can confine ourselves to a mere listing of the types of predictions of thermostatics that are subject to verification. (i) identities among measurable quantities; (ii) the possibility of establishing an absolute temperature scale that is independent of the properties of any particular substance; and (iii) the behavior of systems in the space of intensities, in particular, under conditions of special interest, such as phase equilibrium, critical points, and the vicinity of absolute zero.

The experimental verification is overwhelming on all three counts and for all substances investigated. In particular, the present theoretical framework is wide enough to describe the so-called λ-phenomenon or second order transition which did not fit into the classical theory.

The discussion of the λ-phenomenon is a convenient point of departure for bringing out the characteristic new aspects of the present theory.

From the point of view of the classical phase theory of Gibbs, λ-points have contradictory properties. On the one hand, they are like critical points, the specific heat c_p and the other compliance coefficients exhibit characteristic singularities. On the other hand, instead of appearing in isolated points in the $p - T$ plane, the λ-points of one-component systems form λ-lines in that plane. In this respect, they are more like lines of heterogeneous equilibrium. Ehrenfest sought to solve this difficulty simply by creating new categories which he called phase transitions of higher order. Quite apart from the fact, whether or not these categories are complete enough, this procedure can be accepted as satisfactory only if one considers thermodynamics merely as a taxonomical discipline in which experimental facts are characterized according to their empirical features.

In the present theory we return to the point of view implicit in Gibbs' work according to which taxonomical conclusions are derived from fundamental principles. The failure of the classical theory to account for λ-lines can be traced to the fact that the specification of homogeneous phases or modifications in terms of extensive variables is incomplete, and has to be supplemented by nonclassical variables that account for their symmetry properties. The incorporation of the symmetry concept into the fundamental principles of thermostatics is among the most characteristic features of our generalization of thermostatics. We now proceed to assess the merits of this innovation.

In the first place, the new definition of phase leads to a generalized version of the phase rules. For one-component systems the theory predicts the possibility of λ-lines in the p, T plane, along which the compliance matrix is singular, provided the following situation prevails: the λ-line separates two regions of the p, T plane that correspond to what we call the high and low forms, respectively. The former has the higher symmetry, and its group contains at least one element R which is no longer in the symmetry group of the low form. Assuming that R^2 is an element of the symmetry group of the low form we conclude that the low form consists of doublet modifications which are transformed into each other by the operation R[37]. In the ideal case the low form appears as one or the other of the two modifications forming the doublet. Practically, however, one usually has twinning and the two modifications appear as different domains of the same sample.

The equivalence under the operation R assures the identity of the modifications from the energetic point of view. At the same time the distinctness of the two numbers of the doublet clearly requires the use of nonclassical variables that are not invariant under R. The value of these variables goes gradually to zero as the λ-line is approached.

This situation is to be compared with the classical gas-liquid critical points which are isolated singular points in the p, T plane and where two modifications become identical that differ from each in classical variables, rather than forming a doublet.

We draw attention to the unusual logic of the argument: we assume that the singularity of the compliance matrix is given from experiment and we infer the existence of modification doublets. In statistical mechanics the usual procedure is the opposite. We start from a hypothetical model; the symmetry properties, such as the occurrence of modification doublets, are known from the outset, and the problem is to establish the thermodynamic functions, in particular the singularity of the compliance matrix. This is only one of many features for which the thermodynamic and the mechanical procedures appear complementary to each other.

The comparison of our theoretical conclusions with experiment raises another methodological problem. Apart from limiting cases falling within the scope of classical crystallography, the symmetry properties of thermodynamic systems cannot be discussed in macroscopic terms. Thus the injection of symmetry into thermostatics calls at the same time for an extension of the theory in the microscopic direction.

In Sections X and XI we have made a start toward a microscopic discussion of symmetry. However, we went only as far as qualitative arguments would allow us to go. In such a way we arrived at a classification of the λ-transitions into

(i) displacive transitions,
(ii) transitions involving configurational ordering,
(iii) transitions involving momentum ordering.

"Momentum" is understood here in the generic sense and includes angular momentum and in particular spin. All of these dynamic quantities change sign on time reversal.

The characteristic symmetry operation R is defined for class i transitions in geometrical terms, for class ii in terms of probabilities of configurations, and for class iii in terms of quantum mechanical probability amplitudes, the operation R can be chosen in this case as the operation of time reversal.

We conclude from the survey of the experimental material that whenever the structural properties of a substance are sufficiently understood, the λ-phenomenon conforms to the conditions set by the generalized phase rule.

The case of liquid helium is in a special category. The theory leads to the inference that the ground state wave function of the superfluid is at least doubly degenerate. The operation R that transforms the two wave functions into each other is presumably the operation of time reversal.

This conclusion is at variance with the existing theories of superfluidity, and does not seem to be amenable to any direct experimental check. Under these conditions we are bound to ask: how reliable are the conclusions reached and what is the outlook for providing a quantum mechanical substantiation for them?

Concerning the first question we point out that the logic of MTE is considerably more rigorous than that of traditional thermostatics. In fact the cumulative effect of many apparently minor changes brings about a logical structure that forms the basis for a reconsideration of the traditional relation of thermodynamics and mechanics, an idea that will be developed elsewhere.

Among other features, we do not formulate the second law in the usual weak form, merely excluding the possibility of processes away from equilibrium. We rather postulate that equilibrium is reached, and the proper entropy is attained. The results are not universally valid for actual systems, but represent the *normal equilibrium behavior*, a practical criterion of equilibrium for actual systems.

Our strong formulation of the entropy maximum postulate yields a theorem which we have called the *principle of thermostatic determinism*. Suppose that an infinite reservoir is in equilibrium with a simple system for which it provides the surroundings. The principle of determinism states that the intensities of the surroundings uniquely determine the densities of the systems, provided condition (5.14) is satisfied. Conversely, the densities uniquely determine the intensities, if condition (5.17) holds. The inferences leading from the properties of the surroundings to that of the system and vice versa, correspond to basic experimental situations. Hence the special designation of the uniqueness theorem seems justified.

The principle of determinism has its limitations. Condition (5.14) breaks down at the critical point, and (5.17) fails near absolute zero. These limitations are real.

In a surroundings having the intensities of the critical point, the densities of the system are indeed subject to large uncertainties which originate in extremely large fluctuations.

Near absolute zero, where the entropy is practically zero, the system becomes insensitive to temperature variations. Physically, this effect comes from long-range space-time correlations in the system, which, within limits, renders it insensitive to outside influences.

The actual occurrence of these extreme situations makes it evident that a correct thermodynamics should include both the elements of randomness and of space-time correlations. The difficulties of developing such a theory are at present very great; however a step by step approximation seems feasible.

First, there is a purely statistical theory in which space-time correlations are assumed to be absent. A statistical theory of this sort, which is applicable to fluctuations around equilibrium, was developed in collaboration with Quay (*14*). This theory, the statistical thermodynamics of equilibrium (STE), will be published as the second paper of this series.

This theory appears as a very natural generalization of MTE. In MTE the virtual states of a composite system have only a formal meaning, they are comparison states for the extremum problem that determines equilibrium. In STE equilibrium is conceived as a statistical distribution over virtual states, and these assume physical reality as fluctuation states. The additive invariants now become additive random variables. Their distributions and other relevant statistical questions can be derived by using methods which had been advanced by Szilard (*61*) and Mandelbrot (*62*). By and large STE turns out to be an axiomalization of Gibbs' theory of ensembles (*63*).

The next step should be a quantum mechanical theory of space-time correlations. A satisfactory solution of this problem seems to be the prerequisite for substantiating our conclusions concerning the nature of the superfluid ground state.

APPENDIX A. QUASI-THERMODYNAMIC VARIABLES

We proceed to discuss a few ideas concerning the definition of quasi-thermodynamic variables.

We divide a thermodynamic system into M^3 volume elements (cells) of equal size and shape, where M is a large number. For crystals the cell may coincide with the unit cell. For fluids it is conveniently a cubic cell that contains a moderately large number of molecules.

The cells are numbered, as usual by vectors **n**, with integral components running from 0 to $M - 1$. Let the system be specified by a set of numbers $z_{\mathbf{n}}$, which denote the value of the quantity z in the cell **n**. The physical nature of the quantity z is left open at first. The fundamental domain of M^3 cells may be extended by means of periodic boundary conditions, and the $z_{\mathbf{n}}$ expressed in terms of normal coordinates.

Assuming, for simplicity a simple cubic lattice, we define

$$\xi_{\mathbf{k}} = \sum_{\mathbf{n}} z_{\mathbf{n}} e^{-2\pi i \mathbf{k}\cdot\mathbf{n}}, \tag{A1}$$

where the components of k are $0, 1/M, \cdots, M - 1/M$. The zero-wave vector normal coordinate

$$\xi_0 = \sum_{\mathbf{n}} z_{\mathbf{n}} \tag{A2}$$

is a translation invariant. Moreover

$$\xi_{1/2,1/2,1/2} = \sum_{\mathbf{n}} z_{\mathbf{n}}(-1)^{n_1+n_2+n_3} \tag{A3}$$

is an invariant with respect to the cubic group with a lattice constant of two units. We define the coordinates (2) and (3) as macrovariables, that are appropriate for the description of course-grained homogeneous phases. These conclusions were reached also by Landau and Lifshitz from more elaborate considerations (Chapter XIV of Ref. *20*).

The coordinates ξ_0 include the additive invariants of *P a3*. The normal coordinates $\xi_{\mathbf{k}}$ for which the components of the wave vector are not all zero or ½, are microvariables. They multiply by a phase factor $e^{2\pi i \mathbf{k}\cdot\Delta\mathbf{n}}$ as the lattice is displaced by $\Delta\mathbf{n}$. These variables are used for the description of instantaneous spatial fluctuations.

In addition to these thermodynamic variables there are translational invariants, and these are the ones that are admissible as quasi-thermodynamic variables η. If z is the coordinate of an atom in the unit cell, ξ_0 represents the ideal lattice of the average positions of the equivalent atoms. The corresponding microvariables are phonons.

If an atom or a group of atoms has two equivalent positions in a unit cell, we may assign the values $z = +1$ and -1 to cells in which one or the other position is actually occupied. In this case $\xi_0 = N^+ - N^-$, where N^+ and N^- are the number of atoms in the two positions, respectively. Since $N^+ + N^- = M^3$, the quantity

$$\frac{\xi_0}{M^3} = \frac{N^+ - N^-}{N^+ + N^-} \tag{A4}$$

is the well-known parameter of long-range order, while the variables with $k \neq 0$ describe the microdistribution or short-range order.

The nature of the λ-phenomenon depends essentially on the coordinate z, or ξ_0. In most cases, this parameter is of type (4). However, in some instances, as in Rochelle salt and barium titanate (*11*, *36*), the parameter η_0 relates to a small distortion of the crystal lattice from a higher to a lower symmetry form (orthorhombic to monoclinic in Rochelle salt). This case is sometimes called a displacive transition.

From the quantum mechanical point of view, in both of the cases mentioned above, ξ_0 corresponds to the absolute square of the wave function. However, it is very likely that in some cases, and particularly in helium, ξ_0 is to be defined in terms of the wave function itself.

It is apparent from the discussion of the text that not all parameters ξ_0 that are admissible on the basis of the foregoing considerations will be actually used as quasi-thermodynamic parameters η. The fundamental equation in, say, the energy scheme can be unconditionally minimized with respect to the $\xi_0 = \eta$. Thus

$$\partial u / \partial \xi_0 = 0. \tag{A5}$$

As a rule, this equation yields a unique solution for ξ_0. Only in those cases in which this solution is not single-valued will the parameter η be significant in accounting for a λ-phenomenon.

APPENDIX B. THE DIAGONALIZATION OF QUADRATIC FORMS

Consider the quadratic form

$$w_r = \tfrac{1}{2} \sum u_{ik} \xi_i \xi_k , \tag{B1}$$

where

$$u_{ik} = u_{ki} . \tag{B2}$$

We assume at first that the form is nonsingular, that is, its determinant

$$D_r = | u_{ik} | \neq 0. \tag{B3}$$

There are infinitely many linear (affine) transformations $\eta = T\xi$ that diagonalize the form (1). In particular we may obtain a diagonal form with the diagonal elements ± 1 and 0. The restriction to length-preserving, orthogonal transformations leads to the well known eigenvalues which in this case turn out to be physically meaningless. The physically significant diagonal form is obtained from the requirement that the transformation be unimodular, of determinant unity, or, in geometrical language, volume preserving.

We assume $u_{11} \neq 0$ and proceed by "completing the square":

$$w_r = \frac{1}{2} u_{11} \left(\xi_1 + \frac{1}{u_{11}} \sum_2^r u_{ik} \xi_k \right)^2 - \frac{1}{2u_{11}} \left(\sum_2^r u_{ik} \xi_k \right)^2 + \frac{1}{2} \sum_2^r u_{ik} \xi_i \xi_k . \tag{B4}$$

Substituting

$$\eta_1 = \xi_1 + \frac{1}{u_{11}} \sum_2^r u_{1k} \xi_k , \tag{B5}$$

we obtain

$$w_r(\xi_1 , \xi_2 , \cdots , \xi_r) = \tfrac{1}{2} u_{11} \eta_1^2 + w_{r-1}(\xi_2 , \xi_3 , \cdots , \xi_r), \tag{B6}$$

where the residual form w_{r-1} is of only $r - 1$ variables. This procedure can be applied to w_{r-1} and then continued until the diagonalization is completed, provided that each residual form contains at least one nonvanishing diagonal element.

If all diagonal elements vanish, the matrix of the form has a segment such as

$$\begin{Vmatrix} 0 & u_{jl} \\ u_{lj} & 0 \end{Vmatrix} . \tag{B7}$$

In this case, we substitute

$$\xi_j = \sqrt{\tfrac{1}{2}}(\eta_j - \eta_l) \tag{B8}$$

and

$$\xi_l = \sqrt{\tfrac{1}{2}}(\eta_j + \eta_l),$$

which yields

$$2u_{jl}\xi_j\xi_l = u_{jl}(\eta_j^2 - \eta_l^2). \tag{B9}$$

The combination of substitutions such as (5) and (8) always allows us to complete the diagonalization:

$$w_r = \tfrac{1}{2}\sum_1^r \lambda_i \eta_i^2. \tag{B10}$$

We note that transformation (8) yields diagonal elements of opposite sign (hyperbolic form). Such situations are of no physical interest in thermostatics, hence we confine ourselves to the case in which the transformation connecting the ξ_i and the η_i is of the "triangular" form:

$$\begin{aligned} \eta_i &= \xi_i + a_{12}\xi_2 + a_{13}\xi_3 + \cdots + a_{1r}\xi_r\,, \\ \eta_2 &= \qquad\quad \xi_2 + a_{23}\xi_3 + \cdots + a_{2r}\xi_r\,, \\ \eta_r &= \qquad\qquad\qquad\qquad\qquad\quad \xi_r\,. \end{aligned} \tag{B11}$$

If the form (1) is singular and the corresponding matrix is of rank $(r - 1)$, the last diagonal element vanishes, and we have $\lambda_r = 0$. In general, more than one diagonal element may vanish, but, thus far, such cases have not been found of interest in thermostatics.

The coefficients λ_i can be calculated in a straightforward manner by a step-by-step construction of the transformation (11). However, it is easier to proceed as follows:

The transformation (11) is unimodular, hence the discriminant of (1) is invariant.

$$\lambda_1\lambda_2 \cdots \lambda_r = D_r\,. \tag{B12}$$

This is seen also from the fact that (11) involves merely the subtraction of the multiple of one row of the determinant (3) from another while similar operations are performed on the columns.

Similar relations hold for the principal minors

$$\lambda_1\lambda_2 \cdots \lambda_k = D_k\,, \tag{B13}$$

where

$$\begin{aligned} D_k &= \begin{vmatrix} u_{11}\, u_{12} & \cdots & u_{1k} \\ u_{21}\, u_{22} & \cdots & u_{2k} \\ \cdots & \cdots & \cdots \\ u_{k1}\, u_{k2} & \cdots & u_{kk} \end{vmatrix} = \frac{\partial(P_1, P_2, \cdots, P_k)}{\partial(x_1, x_2, \cdots, x_k)} \\ &= \frac{\partial(P_1, P_2, \cdots, P_k, x_{k+1} \cdots x_r)}{\partial(x_1, x_2, \cdots, x_r)}\,. \end{aligned} \tag{B14}$$

From (13), we have

$$\lambda_k = D_k/D_{k-1} \tag{B15}$$

and from (14),

$$\lambda_k = \frac{\partial(P_1, P_2, \cdots, P_k, x_{k+1} \cdots x_r)}{\partial(P_1, P_2, \cdots, P_{k-1} \cdots x_r)} = (\partial P_k/\partial x_k)_{P_1 P_2 \cdots P_{k-1} x_{k+1} \cdots x_r}. \tag{B16}$$

While different unimodular transformations lead to different sets of λ_i, the product (12) is a unimodular invariant.

APPENDIX C. DERIVATION OF THE PIPPARD RELATIONS FOR THE λ-LINES OF ONE-COMPONENT SYSTEMS.

We consider the implications of the working hypothesis that in the neighborhood of the λ-line the Gibbs function can be approximately expressed as

$$G(p, T) = f(t) + tg(p) + h(p), \tag{C1}$$

where

$$t = T - T_\lambda(p) \tag{C2}$$

and

$$\xi = (dp/dT)_\lambda = (dT_\lambda/dp)^{-1} \tag{C3}$$

is the slope of the λ-line and is assumed to be constant; $f(t)$ is a singular function with the properties:

$$\lim_{t\to 0} f(t) = 0, \qquad \lim_{t\to 0} f'(t) = \text{finite}, \qquad \lim_{t\to 0} f''(t) = -\infty,$$

and $g(p)$ and $h(p)$ are slowly varying functions of the pressure. The primes denote differentiation.

The meaning of this assumption is that the parabolic singularity does not vary appreciably as we move along the λ-line.

We obtain through differentiation

$$-(\partial G/\partial T)_p = S = -f' - g \tag{C4}$$

$$(\partial G/\partial p)_T = V = -f'\xi^{-1} - g\xi^{-1} + tg' + h' \tag{C5}$$

$$-(\partial^2 G/\partial T^2)_p = C_p/T = -f'' \tag{C6}$$

$$\partial^2 G/\partial T \partial p = V\alpha = -f''\xi^{-1} + g' \tag{C7}$$

$$-(\partial^2 G/\partial p^2)_T = V\kappa_T = -f''\xi^{-2} + 2\xi^{-1}g' - tg'' - h''. \tag{C8}$$

From here

$$\alpha = C_p/VT\xi + \alpha_0, \tag{C9}$$

$$\kappa_T = \alpha\xi^{-1} + \kappa_0. \tag{C10}$$

The intercepts in these relations are

$$V\alpha_0 = -(\partial S/\partial p)_\lambda = g', \tag{C11}$$

$$V\kappa_0 = -(\partial V/\partial p)_\lambda = \xi^{-1}g' - h'' - tg''. \tag{C12}$$

The determinant of the compliance matrix is

$$\begin{aligned} D^{-1} &= C_p V\kappa_{T/T} - (V\alpha)^2 \\ &= -C_p V\alpha_{0/T\xi} + C_p V\kappa_{0/T} - (V\alpha_0)^2, \end{aligned} \tag{C13}$$

where D is the stiffness matrix.

For the sound velocity μ_s we have

$$\begin{aligned} \mu_s^2 &= -V^2\left(\frac{\partial p}{\partial V}\right)_s = -\frac{V^2 C_p D}{T} \\ &= V^2(\xi^{-1}V\alpha_0 + V\kappa_0 - T(V\alpha_0)^2/C_p)^{-1}. \end{aligned} \tag{C14}$$

This expression is finite for $C_p \to \infty$. The same is true of the specific heat at constant volume $C_V = T(V\kappa_T D)^{-1}$ which tends to a finite maximum at the λ-line:

$$C_{V\lambda} = T\xi V(\xi\kappa_0 - \alpha_0). \tag{C15}$$

Setting $T \approx T_\lambda$ and $g'' \approx 0$, we obtain the relations of Pippard (p. 143 of Ref. *10*) and those of Chase (*36*).

Received: February 26, 1960

REFERENCES

1. L. Onsager, *Phys. Rev.* **37,** 405 (1931); **38,** 2265 (1931).
2. J. W. Gibbs, "Collected Works," Vol. I, "*Thermodynamics*." Yale Univ. Press, New Haven, 1948.
3. C. Carathéodory, *Math. Ann.* **67,** 355 (1909); *Sitzber. preuss. Akad. Wiss.* p. 39 (1925).
4. J. D. van der Waals, "Lehrbuch der Thermostatik." J. A. Barth, Leipzig, 1927.
5. P. T. Landsberg, *Revs. Modern Phys.* **28,** 363 (1956).
6. P. Ehrenfest. *Z. Phys. Chem.* **77,** 227 (1911).
7. W. Blaschke, "Vorlesungen über Differential Geometrie," Vol. II. "Affine Differential Geometrie," Chapter 4. Springer, Berlin, 1923.
8. W. H. Keesom, *Comm. Kamerlingh Onnes Lab. Leiden* Suppl. 75a and *Proc. Koninkl. Akad. Wetenschap. Amsterdam* **36,** 147 (1933).
9. P. Ehrenfest, *Comm. Kamerlingh Onnes Lab. Leiden* Suppl. 75b and *Proc. Koninkl. Akad. Wetenschap. Amsterdam* **36,** 153 (1933).
10. A. B. Pippard, "Elements of Classical Thermodynamics for Advanced Students of Physics, Cambridge Univ. Press, Cambridge, 1957.
11. L. Tisza, "Phase Transformations in Solids," Smoluchowski, Mayer, and Weyl, eds., Chapter 1. Wiley, New York, 1951.
12. M. J. Klein and L. Tisza, *Phys. Rev.* **76,** 1861 (1949).

13. L. Tisza, *Phys. Rev.* **110,** 587 (1958).

14. P. M. Quay and L. Tisza (to be published).

15. F. Simon, Ergeb. exakt. Naturw. **9,** 222 (1930);
L. Pauling, *J. Am. Chem. Soc.* **57,** 2680 (1935);
K. Clusius, L. Popp, and A. Frank, *Physica* **4,** 1105 (1937);
J. O. Clayton and W. F. Giauque, *J. Am. Chem. Soc.* **54,** 2610 (1932).

16. R. H. Fowler and E. A. Guggenheim, "Statistical Thermodynamics," pp. 535–39. Cambridge Univ. Press, London and New York, 1939.

17. W. F. Giauque and J. W. Stout, *J. Am. Chem. Soc.* **58,** 1144 (1936).

18. See, e.g., M. Planck, *Physica* **2,** 1029 (1935).

19. See C. Kittel, "Introduction to Solid State Physics," 2nd ed., p. 414. Wiley, New York, 1956.

20. L. D. Landau and E. M. Lifshitz, "Statistical Physics," pp. 33, 215. Pergamon Press, London and Paris, 1958.

21. N. F. Ramsey, *Phys. Rev.* **103,** 20 (1956).

22. M. J. Klein, *Phys. Rev.* **104,** 589 (1956).

23. T. Ehrenfest-Afanassjewa, *Z. Physik* **33,** 933, (1925); **34,** 638 (1925); Also "Die Grundlagen der Thermodynamik," p. 4. E. J. Brill, Leiden, 1956.

24. F. London, "Superfluids," Vol. II, p. 72. Wiley, New York, 1954.

25. K. R. Atkins, "*Liquid Helium.*" Cambridge Univ. Press, London and New York, 1959.

26. S. R. deGroot, "Thermodynamics of Irreversible Processes." North Holland Publishing Co., Amsterdam, 1952.

27. See J. E. Mayer and M. G. Mayer, "Statistical Mechanics." Wiley, New York, 1940.

28. J. A. Beattie, lecture notes (unpublished).

29. B. H. Zimm, *J. Chem. Phys.* **19,** 1019 (1951).

30. See G. H. Wannier, *Revs. Modern Phys.* **17,** 50 (1945).

31. F. Seitz, "The Modern Theory of Solids," pp. 37, 57, 511. McGraw-Hill, New York, 1940.

32. A. Eucken, *Z. Elektrochem.* **45,** 126 (1939) (review paper);
C. P. Smyth, *Chem. Rev.* **19,** 329 (1936) (review paper).

33. W. M. Fairbank, M. J. Buckingham, and C. F. Kellers, *in* "Proceedings of the Fifth International Conference on Low Temperature Physics and Chemistry, p. 50. Madison, Wisconsin, 1958.

34. L. D. Landau, *Physik. Z. Sowjetunion* **11,** 26 and 545 (1937) and Chapter XIV of Ref. *20*.

35. L. Onsager, *Phys. Rev.* **65,** 117 (1944);
G. F. Newell and E. W. Montroll, *Revs. Modern Phys.* **25,** 353 (1953).

36. A. B. Pippard, *Phil. Mag.* [8], **8,** 473 (1956) and Chap. 9 of Ref. *10*;
C. E. Chase, *Phys. Rev. Letters* **2,** 197 (1959).

37. A. I. Khinchin, "Mathematical Foundations of Statistical Mechanics," translated by G. Gamow. Dover, New York, 1949.

38. E. Justi and M. V. Laue, *Physik. Z.* **35,** 945 (1934).

39. L. Pauling, *Phys. Rev.* **36,** 430 (1930).

40. R. H. Fowler, *Statistical Mechanics*, 2nd ed. Cambridge Univ. Press, Cambridge, 1936.

41. N. Bloembergen, E. M. Purcell, and R. V. Pound, *Phys. Rev.* **73,** 679 (1948).

42. N. L. Alpert, *Phys. Rev.* **75,** 398 (1949).

43. J. T. Thomas, N. L. Alpert, and H. C. Torrey, *J. Chem. Phys.* **18,** 1511 (1950).

44. T. Nagamiya, *Progr. Theoret. Phys. Japan* **6,** 702 (1951).

45. K. Tomita, *Phys. Rev.* **89,** 429 (1953).

46. B. Matthias, Chapter XII of Ref. *11*.

47. J. C. Slater, *J. Chem. Phys.* **9,** 16 (1941).
48. M. J. Buerger, *J. Chem. Phys.* **15,** 1 (1947).
49. W. H. Bragg and R. E. Gibbs, *Proc. Roy. Soc.* **A109,** 405 (1925);
R. E. Gibbs, *Proc. Roy. Soc.* **A110,** 443 (1926).
50. H. Moser, *Physik. Z.* **37,** 737 (1936).
51. C. G. Shull and J. S. Smart, *Phys. Rev.* **76,** 1256 (1949);
C. G. Shull, W. A. Strauser, and E. O. Wollan, *Phys. Rev.* **83,** 333 (1951).
C. G. Shull, E. O. Wollan, and W. C. Koehler, *Phys. Rev.* **84,** 912 (1951).
52. E. P. Wigner, "Group Theory," Translated by J. J. Griffin, Chapter 26. Academic Press, New York and London, 1959;
H. Zocher and C. Torok, *Proc. Natl. Acad. Sci. U.S.* **39,** 681 (1953).
53. R. Stevenson, *J. Chem. Phys.* **27,** 656 (1957).
54. H. M. James and T. A. Keenan, *J. Chem. Phys.* **31,** 12 (1959).
55. W. H. Keesom and A. P. Keesom, *Proc. Koninkl. Akad. Wetenschap. Amsterdam* **35,** 786 (1932).
56. T. D. Lee and C. N. Yang, *Phys. Rev.* **87,** 910 (1952);
C. N. Yang and T. D. Lee, *Phys. Rev.* **87,** 404 (1952);
F. Cernuschi and H. Eyring, *J. Chem. Phys.* **1,** 574 (1939).
57. See J. O. Hirschfelder, C. F. Curtis, and R. B. Bird, "Molecular Theory of Gases and Liquids," p. 362. Wiley, New York, 1954.
58. H. Fröhlich, *Physica* **4,** 639 (1937).
59. H. A. Kramers, *Koninkl. Ned. Akad. Wetenschap, Proc.* **33,** 959 (1930).
60. A. Sommerfeld and H. Bethe, *Handbuch der Physik,* Vol. 24, Part 2, p. 607. Springer, Berlin, 1933.
61. L. Szilard, *Z. Physik* **32,** 735 (1925).
62. B. Mandelbrot, I.R.E. Transactions on Information Theory, IT-2, No. 3, September, 1956.
63. J. W. Gibbs, "Collected Works," Vol. II, "Elementary Principles of Statistical Mechanics." Yale Univ. Press, New Haven, 1948.

On the General Theory of Phase Transitions*

LASZLO TISZA
Research Laboratory of Electronics
The Massachusetts Institute of Technology

ABSTRACT

The theory of phase transitions is developed from fundamental principles of thermodynamics. The discussion is based on the so-called fundamental relation of Gibbs, $U = U(X_1, X_2, \cdots, X_{r+1})$, expressing the internal energy as a function of the extensive parameters. The quadratic form occurring in the expansion of this function is responsible for the stability properties of the system. If the form is positive definite, the system is stable; if it is indefinite, it cannot exist in a homogeneous form but breaks up into two or more phases, each of which satisfies the stability condition. States with semidefinite forms are of special interest. They correspond to the generalized critical points of the system. These include, in addition to the liquid-gas critical point, the various λ points occurring in many solids and some liquids. The structure of the above-mentioned quadratic forms is studied by transforming them in a sum of squares. This is achieved not by the eigenvalue method, since the eigenvalues turn out to be devoid of physical meaning, but by the method of completing the square. The diagonal terms appearing in this procedure have been evaluated in terms of quantities permitting a simple physical interpretation. The theory is used for a classification of phase transitions. The discussion of the λ points in ferroelectrics and in the hydrogen halides illustrates the application of the general principles.

It is well known that there are two kinds of phase transitions: the first kind, called also of first order, in which energy, volume, and crystal structure change discontinuously; the second kind, frequently called λ points, Curie points, transitions of higher order, or continuous transitions, in which energy and volume change con-

* Reprinted, with permission, from *Phase Transformations in Solids,* edited by R. Smoluchowski, J. E. Mayer, and W. A. Weyl (John Wiley & Sons, Inc., New York, 1951), pp. 1–37.

tinuously, but the temperature derivatives of these quantities have singularities.

For the transitions of the first kind the phase theory of Gibbs provides a satisfactory thermodynamic framework that can be supplemented by more specific information derived primarily from experiment. In the case of crystals, x-ray and, more recently, neutron-diffraction methods supply detailed information about the geometry of the lattice. Measurements of rates of transitions give some information about the energy (free-energy) barriers separating the different polymorphic modifications. Although an a priori calculation of these quantities is seldom possible, this is due more to practical obstacles than to any difficulties of principle.

The essential thermodynamic features of the transitions of the second kind have proved more elusive. That these transitions reveal common thermodynamic features has been frequently recognized, although the underlying molecular mechanisms were of an unusual diversity. In fact transitions of the second kind mark the onset of ferromagnetism, ferroelectricity, superconductivity, superfluidity in liquid helium, and the establishment of various types of molecular order in alloys and in molecular crystals.

The first attempt to extend the Gibbs phase theory to include the continuous transitions is probably due to Ehrenfest,[1] who tentatively introduced the notion of a transition of the nth order as one having a discontinuity in the nth derivative of the Gibbs function. Ehrenfest derived analogues to the Clausius-Clapeyron relation connecting the slope of the equilibrium curve in the pressure-temperature diagram, dP/dT, with the discontinuities (jumps) of the specific heat, expansion coefficient, and compressibility (ΔC_p, $\Delta\alpha$, $\Delta\kappa$).

Although Ehrenfest's relations are undoubtedly correct, their usefulness depends on whether or not these quantities have a definite meaning. A look at a typical λ curve (Figs. 4 and 6) will reveal that it is not easy to establish a definite experimental value for ΔC_p. This quantity is, moreover, frequently dependent on the experimental techniques, and smaller steps in changing the temperature lead to sharper curves and higher values of ΔC_p. These difficulties were at first considered to be more of a practical kind than difficulties of principle, because a long series of theories seemed to confirm Ehrenfest's ideas. These theories proceeded either on thermodynamic or on statistical mechanical lines. Among the first to be mentioned is Landau's theory.[2] Landau clarified an important feature of the transitions of the second kind. He pointed out that at a λ point in a solid a number of symmetry elements appear or disappear abruptly, whereas the density and more generally the probability distribution in the crystal differs only slightly before and after the transition. In transitions of the first kind, the change of these quantities is, of course, abrupt rather than continuous.

In this connection it is worth noting that in the literature the phenomenon of a rather important structural change taking place within a relatively narrow temperature range (e.g., the onset of rotation of CH_3 groups in high polymers and others) is frequently called a transition. In spite of the relative abruptness of the change there are no sharp temperatures separating states or markedly different properties; and, if these phenomena are called transitions at all (a matter of convention, of course), they should perhaps be called transitions of the third kind in order to avoid confusion with the λ points.

In addition to the above qualitative features, Landau's theory contained also quantitative developments. He concluded that λ points are transitions of the second order (i.e., the specific heat has finite discontinuities ΔC_p). This seemed to corroborate the scheme advanced by Ehrenfest.

The same was true for all the statistical-mechanical theories working with approximate methods.[3,4] Most of these theories refer to the so-called Ising model, in which dipoles capable of two orientations occupy the sites of a rigid lattice. It is assumed that there is an energy difference between the two orientations of a given dipole which depends on the orientations of its nearest neighbors and, in addition, perhaps on an applied field. This model, of course, reminds us of ferromagnetism, but it does not account for the quantum-mechanical nature of the interaction in actual ferromagnets. It can be interpreted more closely as describing the order-disorder phenomena in AB-type alloys.

At any rate the Ising model, in particular the two-dimensional one, seems to be the simplest system revealing the most essential features of a real phase transition of the second kind, and it was turned into a proving ground for various theoretical methods. A rigorous solution even of this simple case offers formidable difficulties only recently overcome by Onsager.[4,5] These calculations brought forth a surprising new turn in the situation.

Onsager obtained the result that the specific heat $C_p(T)$ as a function of the temperature has a logarithmic infinity at the transition point T_0 and has a symmetric shape around the latter. Although both the logarithmic nature of the singularity and its symmetry around T_0 are probably due to the fact that the model is two-dimensional, we see nevertheless that $C_p(T)$ may have singularities not adequately described in terms of a discontinuity:

$$\Delta C_p = C_p^- - C_p^+$$

where

$$C_p^- = \lim_{t \to 0} C_p(T_0 - t)$$

and

$$C_p^+ = \lim_{t \to 0} C_p(T_0 + t)$$

Ehrenfest's scheme is satisfactory if these two limits are finite although different, whereas in the two-dimensional Ising model both are infinite.

It seems particularly puzzling that Landau's theory in its apparent generality fails to include the possibility of a singularity of the kind obtained by Onsager.

The purpose of the present theory is to clarify this difficulty and to develop a theory from the fundamental principles of thermodynamics without any additional *ad hoc* assumptions. The basic ideas of our method can be traced to those of Gibbs, who conceived the problem of phase equilibrium essentially as one of stability.

It may happen that, for an appropriate choice of the external conditions, a system in its homogeneous phase does not satisfy the conditions of thermodynamic stability and breaks up into two heterogeneous parts called phases. The heterogeneous system again satisfies the conditions of stability, and the coexistent phases show no unusual properties. A simple example should help to emphasize this point. Consider ice, which at −1°C is perfectly stable. The stability is not affected by contact with an aqueous salt solution of appropriate concentration. An increase in the salt concentration will result, however, in the melting of the ice. Hence this system is stable with respect to infinitesimal changes of state, although unstable with respect to the formation of an entirely different phase.

An interesting situation may arise in fluid phases. Proceeding along the line of phase equilibrium, the coexisting phases may gradually become more similar and become identical at a terminal point, called also the critical point or critical phase. The critical phases have remarkable properties. They form the limit between the stable states and the states of essential instability, to use Gibbs' terminology. Although stable with respect to infinitesimal changes (they could not otherwise exist in nature) this stability is of a lower order than usual, and it manifests itself in extremely high values of the fluctuations of certain thermodynamic quantities.

The success of thermodynamic methods stems from the fact that the fluctuation of statistical quantities around their thermodynamic averages is extremely small. This no longer holds at the critical points, which thus appear as truly singular states in thermodynamics. The point we wish to make is that the notion of critical point can be generalized in such a way as to include the above-mentioned phase transitions of the second kind. The analogy between gas-liquid critical points and λ points has been noticed already by Onsager and Kaufmann.[5] It is not obvious, however, that this analogy can be made the basis for a quantitative theory. In fact at first we seem to run into difficulty with the phase rule. A coexisting gas-liquid system has one thermodynamic degree of freedom. The end of the equilibrium line, the critical point, is an isolated point in the phase diagram. In contrast a λ point is in reality a λ line, since a transition temperature $T_0(P)$ is obtained for an arbitrary value of the pressure (or stress, etc.). This apparent difficulty can be relatively easily solved for the case of ferroelectricity (and ferromagnetism). Another reason for giving special attention to this problem is that for Rochelle salt the general theory reduces to the well-known interaction theory of Mueller.[6] It may be noted that this theory provided an important clue for the development of our general theory.

The theory can be extended to hold for transitions of the second kind other than those leading to ferroelectricity and ferromagnetism, but only at the price of introducing so-called quasi-thermodynamic parameters.

The next section contains a systematic theory of critical points. Some of the mathematical proofs will be found on page 32. The application of the general theory to fluid systems is given on page 15. A good many of these results can be found in Gibbs' classical work. On pages 18 to 24 the ferroelectric crystals are discussed. On pages 24 to 28 are presented certain quasi-thermodynamic considerations that enable us to subordinate other transitions of the second kind to the general theory of the next section.

Although the theory here presented is essentially thermodynamic, it permits reasonable guesses as to the possible molecular mechanism responsible for various λ points. In addition to the case of ferroelectrics, further examples for this procedure are given on pages 28 to 30. Special attention is given to the transitions in the hydrogen halides and to the question of molecular rotation in crystals.

THE THEORY OF CRITICAL POINTS

All thermodynamic information concerning a system is concisely summarized in a relation which, with Gibbs, we call its fundamental equation:

$$U = U(X_1, X_2, \cdots, X_{r+1}) \tag{1}$$

where U is the internal energy and $X_1, X_2, \cdots, X_{r+1}$ are the generalized coordinates, called also extensive parameters. To each coordinate there belongs a conjugate generalized force

$$P_k = \frac{\partial U}{\partial X_k} \tag{2}$$

The couples of conjugate variables relevant for any particular system are usually chosen from the following set:

$V, -P$ (volume and negative pressure)
S, T (entropy and temperature)

$n_1, n_2, \cdots, n_c; \mu_1, \mu_2, \cdots, \mu_c$ (mole numbers and chemical potentials of the c independent components); and quantities specifying the mechanical, electric, and magnetic strains and stresses.

We shall be particularly concerned with homogeneous systems of such simple geometry that these strains and stresses should also be homogeneous. Under such conditions equation (1) is a first-order homogeneous function of its arguments. It is convenient to introduce the quantities

$$u = \frac{U}{X_{r+1}}, \qquad x_k = \frac{X_k}{X_{r+1}} \tag{3}$$

The fundamental equation appears then with only r independent variables instead of $r + 1$:

$$u = u(x_1, x_2, \cdots, x_r) \tag{1a}$$

We have, of course,

$$P_k = \frac{\partial u}{\partial x_k} \tag{2a}$$

If X_{r+1} is taken to be the volume, the small characters will represent densities of energy, entropy, etc. If it is the total mass or the total number of moles, the small characters will denote so-called specific or molar quantities.

Intrinsic properties of the system (such as its stability) will be determined by a fundamental equation of the form (1*a*). For our general results it is irrelevant whether the symbols in this equation are interpreted as densities or as specific quantities. It may be noted that the various strains are usually described in terms of variables having the nature of densities.

The mechanical strains we shall denote by $S_1, S_2, \cdots, S_6$, which are, in another standard notation, $x_x, y_y, \cdots, x_y$; the corresponding stresses are $T_1, T_2, \cdots, T_6$.

The expression for electromagnetic work

$$\vec{E}\cdot d\vec{D} + \vec{H}\cdot d\vec{B}$$

obtained from Maxwell's equations identifies the electric displacement $\vec{D}$ and the magnetic induction $\vec{B}$ as coordinates, and the fields $\vec{E}$ and $\vec{H}$ as forces.

Let us expand the fundamental equation (1*a*) around a "point" $x_1^0, x_2^0, \cdots, x_r^0$ of the phase diagram:

$$u = u^0 + \sum_{k=1}^{r} P_k^0\, \delta x_k + \tfrac{1}{2} \sum_{1}^{r} u_{ik}\, \delta x_i\, \delta x_k + \cdots \tag{4}$$

The δx_k represent arbitrary small variations of the coordinates. The corresponding variations of the forces are δP_k; hence

$$P_k = \frac{\partial u}{\partial x_k} = P_k^{(0)} + \delta P_k$$

and

$$\delta P_i = \sum_{k=1}^{r} u_{ik}\, \delta x_k \qquad (i = 1, 2, \cdots, r) \tag{5}$$

The coefficients

$$u_{ik} = u_{ki} = \frac{\partial^2 u}{\partial x_i \, \partial x_k} = \frac{\partial P_i}{\partial x_k} = \frac{\partial P_k}{\partial x_i} \tag{6}$$

will be of importance for our investigations. We shall call them the (generalized) stiffness moduli. (Cf. also Zwikker.[7]) They describe the response of the system against an interference tending to change the coordinates. The matrix $\| u_{ik} \|$ will be called the stiffness matrix. The physical meaning of the stiffness moduli is best understood if special variables are inserted for x_k, P_k in equations (5) and (6).

Examples will be discussed later, but it may be noted here that the u_{ik} have familiar physical interpretations, such as adiabatic elastic moduli, electric and magnetic "impermeabilities" (these are reciprocal to the permeabilities), and moduli determining the piezoelectric, electro and magnetocaloric effects and the thermal stresses. If $x_i = s$, $P_i = T$, then $u_{ii} = \partial T/\partial s = T/c_x$, where c_x is the specific heat at no external work (all $x_1, \cdots, x_r$ except x_i constant). In the discussion so far the independent variables have been the coordinates x_k, and it is understood that these are to be kept constant in the differentiation in (6).

It is of great practical interest to interchange the role of coordinates and forces and to consider the latter as independent variables. This can be achieved by solving the set of linear equations (5), provided the determinant

$$D_r = \begin{vmatrix} u_{11} & \cdots & u_{1r} \\ \cdot & \cdot & \cdot \\ u_{r1} & \cdots & u_{rr} \end{vmatrix} \neq 0 \tag{7}$$

We then have

$$\delta x_i = \sum_{k=1}^{r} v_{ik} \, \delta P_k \tag{8}$$

and

$$v_{ik} = \frac{U_{ik}}{D_r} \tag{9}$$

where U_{ik} is the cofactor of the matrix element u_{ik}.

The physical meaning of the v_{ik} is evident from

$$v_{ik} = v_{ki} = \left(\frac{\partial x_i}{\partial P_k}\right)_{P_1, \cdots, P_r} = \left(\frac{\partial x_k}{\partial P_i}\right)_{P_1, \cdots, P_r} \tag{10}$$

It is to be noted that in contrast to (6) it is now the forces that are kept constant while differentiating.

Again borrowing the terminology from the theory of elasticity we call v_{ik} the (generalized) compliance coefficients, $\| v_{ik} \|$ the compliance matrix.

Special cases will be considered later, but it may be noted that the v_{ik} can be isothermal elastic coefficients, electric and magnetic permeabilities, thermal-expansion coefficients, inverse piezoelectric coefficients, and electrocaloric coefficients with appropriate boundary conditions (if the forces are kept constant). For

$$x_i = s, \qquad P_i = T$$

we have

$$v_{ii} = \left(\frac{\partial s}{\partial T}\right)_{P_k} = \frac{c_P}{T}$$

where c_P is the specific heat at constant forces (in particular, constant pressure).

Finally it is frequently of interest to have a mixed set of independent variables, say

$$P_1, P_2, \cdots, P_s, x_{s+1}, \cdots, x_r$$

This case is handled by introducing an appropriate free-energy or thermodynamic potential defined by the relation

$$U[P_1, P_2, \cdots, P_s] = U - \sum_{k=1}^{s} P_k X_k \tag{11}$$

As a simple example we have the Helmholtz free energy $A = U[T] = U - TS$; the Gibbs free energy $G = U[T, -P] = U - TS + PV$; or the potential familiar from the grand canonical ensemble:

$$\Omega = U[T, \mu_1, \mu_2, \cdots] = U - TS - \Sigma \mu_k n_k$$

The well-known relations

$$\frac{\partial U[P_1, \cdots, P_s]}{\partial P_k} = -X_k \qquad (k = 1, 2, \cdots, s) \tag{12a}$$

$$\frac{\partial U[P_1, \cdots, P_s]}{\partial X_k} = P_k \qquad (k = s + 1, \cdots, r + 1) \tag{12b}$$

are verified at once. Equations (12a) can be used to eliminate the variable $X_1, X_2, \cdots, X_s$ from the functions (11). Analogous relations hold for the densities $u[P_1, P_2, \cdots, P_r]$.

It is of interest to consider the last of these potentials in which all the coordinates have been replaced by forces. We claim that this potential vanishes.

$$U[P_1, P_2, \cdots, P_{r+1}] = U - \sum_1^{r+1} P_k X_k = 0 \tag{13}$$

This follows from the assumed homogeneity of the system. U is a first-order homogeneous function of its variables $X_1, X_2, \cdots, X_{r+1}$, and (13) expresses Euler's theorem valid for such functions.

Using (12a) we can write (13) in the differential form:

$$dU[P_1, P_2, \cdots, P_{r+1}] = -\sum_1^{r+1} X_k \, dP_k = 0 \tag{14}$$

which is the well-known Gibbs-Duhem relation. Relation (13) has an important physical meaning. It represents the equation of an r-dimensional surface in the $r + 1$-dimensional space of the variables $P_1, P_2, \cdots, P_{r+1}$. This can be expressed by saying that the number of thermodynamic degrees of freedom of a homogeneous system is $\delta = r$. With electromagnetic forces ignored, this reduces, for a fluid system of c independent components, to $r = \delta = c + 1$. In fact the generalized forces are in this case: $-P$, T, $\mu_1, \cdots, \mu_c$.

We now expand the function $u[P_1, P_2, \cdots, P_s]$ but only with respect to the variables $x_{s+1}, x_{s+2}, \cdots, x_r$ and put $\delta P_1 = \delta P_2 = \cdots = \delta P_s = 0$.

We get

$$u[P_1, P_2, \cdots, P_s] = u^0[P_1, P_2, \cdots, P_s] + \sum_{s+1}^{r} P_k{}^0 \, \delta x_k + \tfrac{1}{2} \sum_{s+1}^{r} w_{ik}{}^{(s)} \, \delta x_i \, \delta x_k \tag{15}$$

Here

$$w_{ik}{}^{(s)} = \frac{\partial^2 u[P_1, \cdots, P_s]}{\partial x_i \, \partial x_k} = \left(\frac{\partial P_i}{\partial x_k}\right)_{P_1, \cdots, P_s} \tag{16}$$

We adopt the convention of indicating only the forces that are to be kept constant and take for granted that the coordinates $x_{s+1}, x_{s+2}, \cdots, x_r$ are held constant too. We are now prepared to discuss equation (4) with a view to establishing the conditions of stable equilibrium.

The point $x_1^{(0)}, \cdots, x_r^{(0)}$ represents a state of equilibrium if the forces $P_1^{(0)}, \cdots, P_r^{(0)}$ are balanced by equal and opposite external forces. This equilibrium is stable if

$$\delta^2 u = \frac{1}{2} \sum_1^r u_{ik}\, \delta x_i\, \delta x_k > 0 \tag{17}$$

for any and all variations of δx_i, δx_k. In fact $\delta^2 u$ is the minimum work that has to be done on the system to achieve a variation δx_k of its state while the forces $P_1^{(0)}, P_2^{(0)}, \cdots, P_r^{(0)}$ are being kept constant.

If a quadratic form satisfies condition (17), it is said to be positive definite.

States corresponding to negative definite or indefinite forms (the latter take positive and negative values) cannot occur in nature. If the external conditions $x_1, x_2, \cdots, x_r$ are such as to lead to, say, an indefinite form, the instability of the system manifests itself in its breaking up into two or more phases, each of which satisfies the condition of stability.

If the number of coexisting phases is p, the number of thermodynamic degrees of freedom reduces to

$$\delta = r + 1 - p \tag{18}$$

since the forces P_k must be constant throughout the system and a Gibbs-Duhem relation, (13) or (14), must be satisfied for each homogeneous region (i.e., each phase separately).*

Equation (18) together with $\delta \geq 0$ expresses a generalized form of the phase rule. For fluid systems of c independent components $r = c + 1$ in the absence of electric or magnetic fields.

Consider now a system of two coexisting phases. It may happen that in the course of the variation of some parameter, as the temperature, the coexisting phases are getting gradually more similar to each other, to become identical at a point of the phase diagram commonly called the critical point. Such points form the border-

* It may be pointed out that for electromagnetic forces the equality of E and H throughout the system is true only for special geometrical conditions that follow from the well-known boundary conditions of the electromagnetic theory.

line between the regions of stability and instability. It is reasonable to assume that the quadratic form determining the stability is also of an intermediate nature, and we tentatively define the critical point by the condition that the quadratic form (17) should be semi-definite positive (i.e., in addition to having positive values it may also be equal to zero).

It is proved in the Appendix to this paper that the quadratic form (17) can be transformed by a non-singular linear transformation in terms of new variables η_k, which need not be mutually orthogonal, to the form

$$\delta^2 u = \tfrac{1}{2}\Sigma\lambda_k\eta_k^{\,2} \tag{19}$$

with

$$\lambda_k = \frac{\partial^2 u[P_1, P_2, \cdots, P_{k-1}]}{\partial x_k^{\,2}} = \left(\frac{\partial P_k}{\partial x_k}\right)_{P_1,\cdots,P_{k-1}} \tag{20}$$

Hence the form is positive definite if, and only if,

$$\lambda_k > 0 \qquad (k = 1, 2, \cdots, r) \tag{21}$$

At a critical point one of the λ_k coefficients vanishes.

It is important to note that the quantities λ_k are not the eigenvalues of the quadratic form (17), since the transformations connecting the x_k with η_k are linear but not orthogonal. Also the λ_k depend on the particular order chosen for the coordinates x_1, x_2, $\cdots$, x_r, and every permutation leads to a new set of stability conditions (21). These various sets are, however, equivalent to each other, and, in particular, the number of positive, negative, and vanishing k's is invariant.[8] The final results are, of course, independent of the special permutation of the variables.

The quantities λ_k can be expressed in terms of the principal minors of the matrix u_{ik}:

$$D_k = \begin{vmatrix} u_{11} & \cdots & u_{1k} \\ \cdot\;\cdot & \cdot\;\cdot\;\cdot & \cdot\;\cdot \\ u_{k1} & \cdots & u_{kk} \end{vmatrix} \tag{22}$$

In fact, we have

$$\lambda_k = \left(\frac{\partial P_k}{\partial x_k}\right)_{P_1,\cdots,P_{k-1}} = \frac{D_k}{D_{k-1}} \tag{23}$$

and the trivial relation

$$\lambda_1 = u_{11} = D_1 \tag{24}$$

In order to prove equation (23), we insert $\delta P_i = \delta P_2 = \cdots \delta P_{k-1} = 0$, $\delta x_{k+1} = \delta x_{k+2} = \cdots \delta x_r = 0$ into the first k equations of the system (5). The solution of this system of k equations for δx_k leads at once to equation (23).

From equations (23) and (24) we obtain that the set of conditions of (21) is equivalent to

$$D_k > 0 \qquad (k = 1, 2, \cdots, r) \tag{25}$$

which is the more generally known form for this criterion.

From equations (23) and (24) we also obtain the relation

$$\lambda_1 \lambda_2 \cdots \lambda_r = D_r \tag{26}$$

Thus in a critical point $D_r = 0$, and from equation (9) we obtain the important result that at such a point the compliance coefficients become infinite, unless of course the corresponding cofactor U_{ik} also vanishes. Thus the critical points appear as truly exceptional points of the system in which the usual thermodynamic formalism becomes singular. The exact nature of the singularity could be established only by the methods of statistical mechanics. Nevertheless two significant results can be obtained within the framework of the present general theory.

The first point concerns the above-mentioned possibility that some compliance coefficients may not become singular because $U_{ik} = 0$. Laying aside the possibility of an accidental vanishing of these quantities, the discussion of which would be beyond the scope of the present method, there are also systematic reasons for such an occurrence. An insight to this problem can be best obtained indirectly by posing the following question: What is the index s for which $\lambda_s = 0$? Since, however, the λ_k still depend on the particular order chosen for the variables $x_1, x_2, \cdots, x_r$, we make s definite by choosing a permutation of the variables for which s has the smallest possible value. We have to distinguish two cases: (1) $s = r$, the normal case; and (2) $s < r$, the exceptional case.

As proved in the Appendix of this paper, the exceptional case can arise only if the variables fall into two classes, $x_1, x_2, \cdots, x_s$ and $x_{s+1}, \cdots, x_r$, so that the "cooperation" or "interaction" of $x_1, x_2, \cdots, x_s$ produces the critical point and the two classes are independent in the sense that

$$\left(\frac{\partial P_s}{\partial x_k}\right)_{P_1, \cdots, P_{s-1}} = 0 \qquad (k = s + 1, s + 2, \cdots, r) \tag{27}$$

in a small but finite r-dimensional domain including the critical point. In the exceptional case only the compliances v_{ik} with $i, k = 1, 2, \cdots, s$ show a singular behavior. The variables $x_1, x_2, \cdots, x_s$ may be called the critical parameters.

In our discussion we have called the definition of the critical point tentative. This definition has to be made more complete. Assuming $\lambda_s = 0$, we can best carry this out in terms of the potential $u[P_1, P_2, \cdots, P_{s-1}]$ which we will briefly denote by u_{s-1}. Its expansion is given by

$$u_{s-1} = u_{s-1}^{(0)} + \sum_{k=s}^{r} P_k^{(0)}\, \delta x_k + \tfrac{1}{2} \sum_{s+1}^{r} w_{ik}^{(s-1)}\, \delta x_i\, \delta x_k + \tfrac{1}{2}\lambda_s(\delta x_s)^2 + A(\delta x_s)^3 + B(\delta x_s)^4 + \cdots \quad (28)$$

Equation (28) is obtained from equation (15) on replacing s by $s - 1$, taking into account the conditions (27), and adding higher terms in δx_s. The work necessary to bring about the virtual change $\delta x_s, \delta x_{s+1}, \cdots, \delta x_r$ is given by the last four terms in (28). The condition of stability requires that this quantity be positive. Thus, in a critical point with $\lambda_s = 0$, we must have also $A = 0$, $B > 0$. Hence we obtain the complete definition of a critical point:

$$\lambda_s = \frac{\partial^2 u_{s-1}}{\partial x_s^{\,2}} = 0, \qquad \lambda_k > 0 \qquad (k \neq s) \qquad (29a)$$

$$\frac{\partial^3 u_{s-1}}{\partial x_s^{\,3}} = 0, \qquad \frac{\partial^4 u_{s-1}}{\partial x_s^{\,4}} > 0 \qquad (29b)$$

$$\frac{\partial^2 u_{s-1}}{\partial x_s \partial x_k} = 0 \qquad (k = s + 1, s + 2, \cdots r) \qquad (29c)$$

In the normal case we put $s = r$ in equations (29*a*) and (29*b*) and omit (29*c*) altogether.

Finally we take up the questions of the mathematical nature of the singularity of the compliance coefficients. To fix our ideas, let us consider the dielectric constant of an unstressed crystal:

$$\varepsilon_P = \left(\frac{\partial D}{\partial E}\right)_P \qquad (30)$$

where P stands for all the external forces such as pressure and stress. We ignore the tensor properties of ε_P by assuming that E and D point in an appropriate crystallographic direction.

Let us suppose that the critical point is reached by means of varying the temperature while the external stresses vanish. We inquire into the nature of the singularity of the function $\mathcal{E}_P(T)$ at $T = T_0$. According to equation (9), $\mathcal{E}_P(T) \sim D_r^{-1}$. Assuming tentatively $D_r \sim (T - T_0)$ we would obtain a Curie-Weiss law for $\mathcal{E}_P$, which is approximately correct for ferroelectric crystals.[6] Nevertheless such a temperature-dependence cannot hold rigorously. In fact, $\int \mathcal{E}_P \, dT$ must be finite even if the integration includes the critical temperature, since physically it essentially represents differences of electric displacements (due mainly to polarization). This has the important implication that the determinant D_r vanishes more slowly than the first power of $T - T_0$; hence it cannot be expanded into a Taylor series around T_0. Since D_r is a rational function of the stiffness moduli u_{ik} of the expansion in equation (4), we come to the following conclusion: at least one of the moduli u_{ik} cannot be an analytic function of the temperature (e.g., it must have a branch point). This circumstance does not invalidate the expansion in equation (4) which proceeds in powers of the coordinates x_k rather than of the temperature.

Concerning the physical interpretation of the singularities of the compliance coefficients a word of caution may be in order. The above theory merely states that these coefficients have abnormally large values when approaching the critical point. At the critical point itself the thermodynamic theory ceases to be valid, and so it cannot be decided with certainty whether these coefficients become actually infinite or only very large. A decision could be obtained only by a rigorous statistical calculation. From the experimental point of view this does not seem to be a grave shortcoming, since this question is hardly of experimental significance. All we intend to state is that the singularities of the compliance coefficients obtained above are the theoretical counterparts of such experimental anomalies as are exhibited in Figs. 2, 3, 6, and 7.

CRITICAL POINTS IN FLUIDS

The well-known properties of critical points of fluid systems will be discussed in some detail in order to illustrate the use of the general theory of the preceding section.

Consider a one-component system with the fundamental equation

$$U = U(S, V, N) \tag{31}$$

in which the variables are entropy, volume, and mole number. In terms of molar quantities the fundamental equation will be

$$u = u(s, v) \tag{32}$$

The stiffness matrix is now

$$\left\| \begin{matrix} \left(\frac{\partial T}{\partial s}\right)_v & \left(\frac{\partial T}{\partial v}\right)_s \\ -\left(\frac{\partial P}{\partial s}\right)_v & -\left(\frac{\partial P}{\partial v}\right)_s \end{matrix} \right\| = \left\| \begin{matrix} \frac{T}{c_v} & -\frac{T\alpha}{\kappa c_v} \\ -\frac{T\alpha}{\kappa c_v} & \frac{c_P}{c_v \kappa v} \end{matrix} \right\| \tag{33}$$

where c_P, c_v, α, κ are the molar heat capacity at constant pressure and volume, the volume expansion coefficient, and the compressibility respectively.

The compliance matrix

$$\left\| \begin{matrix} \left(\frac{\partial s}{\partial T}\right)_P & -\left(\frac{\partial s}{\partial P}\right)_T \\ \left(\frac{\partial v}{\partial T}\right)_P & -\left(\frac{\partial v}{\partial P}\right)_T \end{matrix} \right\| = \left\| \begin{matrix} \frac{c_P}{T} & \alpha \\ \alpha & v\kappa \end{matrix} \right\| \tag{34}$$

Furthermore

$$\lambda_1 = \frac{\partial^2 u}{\partial s^2} = \frac{\partial T}{\partial s} = \frac{T}{c_v} \tag{35}$$

$$\lambda_2 = \frac{\partial^2 u[T]}{\partial v^2} = \frac{\partial^2 a}{\partial v^2} = -\left(\frac{\partial P}{\partial v}\right)_T \tag{36}$$

where $u[T] = a = u - Ts$ is the Helmholtz function per mole.

The critical point is determined by $\lambda_2 = 0$. Or, according to (29),

$$\left(\frac{\partial P}{\partial v}\right)_T = -\left(\frac{\partial^2 P}{\partial v^2}\right)_T = 0, \qquad -\left(\frac{\partial^3 P}{\partial v^3}\right)_T > 0 \tag{37}$$

At the same time all the elements of the compliance matrix become infinite:

$$c_P = \infty, \qquad \alpha = \infty, \qquad \kappa = \infty \tag{38}$$

If we take a different order for the variables, namely $x_1 = v$, $x_2 = s$, the critical point obtains in the form

$$\lambda_2 = \frac{\partial^2 u[P]}{\partial s^2} = \frac{\partial^2 h}{\partial s^2} = \frac{T}{c_P} = 0 \qquad (39)$$

where $u[P] = h = u + Pv$ is the enthalpy per mole. The reader can easily convince himself that the final results are identical with those obtained above.

The graphical presentation of critical points is instructive. Figures 1*a* and 1*b* respectively give one in the *P-T*, *a-v* diagrams. It

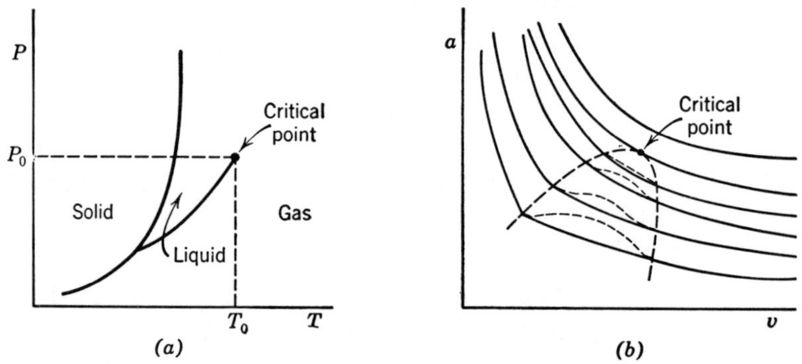

FIG. 1. Schematic representation of gas-liquid critical point: (*a*) in *P-T* diagram; (*b*) in *a-v* diagram.

is clear from the figures that we have here a single, isolated critical point. Formally this follows from the fact that it is defined by two equations (37), each corresponding to a curve in the *P-T* plane.

We now consider briefly two component systems. We have here an additional couple of variables, $x_3 = c = n_1/n_2$, $P_3 = \mu$, where c and μ are mole concentration and chemical potential of the first component respectively.

The critical point is defined by two equations:

$$\left(\frac{\partial \mu}{\partial c}\right)_{P,T} = 0, \qquad \left(\frac{\partial^2 \mu}{\partial c^2}\right)_{P,T} = 0, \qquad \left(\frac{\partial^3 \mu}{\partial c^3}\right)_{P,T} > 0 \qquad (40)$$

Since the number of independent variables has increased, we have now a "critical line" marking the boundary of the surface of phase equilibrium in the P, T, μ space. It may be added that a two-component system may have two critical lines. For the first one the coexisting phases are vapor and liquid of different concentrations; for the second, occurring only in liquids of limited miscibility, we have two coexisting liquid phases of different concentrations.

The discussion of three and more components could be carried out on the same line; the addition of each new component will increase by one the number of dimensions of the manifold of critical points.

CRITICAL POINTS IN SOLIDS: FERROELECTRICS

The identification of λ points in crystals with critical points seems at first to meet with difficulty. We saw in the preceding section that one-component systems exhibit only isolated critical points; thus not only the temperature but also the pressure should have a definite critical value. In contrast the λ points and Curie points obtain for arbitrary values of the pressure and all other external forces. Thus we may take an unstressed crystal and obtain the λ point by simply varying the temperature in the absence of any electric or magnetic field.

The solution of this apparent difficulty is quite simple in the case of ferromagnetic and ferroelectric Curie points. The latter will be our concern in this section, and more general λ points will be discussed in the next.

It is a frequent occurrence that a crystal can exist in twin modifications which are transformed into each other by a mirroring, an inversion, or a 180° rotation. In the case of ferromagnetism and ferroelectricity, these twin modifications are characterized by opposite orientations of the magnetic or electric polarization respectively. It is of course assured by symmetry that the twin modifications have identical stability properties. They may actually coexist in the same crystal as so-called domains or at least be stable for the same values of the hydrostatic pressure and the temperature.† In contrast to the vapor-liquid equilibrium, this coexistence does not reduce the number of thermodynamic degrees of freedom. In some cases the twinning will persist throughout the whole range of stability of the crystal. In particular this will happen in tourmaline or other pyroelectric crystals. On the other hand, it may occur that a change in a parameter, say the tempera-

† Certain stress and field components would, of course, favor one twin modification over the other.

ture, will minimize the difference between the twin modifications so that this difference disappears at a temperature T_0 which then can be considered a critical temperature in the sense discussed on page 11. Beyond (usually above) this temperature, the crystal is symmetric with respect to the symmetry operation connecting the twins.

A discussion of Rochelle salt will help to illustrate both the above considerations and the general theory. The relevant pairs of variables are now

$$\begin{array}{lll} S; & S_1, S_2, \cdots, S_6; & D_x, D_y, D_z \\ T; & T_1, T_2, \cdots, T_6; & E_x, E_y, E_z \end{array}$$

(entropy, temperature, strain, stress, electric displacement and field). It is assumed that both strain and field are homogeneous.

For an account of the experimental results on Rochelle salt we refer to Chapters 20 to 25 of reference 9. A concise summary of the relevant facts follows.

Rochelle salt has two Curie points, T_u and T_l. Above the upper and below the lower Curie points ($T < T_l$ and $T > T_u$), it is orthorhombic. Between these two points ($T_l < T < T_u$) it is monoclinic and possesses a permanent electric moment (displacement) along the monoclinic axis even in the absence of an external electric field.

One of the factors that make Rochelle salt ideally suited for a quantitative study is that its critical points result from the interaction of only two variables. In the terminology of page 13 we deal with an exceptional case.

The critical parameters are: (*a*) the electrical displacement along the monoclinic axis D_x and (*b*) the strain component $S_4(= y_z)$. In the orthorhombic state $S_4 = 0$, $T_4 = 0$, $D_x = 0$, $E_x = 0$. Assuming that deviations from the orthorhombic state are small ($\delta S_4 = S_4$, etc.), we obtain the following simple description of the system in terms of the stiffness moduli:

$$\begin{aligned} T_4 &= c_{44}S_4 + hD_x \\ E_x &= hS_4 + \beta D_x \end{aligned} \tag{41}$$

The inverse relations in terms of the compliance coefficients are

$$\begin{aligned} S_4 &= s_{44}T_4 + dE_x \\ D_x &= dT_4 + \epsilon E_x \end{aligned} \tag{42}$$

Here c_{44} and s_{44} are the elastic modulus and constant; ϵ and β are the electric permittivity and impermeability respectively; d and h are the direct and converse piezoelectric coefficients.

Applying the general methods of page 12 we get

$$\lambda_1 = \frac{\partial^2 u}{\partial S_4{}^2} = \left(\frac{\partial T_4}{\partial S_4}\right)_{D_x} = c_{44} \tag{43}$$

$$\lambda_2 = \frac{\partial^2 u[T_4]}{\partial D_x{}^2} = \left(\frac{\partial E_x}{\partial D_x}\right)_{T_4} = \frac{1}{c_{44}} \begin{vmatrix} c_{44} & h \\ h & \beta \end{vmatrix} = \beta - \frac{h^2}{c_{44}} \tag{44}$$

The critical points are defined by the condition $\lambda_2 = 0$ resulting in infinite values for the compliance coefficient s_{44}, d, ϵ. It is noteworthy that at the critical point the dielectric constant of the clamped crystal is finite:

$$\left(\frac{\partial D_x}{\partial E_x}\right)_{S_4} = \frac{1}{\beta} = \frac{c_{44}}{h^2} \tag{45}$$

whereas the dielectric constant of the free crystal is infinite:

$$\left(\frac{\partial D_x}{\partial E_x}\right)_{T_4} = \frac{1}{\lambda_2} = \infty \tag{46}$$

Formulas (41) to (46) form the backbone of Mueller's "interaction" theory of Rochelle salt, and we refer to his papers [6] and to Cady [9] for a detailed discussion of the experimental material supporting this theory. From the point of view of our general theory we can add that the above discussion can be correct only if the criterion (29*c*) for an exceptional critical point is satisfied. Applied to the present case this means

$$\frac{\partial^2 u[T_4]}{\partial D_x \partial \xi} = \left(\frac{\partial E_x}{\partial \xi}\right)_{T_4, D_x} = 0 \tag{47}$$

where ξ stands for any of the variables D_y, D_z, S_1, S_2, S_3, S_5, S_6, and S. All these relations, except the last one when ξ is the entropy, are satisfied because of the rhombic symmetry of the crystal. Hence the only significant one of the conditions (47) is

$$\left(\frac{\partial E_x}{\partial S}\right)_{T_4, D_x} = \left(\frac{\partial T}{\partial D_x}\right)_{T_4, S} = 0 \tag{48}$$

and the specific heat C_p now has a λ anomaly (Fig. 3), whereas no singularity or even any measurable hump has been found in Rochelle salt.[11] In place of equations (41) and (42) we must deal now with systems of the ten pairs of variables enumerated at the beginning of this section. Closer investigation may show that a certain amount of approximate reduction of the number of relevant

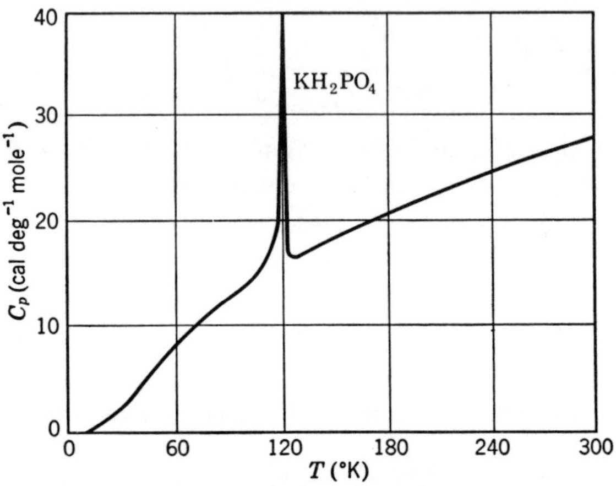

FIG. 3. Heat capacity of KH_2PO_4. [From C. C. Stephenson and J. G. Hooley, *J. Am. Chem. Soc.*, **66**, 1397 (1944).]

critical variables is also possible in this case, but the situation is bound to be more complex than that presented by Rochelle salt.

These results are noteworthy from the following point of view. Our thermodynamic method is not competent to make any detailed statements concerning the molecular mechanisms involved. Nevertheless we have arrived at a significant restriction of the possible molecular mechanisms. Since entropy is a measure of molecular disorder, we have to conclude that the permanent polarization obtains in KDP by means of the ordering of dipoles or rather hydrogen bonds,[12] whereas no such mechanism can be operative in Rochelle salt. In the latter the polarization should result rather from a small shift of the atomic parameters.

If the temperature falls below the upper Curie point, the polarization increases gradually from zero. In Rochelle salt this is uniformly true of every unit cell. In KDP the dipole moment of every unit cell is large and essentially constant. It is the surplus of those pointing in one direction which increases gradually.

The second form is obtained by interchanging the order of the differentiation in (47).

Condition (48) has a simple physical meaning: it states that the electrocaloric effect vanishes (i.e., the adiabatic polarization of the free crystal involves no change in temperature). It should be noted that in the rhombic (generally the non-ferroelectric) state the

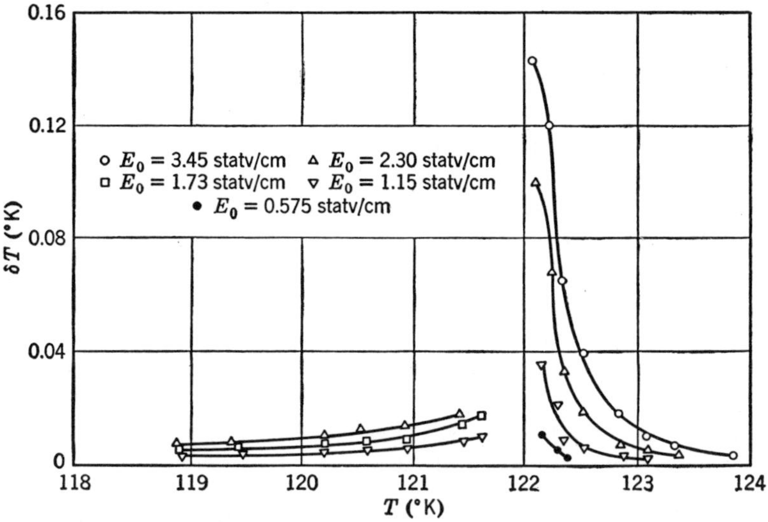

FIG. 2. Electrocaloric effect for KH_2PO_4; Curie temperature, 122.0°K. [From Hutzenlaub.[10]]

electrocaloric effect is by symmetry quadratic in the field and vanishes in zero field. Condition (48) must be satisfied also for non-vanishing fields. This condition is actually satisfied for Rochelle salt, where there is no observable quadratic electrocaloric effect for $T > T_u$.‡ The slight linear electrocaloric effect for $T < T_u$ [10] is irrelevant in this connection since the above formalism does not directly apply to the less symmetric monoclinic state.

The significance of these results becomes apparent through comparison with the case of the ferroelectric phosphates and arsenates (cf. Chapter 27 of reference 9). We take the much studied KH_2PO_4 (briefly KDP) as a representative of this group. In this case the electrocaloric effect does not vanish as indicated in Fig. 2 (taken from Hutzenlaub [10]). Accordingly the entropy is a critical variable,

‡ Private communication from Professor Hans Mueller.

Let N_+ and N_- be the numbers of unit cells with, respectively, positive, negative dipole moments. The resultant moment will be proportional to the quantity §

$$\eta = \frac{N_+ - N_-}{N_+ + N_-} \tag{49}$$

Averaging over a canonical ensemble, η becomes a function of the temperature. For $T > T_0$, $\eta = 0$; for $T < T_0$, $0 \leq |\eta| \leq 1$. The twin modifications are characterized by positive and negative values of η.

Another support for the above views is found in the fact that the spontaneous polarization of Rochelle salt disappears below a lower Curie point. Persistent efforts to locate such a lower Curie point in KDP have proved futile. It is obvious from our point of view that a lower Curie point in KDP would violate thermodynamic principles, since cooling should give rise to an increase in disorder. No such reasons preclude a lower Curie point in Rochelle salt.

It may be noted that the vanishing of the electrocaloric effect in Rochelle salt cannot be quite rigorous, since the susceptibility is a function of the temperature. The essential point is that it will be of a lower order of magnitude than in a case where the polarization is a result of molecular order.‖

We shall consider briefly the case of $BaTiO_3$. (Cf. Paper 13.) $BaTiO_3$ has a slight hump in the specific heat (Fig. 4), but definitely no λ anomaly. Thus the entropy is not a critical variable, and again we are not dealing with an order-disorder phenomenon. Above T_0 the crystal is cubic; below, tetragonal and further tran-

§ The actual situation in KDP is somewhat more complex since the dipoles may also be oriented perpendicularly to the preferred axis;[12] but this circumstance does not significantly modify our general statements.

‖ From the mathematical point of view the distinction between the normal and the exceptional case is so abrupt that we might wonder whether the *approximate* validity of equation (48) could warrant that the case is *approximately* exceptional. A more careful investigation has been carried out and will be published by R. J. Harrison. The result of this is, briefly: in evaluating the analytical properties of the experimental curves representing the specific heats, electrocaloric effects, and the like, we must bear in mind the unavoidable smoothing-out process inherent in experiment. This smoothing out tends to compensate for the deviations from equation (48). In final count the more sophisticated discussion substantiates the simple picture in the text.

sitions occur at lower temperatures. The electric moment can be oriented in the $\pm x$, $\pm y$, $\pm z$ directions, thus making the domain structure more complex than in Rochelle salt.

It would be convenient to have a terminology distinguishing between the KDP type and the Rochelle salt and $BaTiO_3$ type of transitions. Previously [13] the latter were called "displacive" transitions to be distinguished from "order-disorder" transitions. This term was used before by Buerger. Mueller called the Rochelle salt transition a "morphic" one. The term "displacive" transition is

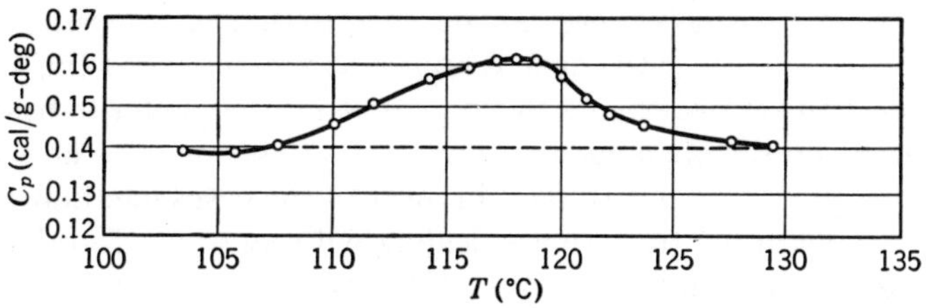

FIG. 4. Specific heat of $BaTiO_3$. [From H. Blattner and W. Merz, *Helv. Phys. Acta*, **21**, 210 (1948).]

more descriptive, but unfortunately it has been used for the α-β transition of quartz which is actually of the order-disorder type as indicated by its λ anomaly of the specific heat.

QUASI-THERMODYNAMICS

A number of phase transitions can be handled essentially on the lines of the discussion of the preceding section. Let us think first of the ferromagnetic Curie points, where magnetic induction, entropy, and possibly strain components (magnetostriction) will be among the critical parameters. We may note that λ anomalies of the specific heat seem to be somewhat less sharp in the ferromagnetic case than usual. This may be due to the fact that the quantum-mechanical nature of the interaction (spin waves) provides more low-energy disordered states than the Ising-type interaction. An elaboration of this point is, however, beyond the scope of the present theory.

In addition to ferroelectricity and ferromagnetism there are other types of electric and magnetic transitions. Thus the dipoles may be arranged in an antiparallel ordered pattern below the transition point. These phenomena can probably be discussed on the purely thermodynamic lines of our theory, although each case may need a special study.

Our main concern in this section is the discussion of λ points for which the interpretation as a critical point runs into difficulties because the transition is not connected with electromagnetic effects, and there seems to be no bona fide thermodynamic variable to take the place of D and B. Such cases are notably the λ points in substitutional alloys, in molecular crystals, and in liquid helium. The failure of the purely thermodynamic theory is most complete for helium, in which there can be no question of coexisting twin modifications. Since too little is known at present about the molecular mechanism underlying this phenomenon, we prefer to base our discussion on the phenomena for which there seems to be no serious doubt in that respect. In fact, in most of the cases mentioned it is generally accepted that we are dealing with an order-disorder phenomenon of one kind or another. The analogy between these cases and, say, the Curie point of KDP is so close that we should hardly be willing to give up the idea of having a theory including both groups of cases.

In fact, the well-known work of Bragg and Williams [3] on order-disorder transformations in alloys was based on the analogy with ferromagnetism. The essential idea was the introduction of a parameter of long-range order, η, playing a role similar to magnetization. The quantitative results of Bragg and Williams were impaired by their assumption that the configurational energy was a function of η alone (inner field assumption). This should not overshadow the fact that the parameter η has a perfectly precise meaning, and the definition to be given below could be easily translated into the terminology of Onsager's rigorous theory.[5]

We define η again by equation (49) but admit a broader physical interpretation for the quantities N_+ and N_-. We shall accept any token by which two tags + and − can be assigned to every unit cell (N_+ and N_- denoting the number of unit cells of each kind). The two assignments, for example, may refer to the two possible orientations of molecules or radicals such as CH_4, NH_4, NO_3. (The tetrahedra may occupy the even or odd corners of a cube; the NO_3 triangles may occupy the even or odd corners of a hexagon.)

In an alloy of the AB type the following situations may arise. The lattice sites are divided into two groups, as an α and a β lattice.

We assign the + to each unit cell in which the α lattice is occupied by an A atom, the $-$ if it is occupied by a B atom. A concrete example is β-brass, CuZn. The lattices α, β are simple cubic; one is in the body center of the other. The completely ordered structure has a CsCl structure.

The variable η thus defined can be introduced into the fundamental equation

$$u = u(x_1, x_2, \cdots, x_r, \eta) \tag{50}$$

We call the variable η a quasi-thermodynamic one. This name should express the fact that in some respects it can be used as another thermodynamic variable, whereas characteristic differences should also be noted. Thus equation (50) includes states that are not realized in nature. The actually realized states can be obtained from the condition

$$\frac{\partial u}{\partial \eta} = 0 \tag{51}$$

This relation can be interpreted as stating that there is no force conjugate to η by the application of which we could directly influence η in such a way as a magnetic field would influence magnetization.

Equation (51) can be solved for the equilibrium value η_0 of η. Inserted into equation (50) this leads us back to the usual fundamental equation. The more explicit form (50) is useful, however, for a qualitative understanding of the molecular mechanism underlying the transitions.

It may happen that u is an even function of η, and positive or negative solutions for η_0 of the same absolute value will correspond to twin modifications and hence may lead to critical points exactly as in the case of ferroelectricity.

It may be worth pointing out that in cases such as the Cu_3-Au alloy there is no symmetry between positive and negative values of the order parameter, and accordingly there is no critical point; but the order-disorder transition is of the first kind.

The use of quasi-thermodynamic variables brings up a question of a rather fundamental nature. Is it possible to give a more formal characterization of the variables that qualify as quasi-thermodynamic ones and can give rise to λ points? This question is closely connected with the ergodic problem of statistical mechanics. What are the macrovariables which are left as constants in the partition sum after the statistical averaging has been carried out over the microvariables? The usual assumption is that the macrovariables are only the energy, the additive integrals of the motion (momentum, angular momentum), and the quantities fixing the external conditions of the system. Our above considerations imply that other macrovariables are possible.

In connection with a calculation of the critical fluctuations carried out in collaboration with Dr. M. J. Klein, [14] we were led to a tentative formulation of a criterion for macrovariables. We shall give a brief and not quite precise summary of the results obtained.

Let us divide the system into N equal cells, which may be macroscopic or may also coincide with the unit cells of the crystal. Consider the coordinates or quantum numbers $z_1, \cdots, z_\nu$ which specify the single cells. In order to specify the whole system, we need N_ν parameters. We can choose these in the form of normal coordinates having definite phase differences for neighboring cells. According to the criterion in question, only normal coordinates with zero phase differences (infinite wavelength) can qualify for inclusion into the fundamental equation. These special normal coordinates are invariants of the translational group of the crystal.

This definition includes the additive integrals of motion but in addition admits, as macroparameters, atomic parameters specifying the equilibrium positions of the atoms in the unit cell and the parameters of long-range order. The corresponding statistical parameters would be the lattice vibrations (phonons) and the short-range order. It will be noted that the macroparameters give rise to coherent x-ray scattering, the statistical parameters to diffuse scattering. Still other parameters can be introduced in particular cases. These macroscopic parameters are subdivided into thermodynamic and quasi-thermodynamic ones. The former have conjugate forces corresponding to them; the latter have none, as explained above in connection with the order parameter.

Finally it is of interest to compare our methods with that of Landau,[2] who made an earlier attempt to develop a quasi-thermodynamic theory of the transitions of the second kind, especially since some features of Landau's theory have been incorporated in this paper.

Let us consider a system with the coordinates s, v, η, for which our condition for the critical point reduces to that of Landau:

$$\lambda_3 = \frac{\partial^2 u[T, P]}{\partial \eta^2} = \frac{\partial^2 g}{\partial \eta^2} = 0 \tag{52}$$

where $u[T, -P] = g = u - Ts + Pv$ is the Gibbs function per mole. Landau computes the specific heat c_P and finds that to the specific heat of the disordered phase c_{P0} there should be added an extra term for the ordered phase alone. Landau tacitly assumes that c_{P0} is a smooth function and obtains a definite Δc_P (transition of the second order). Actually Landau failed to realize that because of the "interaction" of the thermodynamic parameters the condition (52) implies that the compliance matrix and, in particular, c_{P0} become singular at the critical point.

THE DIAGNOSIS OF CRITICAL POINTS

The diagnosis of λ points aims at the establishment of the molecular mechanism responsible for the observed anomalies of the various compliance coefficients. The discussion of ferroelectric phe-

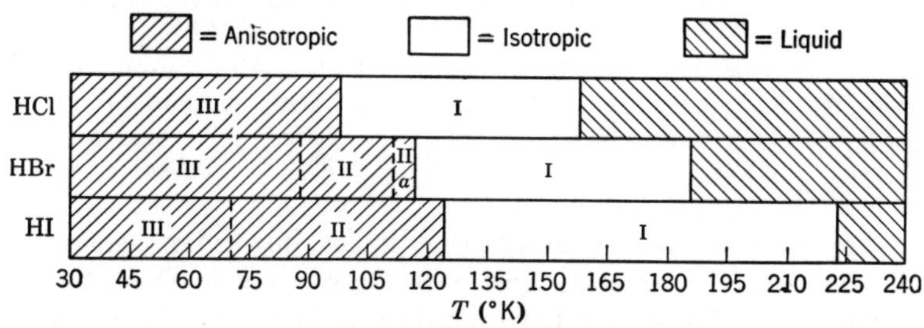

FIG. 5. Transitions in the hydrogen halides. [From A. Kruis and R. Kaischew, *Z. physik. Chem.*, **B41**, 427 (1938).]

nomena has shown that considerable insight into the molecular mechanism may be gained from a purely qualitative discussion of the experimental findings. A few more examples should be discussed in this section.

An instructive case is provided by the hydrogen halides, which have a large number of transitions all but one of which we were able to interpret in molecular terms. Figure 5 gives a schematic representation of the positions of the transitions plotted against temperature. The transition in HCl is of the first kind. All the others are of the second kind. This figure contains also the result of measurements with the polarization microscope clearly identifying the isotropic and anisotropic phases.

Figure 6 gives the specific heat C_p and Fig. 7 the dielectric constant of HBr. The typical singularities in C_p suggest order-disorder effects for all three transitions. The anomaly in $\mathcal{E}$ indicates that an ordering of dipoles is involved in the II-III transitions.

We suggest the following picture. In phase III the dipoles are ordered, probably in an antiparallel pattern. Phase II is disordered

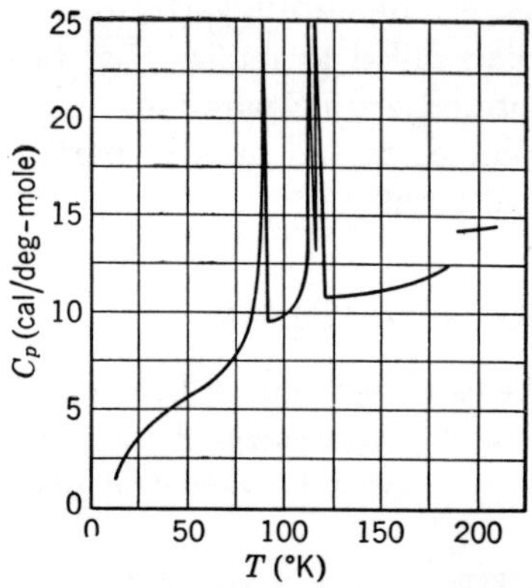

FIG. 6. Specific heat of HBr. [From W. F. Giauque and R. Wieke, *J. Am. Chem. Soc.*, **50**, 2193 (1928).]

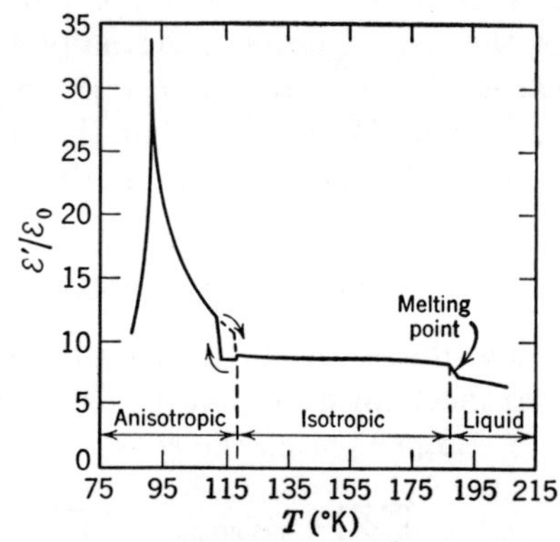

FIG. 7. Dielectric constant *vs.* temperature: solid HBr. [From C. P. Smyth and C. S. Hitchcock, *J. Am. Chem. Soc.*, **55**, 1830 (1933).]

with regard to the orientation of the dipoles, but the molecules are still lined up with their longest axis (anisotropy of this phase!). In phase I the dipoles are distributed at random over the six directions $\pm x$, $\pm y$, $\pm z$. We have not succeeded in finding any interpretation for phase II*a*. If this picture is correct, the II-III transition is brought about by dipolar forces, the I-II transition by van der Waals forces. This conclusion is supported by the comparison of the corresponding transition temperatures in the three halides (Fig. 5). In fact the dipole moments μ for these molecules are [15]

$$\mu \times 10^{-18} \text{ esu}$$

$$\text{HCl} \sim 1$$

$$\text{HBr} \sim 0.8$$

$$\text{HI} \sim 0.4$$

whereas the van der Waals forces increase with the size of the halogen.

The data presented do not exclude the possibility that phase III is ferroelectric. However, no permanent polarization has been reported, and according to the calculations of Luttinger and Tisza [16] an antiparallel ordered dipole array is more favorable in this case. The same calculations indicate the possibility for permanent polarization if a long cylindrical sample is cooled in a longitudinal electric field. An experimental investigation of this question would be desirable.

Another point of interest is that we have interpreted the cubic symmetry (isotropy) of phase I in terms of an orientational disorder. This is in line with ideas usually connected with Frenkel and Landau in contrast to those of Pauling and Fowler, who advanced the hypothesis of transitions from hindered to free rotation to explain these λ points. There has been an argument of long standing between these viewpoints with no conclusive solution until recently. The main difficulty is that in most cases the rotation involves protons, which cannot be located by x-rays.

It is therefore of interest that recently Purcell and collaborators [17] have shown that the method of paramagnetic nuclear resonances is capable of revealing whether or not the paramagnetic nuclei (protons) perform rapid motions as required in the case of free rotation. This method has been applied by Alpert [18] to a certain number of crystals, the hydrogen halides among them. These experiments showed without ambiguity that a free rotation does not set in at the II-I transition. The experiments do indicate, however, a gradual increase of the flipping frequency of the molecules within phase I. This process is not accompanied by any abrupt thermodynamic effects. Probably the only case where a rotational transition has not yet been excluded is that of light and heavy methane, a further study of which would be of interest, possibly with the neutron-diffraction method.

It might be added in this connection that the rotation of NO_3 groups in $NaNO_3$ crystals has been reinvestigated by Siegel.[19] In contrast to earlier reports his measurements of the intensity distribution of the background of the x-ray scattering seems to preclude the possibility of a free rotation.

SUMMARY

A thermodynamic theory of phase transitions has been developed on the basis of the fundamental equation of Gibbs expressing the internal energy as a function of the extensive parameters. In the expansion of the fundamental equation we consider the quadratic form representing the second-order terms. The coefficients of this form are the so-called stiffness moduli providing a specification of the observable thermodynamic properties of the system. This specification can be equally well obtained in terms of the compliance coefficients. These are derivatives in which the forces (intensive parameters) are the independent variables. The compliance matrix is the inverse of the stiffness matrix.

Thermodynamic stability requires that the above-mentioned quadratic form be positive definite. States with indefinite forms cannot exist in nature, and they break up into one or more phases. Critical points correspond to the limiting case of semidefinite quadratic forms. In this case the stiffness matrix becomes singular (its determinant vanishes), and the elements of the compliance matrix become infinite.

It was shown that the ferroelectric and ferromagnetic Curie points can be considered special cases of critical points. Various λ points that are not connected with electric and magnetic anomalies can also be interpreted as critical points, but only if so-called quasi-thermodynamic variables are introduced in addition to the ordinary thermodynamic ones.

A classification of critical points (transitions of the second kind) is obtained by establishing the nature of the variables giving rise to the critical point. The procedure is demonstrated by the detailed discussion of the transitions in ferroelectric crystals and in the hydrogen halides.

Our term "transition of the second kind" should be distinguished from the usual "transition of second order." The latter term implies that the compliance coefficients such as specific heat have discontinuities at T_0 but have regular finite values within both phases. In contrast, at the transitions of the second kind, C_p or possibly other compliance coefficients have singularities. These indicate that the present thermodynamic formalism is stretched to its limits at the critical point. The nature of the singularity can be established only by statistical methods. In contrast to earlier thermodynamic theories, our conclusions are compatible with Onsager's rigorous theory of the two-dimensional Ising model.

APPENDIX

Consider the quadratic form

$$\delta^2 u = \tfrac{1}{2} \sum_{1}^{r} u_{ik}\xi_i\xi_k \tag{1}$$

where $\xi_k = \delta x_k$ establishes the connection with the notation of page 11.

Quadratic forms are most conveniently discussed by reduction to the diagonal form:

$$\delta^2 u = \tfrac{1}{2} \sum_{1}^{r} \lambda_k \eta_k^2 \tag{2}$$

This is achieved by a non-singular linear transformation of the variables. It should be noted that this transformation is not uniquely determined, and accordingly we obtain different sets of diagonal elements λ_k. But according to Sylvester's law of inertia [8] the number of positive, negative, and zero terms among the coefficients λ_k is independent of the particular choice of the transformation. These numbers completely determine the nature of the form (positive, negative definite, semidefinite, and indefinite) and lead to the criteria stated on page 14.

The most frequently used procedure for the diagonalization of equation (1) is the so-called eigenvalue method in which the transformation connecting the ξ_k and η_k is orthogonal. This method is inappropriate in our case, since the eigenvalues of equation (1) turn out to be devoid of physical meaning. We shall use an alternative procedure [8] which will lead to the expressions for λ_k given on page 12.

We assume at first that the matrix u_{ik} is of rank r (i.e., $\delta^2 u \neq 0$) except for the trivial case $\xi_1 = \xi_2 = \cdots = \xi_r = 0$. This implies that $u_{kk} \neq 0$ for all k.

We apply the familiar procedure of completing the square:

$$\delta^2 u = \tfrac{1}{2} u_{11} \left(\xi_1 + \frac{1}{u_{11}} \sum_{2}^{r} u_{1k}\xi_k \right)^2 - \frac{1}{2u_{11}} \left(\sum_{2}^{r} u_{1k}\xi_k \right)^2 + \tfrac{1}{2} \sum_{2}^{r} u_{ik}\xi_i\xi_k$$

Introducing the new variable

$$\eta_1 = \xi_1 + \frac{1}{u_{11}} \sum_{2}^{r} u_{1k}\xi_k \tag{3}$$

and the notation

$$u_{ik}{}^{(1)} = u_{ik} - \frac{1}{u_{11}} u_{1i} u_{1k} = \frac{1}{u_{11}} \begin{vmatrix} u_{11} & u_{1k} \\ u_{1i} & u_{ik} \end{vmatrix} \tag{4}$$

we obtain

$$\delta^2 u = \tfrac{1}{2} u_{11} \eta_1{}^2 + \tfrac{1}{2} \sum_2^r u_{ik}{}^{(1)} \xi_i \xi_k \tag{5}$$

The residual form contains only $r - 1$ variables. The assumed non-singular nature of $\delta^2 u$ implies that $u_{kk}{}^{(1)} \neq 0$, $k = 2, 3, \cdots, r$; hence the procedure can be continued. Introducing

$$\eta_2 = \xi_2 + \frac{1}{u_{22}{}^{(1)}} \sum_3^r u_{2k}{}^{(1)} \xi_k \tag{6}$$

we obtain

$$\delta^2 u = \tfrac{1}{2} u_{11} \eta_1{}^2 + \tfrac{1}{2} u_{22}{}^{(1)} \eta_2{}^2 + \tfrac{1}{2} \sum_3^r u_{ik}{}^{(2)} \xi_i \xi_k \tag{7}$$

or finally the reduced form (2) with

$$\lambda_k = u_{kk}{}^{(k-1)} \tag{8}$$

We now consider the case of a singular matrix. Here the procedure comes to a stop because one of the diagonal elements vanishes, as

$$u_{ss}{}^{(s-1)} = 0 \tag{9}$$

We claim that in this case $\delta^2 u$ is indefinite (takes positive and negative values) unless

$$u_{sk}{}^{(s-1)} = 0 \qquad (k = s + 1, s + 2, \cdots, r) \cdot \tag{10}$$

In fact we assume that $u_{st}{}^{(s-1)} \neq 0$. Then, for $\eta_1 = \eta_2 = \cdots = \eta_{s-1} = 0$ and $\xi_k = 0$ for all $k = s + 1, s + 2, \cdots, r$ except $k = t$, we have

$$\delta^2 u = u_{st}{}^{(s-1)} \xi_s \xi_t + \tfrac{1}{2} u_{tt}{}^{(s-1)} \xi_t{}^2$$

This is indeed indefinite, since for large enough $|\xi_s|$ the first term will determine the sign of the form.

An important corollary, used on page 13, is that for semidefinite forms equation (9) implies equation (10).

All that remains to be done is to compute the matrix elements of the residual form $u_{ik}{}^{(s)}$. This could be done by iterating the procedure leading to equation (4); the result can also be obtained in one stroke with the use of the Legendre transforms $u[P_1, P_2, \cdots, P_s]$. In fact the following simple relation is valid:

$$u_{ik}^{(s)} = w_{ik}^{(s)} \tag{11}$$

where the $w_{ik}^{(s)}$ are the coefficients of expansion given in equation (15) of the main text. Together with equation (16) this leads to the desired interpretation of $u_{ik}^{(s)}$ and in particular to

$$\lambda_k = u_{kk}^{(k-1)} = \left(\frac{\delta P_k}{\delta x_k}\right)_{P_1,\cdots,P_{k-1}}$$

The proof of equation (11) is obtained by a simple step-by-step calculation. We shall briefly indicate the first two steps. We have

$$u[P_1] = u - P_1 x_1$$

and

$$P_1 x_1 = (P_1^{(0)} + \delta P_1)(x_1^{(0)} + \delta x_1)$$

Since we are interested only in the quadratic expansion terms, only $\delta P_1\, \delta x_1$ need to be considered. From equation (3) we have $\delta P_1 = u_{11}\eta_1$; hence

$$\delta P_1\, \delta x_1 = u_{11}\eta_1\xi_1 = u_{11}\eta_1\left(\eta_1 - \frac{1}{u_{11}}\sum_2^r u_{1k}\xi_k\right)$$

The subtraction of this expression from equation (5) leaves the residual form unaffected; hence

$$u_{ik}^{(1)} = w_{ik}^{(1)}$$

Next we have

$$u[P_1, P_2] = u[P_1] - P_2 x_2$$

Again only $\delta P_2\, \delta x_2 = \delta P_2 \xi_2$ is of interest. Using

$$P_2 = \frac{\partial u[P_1]}{\partial x_2}$$

we obtain

$$\delta P_2 = a\eta_1 + \sum_2^r u_{2k}^{(1)}\xi_k = a\eta_1 + u_{22}^{(1)}\eta_2 \tag{12}$$

where a is a constant of no interest to us. Using equations (12) and (6) we obtain

$$\xi_2\, \delta P_2 = \left(\eta_2 - \frac{1}{u_{22}^{(1)}}\sum_3^r u_{2k}\xi_k\right)(a\eta_1 + u_{22}^{(1)}\eta_2) \tag{13}$$

The subtraction of equation (13) from equation (7) leaves the residual form unchanged, and we obtain

$$u_{ik}^{(2)} = w_{ik}^{(2)}$$

The procedure can obviously be continued.

NOTE

This work was supported in part by The Signal Corps, The Air Materiel Command, and the Office of Naval Research.

REFERENCES

1. P. Ehrenfest, Leiden Comm. Suppl. 75b (1933).
2. L. Landau, *Phys. Z. Sow. Union*, **11**, 26, 545 (1937); L. Landau and E. Lifshitz, *Statistical Physics*, Clarendon Press, Oxford, 1938, Chapter XI.
3. F. C. Nix and W. Shockley, *Rev. Mod. Phys.*, **10**, 1 (1938).
4. G. H. Wannier, *Rev. Mod. Phys.*, **17**, 50 (1945).
5. Lars Onsager, *Phys. Rev.*, **65**, 117; Lars Onsager and Bruria Kaufman, Physical Society Cambridge Conference Report, 1947, p. 127.
6. H. Mueller, *Phys. Rev.*, **57**, 829; **58**, 565, 805 (1940).
7. C. Zwikker, *Physica*, **14**, 35 (1948).
8. W. L. Ferrar, *Algebra*, Clarendon Press, Oxford, 1941, p. 135, for example.
9. W. G. Cady, *Piezoelectricity*, McGraw-Hill Book Co., New York, 1946.
10. J. F. Hutzenlaub, Thesis, M.I.T., 1943.
11. A. J. C. Wilson, *Phys. Rev.*, **54**, 1103 (1938).
12. J. C. Slater, *J. Chem. Phys.*, **9**, 16 (1941).
13. A. von Hippel, R. G. Breckenridge, F. G. Chesley, and L. Tisza, *Ind. Eng. Chem.*, **38**, 1097 (1946).
14. M. J. Klein and L. Tisza, *M.I.T. R.L.E. Tech. Rept.* 111, 1949.
15. C. T. Zahn, *Phys. Rev.*, **24**, 400 (1924).
16. J. M. Luttinger and L. Tisza, *Phys. Rev.*, **70**, 954 (1946).
17. N. Bloembergen, E. M. Purcell, and R. V. Pound, *Phys. Rev.*, **73**, 679 (1948).
18. N. L. Alpert, *Phys. Rev.*, **75**, 398 (1949).
19. L. A. Siegel, Thesis, M.I.T. (1948).

DISCUSSION

G. H. Wannier, P. W. Anderson, J. M. Richardson (Bell Telephone Laboratories):

Our main interest in the paper lies in its application to ferromagnetic and ferroelectric systems, and we find two difficulties in this application:

(*a*) The basic general theory does not seem to apply to the two cases of second-order transitions which can be worked out rigorously, namely the linear and the two-dimensional Ising models.

(*b*) The distinction between normal and exceptional singular points is not, apparently, germane; almost all simple cases are exceptional purely for symmetry reasons, even where it is certain that the concept that thermal properties are decoupled is not appropriate.

In justification of (*a*): The linear Ising model, although not exhibiting a singular point in the range, can be thought of as having a singular point at $T = 0$ and $M = 0$. At this point we have computed some derivatives of $U(S, M)$ and found

$$\left(\frac{\partial^2 U}{\partial S^2}\right)_{M=\mathrm{const}} \sim T^3 e^{\alpha/T} \to \infty \tag{1}$$

which violates the basic assumption of the paper, that $U(S, M)$ has finite first and second derivatives. We admit that this is a highly artificial case of little interest; still a discussion of the validity of the basic assumption is definitely in order.

We have also worked out the basic determinant

$$\begin{vmatrix} \dfrac{\partial^2 U}{\partial S^2} & \dfrac{\partial^2 U}{\partial S\,\partial M} \\ \dfrac{\partial^2 U}{\partial S\,\partial M} & \dfrac{\partial^2 U}{\partial M^2} \end{vmatrix} \tag{2}$$

at the critical point for the case of the plane Ising model. In this case it turns out that all three derivatives vanish (the cross derivative by symmetry, since U and S cannot know the sign of M). Thus we have here the case of a double zero of the determinant, which is not considered in the paper.

As working hypotheses the basic assumptions of the paper cannot be objected to. However, these two examples directly contradict two of them, and thus it is certainly in order to give an explicit indication that they are working hypotheses, with a discussion of their usefulness and validity.

As for (*b*), we first leave out stress and strain variables (they do not change the picture). Then, allowing the assumption that U can be expanded to second powers in S and M, the basic determinant is (2). The classification of singular points into exceptional and normal then depends on the vanishing or non-vanishing of the off-diagonal element.

$$\frac{\partial^2 U}{\partial S\,\partial M} = \left(\frac{\partial T}{\partial M}\right)_S$$

It is hard to conceive of any ferromagnetic (or ferroelectric, replacing M by P) substances for which this derivative does not vanish for symmetry reasons in the unmagnetized (unpolarized) state. Thus the exceptional case, in which S and M are uncoupled, becomes the only one.

In the paper an attempt is made to make a qualitative distinction between KDP and Rochelle salt. In order to do so it is necessary to examine the behavior of U beyond second powers of S and M, but this is beyond the domain of the phenomenological theory itself.

L. TISZA:

To the above remarks I should like to make the following reply:

(1) The statement that $U(S, M)$ has finite derivatives is not a working hypothesis. It can be shown to be true in every point which is accessible in finite steps. This condition is not satisfied at absolute zero, which indeed has an exceptional character. Thus it is a common occurrence that $\partial^2 U/\partial S^2 = T/C$ should tend to infinity for $T \rightarrow 0$. There is hardly an advantage in considering absolute zero a phase-transition point.

(2) The discussion in the text can be easily extended to the case in which the basic determinant has a multiple zero.

(3) The criterion for the validity of an exceptional critical point should extend over a finite region around the critical point. Thus the distinction between KDP and Rochelle salt is in the *quadratic* electrocaloric effect. This distinction eludes the discussors, who envisage only the case of a vanishing electric field.

(4) The cubic and higher-order expansion terms of the energy are not beyond the domain of the phenomenological theory. Yet, this is irrelevant in the present connection since the distinction between KD'P and Rochelle salt appears in the quadratic expansion terms.

PAPER 5

EQUATION OF STATE OF ^{4}He IN THE CRITICAL REGION

Laszlo Tisza

Department of Physics, Massachusetts Institute of Technology, Cambridge, Massachusetts

and

C. E. Chase

National Magnet Laboratory, Massachusetts Institute of Technology, Cambridge, Massachusetts*

(Received 25 May 1965)

It was pointed out several years ago by one of us [1] that the thermodynamic functions are nonanalytic at the critical point and presumably exhibit branch-point singularities. Recently a great deal of progress has been achieved in providing mathematical descriptions of such singularities.[2] In order to integrate these results with the formalism of thermodynamics, it is necessary to choose a set of "natural variables" that are within the scheme of a single fundamental equation and are adapted to the intrinsic symmetry of the system. This point is too obvious to call for special notice, say, in the case of crystals. There is, however, evidence of a near symmetry in the critical region of fluids, the thermodynamic role of which has not been sufficiently appreciated. This symmetry is masked if the traditionally favored independent variables are used.

In order to demonstrate this thesis we shall analyze the measurement of Edwards and Woodbury [3] (henceforth referred to as EW) who determined the orthobaric densities of ^{4}He to within 36 mdeg of T_c for the liquid and 50 mdeg for the vapor. EW attempted at first to describe

* Supported by the U. S. Air Force Office of Scientific Research.

[1] L. Tisza, in *Phase Transformations in Solids,* edited by R. Smoluchowski, J. E. Mayer, and W. A. Weyl (John Wiley & Sons, Inc., New York, 1951), pp. 1–37; see particularly p. 15.

[2] G. S. Rushbrooke, *J. Chem. Phys.* **39**, 842 (1963); M. E. Fisher, *J. Math. Phys.* **5**, 944 (1964); R. B. Griffiths, *Phys. Rev. Letters* **14**, 623 (1965).

[3] M. H. Edwards and W. C. Woodbury, *Phys. Rev.* **129**, 1911 (1963).

their measurements in terms of the approximate expansion of Landau and Lifshitz [4]

$$N^{-1}(\partial^2 F/\partial v^2)_{T,N} = -(\partial P/\partial v)_T = At + Bv'^2 + \cdots, \tag{1}$$

where $F = F(T, V, N)$ is the Helmholtz free energy, $v = V/N$ is the molar volume, A and B are adjustable positive constants, $t = T - T_c$, and $v' = v - v_c$. Omitting higher order terms, this relation yields the coexistence curve

$$v_G' = v_L' = (-3At/B)^{1/2} \tag{2}$$

According to this relation the molar volumes of the coexisting phases are symmetric around the critical volume, a prediction in conflict with experiment. In order to obtain quantitative agreement with their measurements within about 110 mdeg of T_c, EW added two more terms to the above expansion. In such a way they eliminated the spurious symmetry of Equation 2 with respect to v'. On closer examination, however, it appears that their expression leads to inconsistencies when applied at points away from the coexistence curve; [5] e.g. the critical isotherm (which is identical with that of Landau and Lifshitz) and the coexistence curve cross at a pressure close to P_c.

One might believe that these difficulties are connected with the singular character of the critical point mentioned above. This, however, does not appear to be the case; in the region under consideration, which does not extend into the immediate neighborhood of the critical point, the difficulties can be at least partially overcome by interchanging the role of the variables N and V. Instead of considering a closed system with N fixed and $v = V/N$ variable, we envisage an open system with V fixed and $\rho = N/V$ variable. We obtain a preliminary estimate of the implications of this change of viewpoint by translating the Landau-Lifshitz expansion into the present scheme.

The analog of Equation 1 is

$$V^{-1}(\partial^2 F/\partial \rho^2)_{T,V} = (\partial \mu/\partial \rho)_T = \alpha t + \beta \rho'^2 + \cdots, \tag{3}$$

where α and β are positive constants and $\rho' = \rho - \rho_c$. The slope of the isothermals is now

$$(\partial P/\partial \rho)_T = \rho(\partial \mu/\partial \rho)_T = \rho(\alpha t + \beta \rho'^2) \tag{4}$$

and, by a treatment exactly parallel to that of reference 4, we find

$$\rho_L' = -\rho_G' = (-3\alpha t/\beta)^{1/2} \tag{5}$$

[4] L. D. Landau and E. M. Lifshitz, *Statistical Physics* (Pergamon Press, New York, 1958), p. 259 ff.

[5] We noticed these difficulties in connection with recent measurements of the sound velocity in ^{4}He near the critical point; we needed values of the density in order to calculate the adiabatic compressibility from the relation $\kappa_s^{-1} = \rho u^2$, and in the absence of direct density measurements we used the expression of EW to extrapolate the densities away from the coexistence curve. [C. E. Chase, R. C. Williamson, and L. Tisza, *Phys. Rev. Letters* **13**, 467 (1964).]

The predicted coexistence curve is thus symmetric in the density. This is indeed approximately correct for fluids in general,[6,7] and this equation fits as well with the EW results as their four-parameter expression.[8] According to Equation 5, a plot of $\rho_G'^2$ and $\rho_L'^2$ against $-t$ should yield a single straight line through the origin of slope $3\alpha/\beta$. Such a plot is shown in Figure 1;[9] from the slope of the solid line we find $\alpha/\beta =$

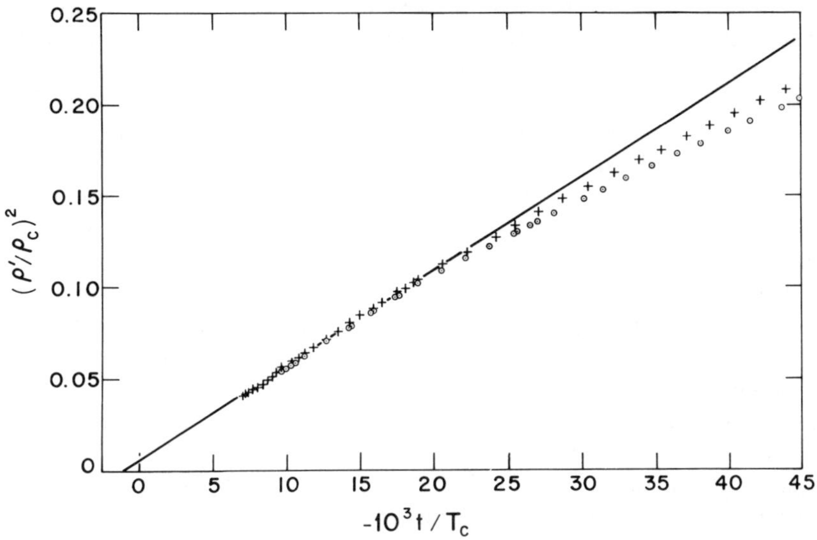

Fig. 1. Graph of $(\rho_L'/\rho_c)^2$ and $(\rho_G'/\rho_c)^2$ against $-t/T_c$ from data of reference 3. Equation 5 predicts a single straight line of slope $3\alpha/\beta$. Crosses, $(\rho_L/\rho_c)^2$; circles, $(\rho_G'/\rho_c)^2$.

$(1.03 \pm 0.02) \times 10^{-4}$ mole^2/cm^6 °K. This diagram should be compared with Figure 3 of EW, which demonstrates the discrepancy between Equation 2 and experiment. The quantity β can be evaluated approximately, as in EW, by comparison with the measured liquid density and isothermal compressibility[10] on the saturation curve at 5°K, since ac-

[6] L. Cailletet and E. Mathias, *Compt. Rend.* **102**, 1202 (1886).

[7] E. A. Guggenheim, *J. Chem. Phys.* **13**, 253 (1945).

[8] It is a curious fact that in ^{4}He the orthobaric densities as measured by EW are indeed proportional to $t^{1/2}$ within the region indicated, although further away the dependence is more nearly $t^{1/3}$ as seems to be the case for most fluids. (See references 2 and 7.) If the former result is real, it might be attributed to the quantum corrections to the law of corresponding states. [J. de Boer, *Physica* **14**, 139, 149 (1948); J. de Boer and R. J. Lunbeck, *Physica* **14**, 509 (1948).] Similar behavior ought to be expected then in ^{3}He, and has in fact recently been observed [R. H. Sherman, *Proceedings of the Conference on Phenomena in the Neighborhood of Critical Points*, National Bureau of Standards, Washington, D. C., 1965 (to be published)].

[9] The straight line in the figure actually has a finite intercept on the abscissa, corresponding to +4.6 mdeg. This may indicate flattening of the coexistence curve close to the critical point, or be the result of errors in the measured temperatures in EW or in the assignment of T_c.

[10] M. H. Edwards and W. C. Woodbury, *Can. J. Phys.* **39**, 1833 (1961).

cording to Equation 4 we have

$$\beta = [\rho^2 \kappa_T (\alpha t/\beta + \rho'^2)]^{-1} \tag{6}$$

The result is $\beta = (6.86 \pm 1.2) \times 10^{13}$ dyne cm^{10}/mole^4; most of the quoted uncertainty is due to a 15 per cent error in the measured value of κ_T. Equations 3 and 4 can be integrated to yield the isotherms in the P, ρ and μ, ρ planes if plausible assumptions are made concerning the temperature-dependent integration constants. Families of such isotherms show none of the inconsistencies associated with the expansion of EW.[11]

The range of usefulness of this simple approximation can be established only by further experiments. A systematic study of the critical region would be indeed most desirable. We wish to emphasize, however, that the case for the pair of conjugate variables ρ, μ depends not on the validity of approximation 3, but rather on the near symmetry of the coexistence curves.[12]

The connection between symmetry and thermodynamic singularity has previously been elaborated by one of us.[13] In that paper critical points were discussed in terms of the diagonalized stiffness matrix. For a one-component system we have two diagonal elements that can be written as $\lambda_1 = T/c_v$, $\lambda_2 = (\partial\mu/\partial\rho)_T = \rho^{-1}(\partial P/\partial\rho)_T$. The classical point is characterized by $\lambda_2 = 0$, a condition that can be satisfied only at an isolated *point* in the P–T diagram. It is the symmetry of the system that renders $\lambda_2 = 0$ possible along a λ *line* in this diagram.

On the basis of the classical theory, which excludes the case $\lambda_1 = \lambda_2 = 0$ as an unlikely accidental degeneracy, c_v is predicted to be finite.[4,13] This *a priori* argument against a singular c_v is removed, however, if the degeneracy is produced by symmetry rather than by accident. An intuitive interpretation of this symmetry is provided by the lattice-gas model of Lee and Yang.[14] At the time this theory was advanced the injection of symmetry seemed to be a fortuitous property of the oversimplified model. It now appears that this symmetry, at least in the immediate neighborhood of T_c, may indeed be a fundamental property of the critical region.

Some aspects of this paper were clarified during the Conference on Phenomena in the Neighborhood of Critical Points, National Bureau of Standards, Washington, D. C., April 5 through 8, 1965.

[11] For a plot of these curves, see C. E. Chase and R. C. Williamson, *Proceedings of the Conference on Phenomena in the Neighborhood of Critical Points*, National Bureau of Standards, Washington, D. C., 1965 (to be published).

[12] The arguments in favor of this scheme far outweigh the slight inconvenience of using the chemical potential which, like the entropy, is measurable only indirectly. The compromise scheme p, ρ turns out to be impractical; this may be connected with the fact that these variables do not form a conjugate pair.

[13] L. Tisza, *Ann. Phys.* (*N. Y.*) **13**, 1 (1961).

[14] T. D. Lee and C. N. Yang, *Phys. Rev.* **87**, 410 (1952); see also C. N. Yang and C. P. Yang, *Phys. Rev. Letters* **13**, 303 (1964).

PAPER 6

THE GEOMETRICAL INTERPRETATION OF THE FORMALISM OF MTE*

1. The Geometrization of Thermodynamics

The major disciplines of theoretical physics are expressed in terms of rigorous mathematical structures, even if the application to empirical situations is only approximate. A conspicuous exception is CK thermodynamics, in which legitimate mathematical methods are hidden behind such devices as the Carnot cycle. Carathéodory and also Born were particularly concerned with removing this shortcoming (see Section 2.5 of Paper 1). As a result of these efforts thermodynamics appears as an instance of differential geometry. In particular the entropy is obtained by integrating Pfaffian differential forms and appears eventually as a surface in a thermodynamic phase space.

A second point of contact between thermodynamics and geometry is brought about by the use of Legendre transformations. However, these two aspects of geometrization are only loosely connected. The main difficulty is that Carathéodory operates in a thermodynamic phase space spanned by operational variables and thus pressure and volume appear on an equal footing, whereas in the Legendre transformation the extensive and intensive variables appear in conjugate pairs. All this constitutes a disharmony rather than an inconsistency. It is nevertheless pleasing to have in MTE a unified theory in which the Gibbs space spanned by the extensive variables is singled out of all thermodynamic phase spaces for special consideration. Other spaces are introduced subsequently by means of a Legendre transformation.

In view of the practical usefulness of Gibbs space, it is natural to ask what, if anything, is the underlying mathematical structure? Does this concept deserve to be admitted among the important abstract spaces of mathematical physics? The latter are usually characterized by a metric

* Some of this work appeared in 1959 and 1960 in the *Quarterly Progress Report* of the Research Laboratory of Electronics, M.I.T., Cambridge, Mass.

based on length and/or the orthogonality of the basis vectors, possibly in the complex domain. Neither of these concepts is definable in Gibbs space spanned by such disparate variables as volume, mole numbers, and energy.

The important fact is that, by means of the entropy, we inject into this unstructured manifold an element of valuation that serves as a "metric," formally based on *volume* rather than *length*, and on *parallelism* rather than *orthogonality*. The result is a specific kind of affine geometry that constitutes a "complementarity" to Euclidean space with its metric based on length: a mathematical counterpart to the complementarity of MTE and CMP (classical mechanics of particles).

Notwithstanding the conceptual difference between the conserved quantities of MTE and the displacement vectors of CMP, the two formalisms have important aspects in common. Thus the variables of the differential equation of a linear harmonic oscillator can be interpreted equally well as the displacement of a point mass or as the charge on a condensor in an *RLC* circuit. This complete identity of the formalisms is fortuitous and stems from the fact that in one dimension there is no difference between length and volume and there is no place for either orthogonality or parallelism. In several dimensions we have the Gibbs space of quantities on the one hand and the configuration space of position vectors on the other, the respective geometries split. The purpose of this paper is to outline the elements of the geometry of Gibbs space.

Finally a note on the emphasis on the geometrical aspects of the formalism. Every geometrical theorem has its analytic-algebraic equivalent, and it is a matter of taste whether or not one insists on the geometrical language. From the algebraic point of view *affine transformation* is merely *linear transformation*. *Volume* spanned by a number of vectors is the *determinant* of the components and *equiaffine* transformations have determinants equal to unity (*unimodular* transformations) and thus leave the volume invariant. A few less obvious points will be added shortly.

In spite of the fact that it is dispensable, some readers may find the geometrical imagery illuminating.

We shall not undertake a systematic development of the affine geometry, but we shall consider three problems to illustrate the role of geometry in MTE.

2. Legendre Transformations

We have considered these transformations in Section 2.3 of Paper 2. We now return to this problem to illustrate it from the geometrical point of view.

For the sake of simplicity we confine ourselves to the case of one independent variable where the geometrical presentation is the most useful.

Geometrical Interpretation

The starting point is the remark that a curve in a plane may be represented in either of two equivalent ways: first as a set of points, second as the set of its tangents, whereby the original curve appears as an envelope (Figures 1*a* and *b*). In order to give a formal demonstration of this statement, we write the equation of a straight line in a form, due to Pluecker, exhibiting the so-called duality between points and lines:

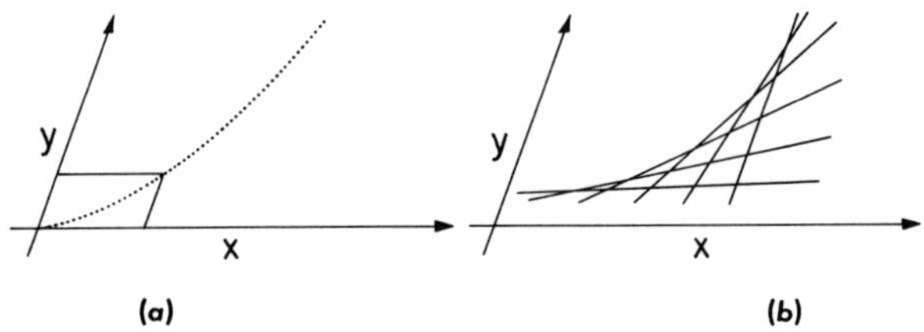

Fig. 1. Representation of a curve in terms of (a) points, (b) lines.

$$y = \xi x - \eta \tag{1}$$

Here x, y are the so-called parallel coordinates of a point whereas ξ, η are called line coordinates inasmuch as any set of constants ξ, η in Equation 1 corresponds to a straight line, ξ being the slope and $-\eta$ the intercept along the y axis. Note that the slope is defined as $\xi = (y_2 - y_1)/(x_2 - x_1)$, but this is not the tangent of an angle, which has no affine invariant meaning.

Let us now consider two continuous differentiable functions:

$$y = y(x) \tag{2}$$

$$\eta = \eta(\xi) \tag{3}$$

Equation 2 represents a set of points, Equation 3 a set of lines. What is the condition that both equations represent the same curve, in other words, that the lines 3 are tangent to the curve 2?

At any particular point x_0, y_0 of the curve 2 the equation of the tangent is

$$y - y_0 = \left(\frac{dy}{dx}\right)_{x=x_0} (x - x_0)$$

The comparison of this equation with Equation 1 provides the required conditions:

$$\xi = \frac{dy}{dx} \tag{4}$$

$$\eta = x\frac{dy}{dx} - y = x\xi - y \tag{5}$$

The "0" subscripts have been omitted; it is to be remembered, however, that x, y have to satisfy the equation of the curve 2. Equation 5 is actually a recipe leading from Equation 2 to 3, provided that x can be eliminated from Equation 4 and inserted in Equation 5. This is possible if $d\xi/dx = d^2y/dx^2 \neq 0$, which shall be assumed.

The transformation $x, y \rightarrow \xi, \eta$ expressed in Equations 4 and 5 is called a Legendre transformation. It has an important dual property inasmuch as it has a form identical to its inverse.

In fact, from Equations 4 and 5 one gets

$$d\eta = x\, d\xi + \xi\, dx - dy = x\, d\xi$$

hence

$$x = \frac{d\eta}{d\xi} \tag{6}$$

and

$$y = x\xi - \eta = \xi \frac{d\eta}{dx} - \eta = \xi^2 \frac{d\eta/\xi}{d\xi} \tag{7}$$

Equation 7 is identical to the original relation $y = y(x)$, provided ξ can be eliminated by solving Equation 6. The condition for this is

$$\frac{dx}{d\xi} = \frac{d^2\eta}{d\xi^2} \neq 0 \tag{8}$$

The proof of the equivalence of Equations 2 and 3 is thus completed, and we have established the connection with the algebraic treatment of Section 2.3 of Paper 2. For the case of several variables we refer to that section.

3. Quadratic Forms

Different types of geometries are characterized by their groups of transformation. The question is to identify the physically meaningful transformations in MTE that can be interpreted within affine geometry.

A transformation group of this sort arises in connection with the specification of the composition of the system in terms of independent components. We have seen in Section 2.3 of Paper 2 that nonsingular linear transformations of sets of independent components lead to other admissible sets of components. Whereas this fact should be remembered as a matter of principle, a case of more practical interest arises in connection with the stiffness form, the fundamental quadratic form of MTE that determines the stability properties of the system (see Sections VII, VIII, and Appendix B of Paper 3; see also the Appendix of Paper 4 for an earlier version).

The relevant properties of quadratic forms are best recognized in the canonical diagonal form. In metric spaces the transformation to such a form is achieved by means of the eigenvalue method. However, this does not apply in Gibbs space, and we have to fall back on another technique operating in terms of "completing the square." Although treated in every text on linear algebra, this method gives rise to some problems not commonly discussed.

Within the group of all linear (affine) transformations, the quadratic form can be brought to a canonical form with the diagonal elements $\lambda = \pm 1$ or 0. This is unsatisfactory for the present purposes since the diagonal elements λ of Equation 8.7 on page 148 are physically significant: they provide a measure of the stability of the system. Although it is easy enough to arrive at the correct coefficients, it is a more subtle problem to give a formal definition of the group of transformations that will ensure this end.

Written in matrix form the elements of this group have a triangular form with 0's under the diagonal and the added restriction that the diagonal contain only 1's. Such matrices form evidently a subgroup of the full linear (affine) group. They are unimodular (their determinant is equal to unity). However, contrary to what is implied on page 189, this condition is insufficient to specify the transformations in question;* we have to require that *the determinants of all principal minors be unity.* We shall refer to the group of these matrices as the *restricted unimodular group*, or in geometrical language the *restricted equiaffine group.* Whereas the equiaffine group is volume-preserving,† the restricted group preserves volume even if the transformation is broken off after the first, second, . . ., step, and the remainder of the matrix is replaced by the unit matrix.

My (by no means exhaustive) search of the literature produced no reference to this group, although the latter is bound to be significant as the orthogonal group's counterpart. Closely related results are found, however, in Gel'fand's book quoted on page 98. He shows (page 164) that the existence of dual spaces ensures a sort of metric, which can be considered as the algebraic core of the duality within MTE.

Gel'fand also considers the reduction of a quadratic form by means of triangular transformations (*op. cit.*), and is concerned with the problem of restricting these in order to arrive at a definite set of diagonal elements. However, the method, credited to Jacobi, yields diagonal elements inverse to ours. This method is not acceptable in MTE although not objectionable as a purely algebraic device.

We turn now to the geometrical interpretation of the diagonalization of forms by means of restricted unimodular matrices for the case of two variables in terms of Figures 2 and 3, which are the equiaffine mapping of each other. Preserving parallelism, but not orthogonality, the di-

* I am grateful to Felix Villars for pointing this out to me.

† W. Blaschke, *Differentialgeometrie,* Vol. II *Affine Differenzialgeometrie* (Julius Springer, Berlin, 1923), p. 70. R. Artzy, *Linear Geometry* (Addison-Wesley, Reading, Mass., 1965).

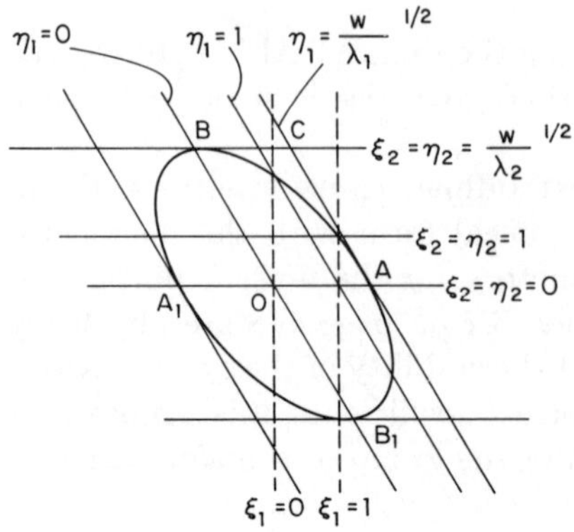

Fig. 2. Equiaffine transformation $\xi_1, \xi_2 \rightarrow \eta_1, \eta_2$ that diagonalizes the quadratic form represented by the ellipse.

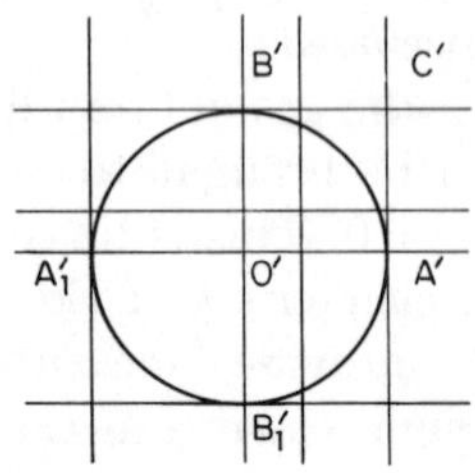

Fig. 3. Equiaffine mapping of Fig. 2 to orthogonal axes.

ameters $A'A_1'$, $B'B_1'$ are mapped into so-called conjugate diameters AA_1, BB_1. The square $O'A'B'C$ is mapped into the parallelogram $OABC$ of equal area.

Figure 2 is to be constructed in the following sequence: Obtain A as the intersection of the conic (ellipse) with the axis ξ_1; construct tangent AC and the parallel diameter BB_1. The tangent in B is parallel to OA.

The generalization to many dimensions is straightforward, provided that the rules of working with oblique coordinates are observed. Such coordinate systems are specified by the families of parallel hyperplanes: ξ_i = constant ($i = 1, 2, \cdots, r$). With orthogonal coordinates, each family of parallel planes is uniquely represented by their normal. This short cut is not available for oblique coordinates. The axis ξ_1 is now defined as the intersection of the planes $\xi_2 = 0, \xi_3 = 0, \cdots, \xi_r = 0$.

Although the line OA represents both the ξ_1 and the η_1 axes, the planes ξ_1 = constant and η_1 = constant are different from each other. If x_1 is interpreted as the entropy, the planes $\xi_1 = 0$ and $\eta_1 = 0$ correspond to processes at constant entropy and temperature, respectively. The peculiarity of the thermodynamic formalism, that the variables to be kept constant have to be especially indicated, appears to be the consequence of the use of oblique coordinate systems. This situation is unaffected by the ingrained habit of drawing the diagrams in Gibbs space with orthogonal axes.

It is evident from Figure 2 that the λ_i are not invariant. (See also page 98.) Thus different coefficients are obtained if the variables are relabeled, for example, $\xi_1 = \delta n$, $\xi_2 = \delta S$, instead of $\xi_1 = \delta S$, $\xi_2 = \delta n$, where n is the number of moles per unit volume.

However, the determinant $D_r = \lambda_1\lambda_2 \cdots \lambda_r$ is an equaffine invariant, a quantitative measure of the stability of the system. Geometrically $(D_r)^{-1/2}$ is the area $OABC$ for the ellipse $w = 1$.

4. Thermostatic Stability and Affine Curvature

It is evident from Papers 2 through 4 that the stiffness form plays a central role in MTE; it determines the stability of thermodynamic systems. It is also apparent that this form should have something to do with the curvature of the Gibbs surface. From the point of view of mathematical theory this conclusion is puzzling, because the theory of curvature relies so heavily on metric expressed in terms of the line element, nonexistent in Gibbs space. This is true both of the elementary theory of curvature and of the Riemannian theory.

It is briefly explained in footnote 4 of Paper 3 that there is a well-rounded, although little-known mathematical structure that can be associated with Gibbs space. Unfortunately, it would be too long-winded to give a detailed background of the differential geometry of surfaces. We confine ourselves to the brief statement that Blaschke's formalism (*op. cit.*, page 239) reduces to that of MTE with the following substitution:

First of all we specialize Blaschke's formalism by choosing the surface parameters $u = x_1$, and $v = x_2$. Next we establish the correspondence between Blaschke's x_1, x_2, and x_3 on the one hand with the thermodynamic variables s, v, and u, respectively. In terms of this definition Blaschke's coefficients of the affine fundamental form become identical to the stiffness moduli of MTE:

$$L = \left(\frac{\partial^2 u}{\partial s^2}\right)_v, \qquad M = \frac{\partial^2 u}{\partial s \partial v}, \qquad N = \left(\frac{\partial^2 u}{\partial v^2}\right)_s$$

This completes the correspondence which can be easily extended to an arbitrary finite number of variables.

We do not enter into this matter in greater detail, because no new physical results have been obtained thus far from the differential geometrical analog. What may be considered noteworthy is that we have a further instance of mathematical complementarity: On the one hand we have Riemannian geometry, the formalism of *universal* gravitation, on the other hand affine geometry, the formalism adapted to *specific* chemical type interactions of matter. To be sure, in MTE this interaction is treated in an entirely macroscopic matter. However, the notion of Gibbs space seems to admit extension to the microscopic domain. Some of this matter will be discussed in Papers 7 and 12. We hope to return to other aspects of the problem elsewhere.

PART B

STATISTICAL THERMODYNAMICS

Reprinted from ANNALS OF PHYSICS, Volume 25, No. 1, October 1963

ANNALS OF PHYSICS: **25,** 48–90 (1963)

The Statistical Thermodynamics of Equilibrium*

LASZLO TISZA† AND PAUL M. QUAY‡

Department of Physics and Research Laboratory of Electronics, Massachusetts Institute of Technology, Cambridge, Massachusetts

A statistical thermodynamics is developed in terms of extensive variables (additive invariants) distributed over a cellular division in space. In general, this distribution is governed by randomness and by correlations. The present theory, however, deals explicitly only with randomness, although correlations are implicit in the so-called fixed variables of the system. Because of this restriction, the theory is valid only for the fluctuations of coupled systems that have reached their equilibrium; hence we call it the statistical thermodynamics of equilibrium, briefly STE. A set of postulates is advanced, the essence of which is the requirement that distribution functions (d.f.) exist for two basic coupling situations. It is implicit that the system has a memory-loss mechanism; and the d.f. does not depend on past history (ergodic property). Such qualitative assumptions are sufficient to derive the Gibbsian d.f.'s in their quantitative form. These d.f.'s describe the coupling of finite systems with infinite environments and can be used to analyze typical situations of measurement by the methods of mathematical statistics. The present point of view sheds some new light on the ergodic problem and on the role of Nernst's law in completing the definition of thermodynamic equilibrium. An attempt is made to clarify the relations between entropy, information, and uncertainty by advancing a generic notion, the dispersal of a d.f., that subsumes these concepts as special cases.

INTRODUCTION

In 1910, Einstein developed a theory of fluctuations by introducing statistical concepts into thermodynamics and without making any explicit reference to the

* This paper, which is a development of a Ph.D. Thesis (to be referred to hereafter as "Thesis") submitted by one of the authors (PMQ) to the Department of Physics, Massachusetts Institute of Technology, August 1958, was supported in part by the U. S. Army Signal Corps, the Air Force Office of Scientific Research, and the Office of Naval Research; and in part by the U. S. Air Force (Office of Scientific Research) under ASD Contract AF 49(638)-95. Reproduction in whole or in part is permitted for any purpose of the United States Government.

† Currently on leave of absence at the Laboratoire de Chimie Physique de la Faculté des Sciences de Paris. John Simon Guggenheim Memorial Fellow.

‡ National Science Foundation Fellow, Massachusetts Institute of Technology, 1956–58. *Present address:* Department of Physics, Case Institute of Technology, Cleveland, Ohio.

coordinates of the particles constituting the system *(1)*. His point of departure was the inversion of Boltzmann's relation $S = k \log W$ connecting entropy and probability *(2)*. Instead of reducing the entropy to probabilities defined in microscopic terms (or, rather, to numbers of microstates), Einstein obtained the relative probabilities of fluctuation states in terms of differences of entropy. He went beyond the Boltzmann theory, inasmuch as he could consider general thermodynamic systems rather than only assemblies of weakly interacting particles. Moreover, he avoided the difficulties connected with the equipartition law without explicitly invoking the quantum theory. Of course, as a counterpart to these improvements, Einstein's theory renounced much of the space-time description that is implicit in Boltzmann's theory and did not provide the framework for the calculation of the thermodynamic functions from molecular parameters.

Einstein's work was independent of Gibbs' statistical mechanics, which had appeared a few years earlier *(3)* and which constitutes the other major generalization of Boltzmann's theory. Both of the generalized theories account for the fluctuations of thermodynamic quantities in systems in equilibrium with their environment. Gibbs' theory is conceptually the more satisfactory, but the quantitative discrepancies between the theories vanish asymptotically for large (macroscopic) systems and the two formalisms are used in the literature interchangeably. The two theories differ considerably, however, in their approach to the foundations. In contrast to Einstein's thermodynamic attitude, Gibbs attempted to derive the statistical theory of equilibrium ensembles from classical mechanics.

The approach taken by the present paper can be considered as a synthesis of selected aspects of those of Einstein and Gibbs. We shall follow the former in choosing a thermodynamic, rather than mechanical point of departure, but we shall sharpen the argument to arrive at a derivation of the more elaborate Gibbsian statistical formalism. This means, of course, that we detach Gibbs' ensemble theory from its connections with classical mechanics.

It is apparent that Gibbs' theory consists of a statistical and of a mechanical part which are only loosely connected with each other. The first aspect of this connection is the simple and important statement that the mechanical phase space provides the so-called sample space for the probability distribution function (d.f.) of the statistical theory. A quantum-mechanical version of this result is incorporated into the present theory, and thus contact is established with the standard calculations of quantum statistics. A second and more deeply reaching aspect of the connection is the derivation of the d.f.'s from purely mechanical principles, handled by Gibbs only through the verbal arguments of the famous Chapter XII. The traditional approach to the foundations required that this loose connection should be tightened. Although the numerous attempts to carry out this program led to interesting developments, there is a consensus of opinion that the problem has not been solved.

In view of this situation, several ideas have been recently offered to put the theory of ensembles on a statistical rather than on a mechanical foundation. Thus, Mandelbrot showed how important parts of statistical thermodynamics can be subsumed under standard procedures of mathematical statistics, without any reference to micromechanics (*4*). Moreover, Jaynes and others have proposed the concepts of information and statistical uncertainty as the sole organizing principles for statistical thermodynamics (*5*). Probability and entropy are interpreted in a purely subjective sense and are supposed to reflect our knowledge of the system only in the sense of information theory (*5a*).

The present paper has certain elements in common with these latter contributions, but departs from them in others. Granted the importance of recognizing the mathematical formalism that is suitable for a physical discipline, it is desirable that the latter should be structured around physical organizing principles. Such an organizing principle is provided by thermodynamics itself if its scope is extended by the introduction of statistical elements of a physical nature (*6*). This is a break with the traditional notion that this discipline is to be strictly macroscopic. If, however, thermodynamics is considered as a theory of measurement, it is fitting for it to catch up with the extension of experimentation into the microscopic domain.

The shortcoming of classical thermodynamics is its lack of space-time detail. This paper is part of a program in which such detail is injected into the theory in successive stages. The theory here developed is called STE, which is short for statistical thermodynamics of equilibrium.[1] This paper was preceded by another (*7*), in which the macroscopic thermodynamics of equilibrium or, briefly, MTE has been developed. We shall refer to (*7*) as (*MTE*).

The relation between MTE and STE is very close, and is explained in terms of their postulational bases in Section I. Both theories deal with the distribution of additive invariants over disjoint regions of space. In MTE, one considers only the equilibrium values of these invariants, collectively denoted by X, as determined from a phenomenological extremum principle. In the deeper theory, STE, the X's are considered as random variables whose averages are the equilibrium values of MTE.

The central problem confronting statistical theories is the interplay of randomness with correlations stemming, say, from a dynamical law. The simplest statistical theory, STE, deals only with pure randomness. This formal specification leads in a natural way to the Gibbsian theory of equilibrium ensembles.

[1] We adhere to the convention of using capitalized abbreviations to denote theories with an explicitly formulated postulational basis. Terms such as "thermostatics" or "thermodynamics" are used in a looser, more conventional sense.

The derivation given in Section II utilizes a fundamental work of Szilard (*8*) which, after some adjustments, fits harmoniously into the present context.[2] An extension to several random variables is found in Section III. Although these results are formally identical to those well known in statistical mechanics, there is a considerable difference in their scope and conceptual interpretation. Some of these questions are discussed in the remainder of the paper. Section IV contains the elements of a theory of measurements based on the concept of parameter estimation (see ref. *4*). This approach yields a reinterpretation of thermodynamic formalism which is applicable even to very small isolated systems. Section V deals with asymptotic theories in which the number of degrees of freedom is allowed to tend to infinity. In Section VI, we discuss the role of the ergodic property in STE and the relation of entropy and information.

The discussion of the approach to equilibrium (H-theorem) is outside the scope of STE since such a discussion requires an explicit handling of correlations. Thus, our approach implies a subdivision of the foundation problem that is quite similar to the one suggested by Schafroth (*9*).

I. DEFINITIONS AND POSTULATES

The postulational basis of STE consists of two parts. That which is summarized in Section I,A is held in common with MTE, and for a fuller discussion we refer to Section I,A of (MTE). Section I,B contains the statistical postulates that replace the entropy-maximum postulate of MTE.

A. Definitions and Postulates from MTE

The independent variables of both MTE and STE are additive, conserved quantities, briefly additive invariants, $X_1, X_2, \cdots$. We shall usually interpret these variables as: internal energy U, mole numbers of chemical components $N_1, N_2, \cdots$, and volume V.[3]

A thermodynamic simple system is a region in space for which particular

[2] Szilard's results were neither developed in nor integrated with the main current of statistical physics. Indeed, his method appears rather *ad hoc* in that context. Recently, however, Mandelbrot showed (*4*) that the method can be considered as an application of the concept of "sufficiency" developed in mathematical statistics. Although we believe that the use of such techniques in statistical thermodynamics is well worth exploring (cf. Section IV), our adaptation of Szilard's method is self-contained and does not rely on specialized statistical knowledge.

[3] This interpretation raises questions of validity: are these variables indeed additive and invariant? Thus, the energy is not exactly additive. Moreover, the identification of the chemical components of the system, hidden innocuously beneath the subscripts of the N's, is not absolute. The identifying properties themselves can vary, depending on the presence or absence of catalysts or anticatalysts, which act, in this way, as constraints on the molecular level. For a further discussion, we refer to pp. 21, 22 of (*MTE*).

values X' of the additive invariants are specified within some short interval of time. Disjoint simple systems can be built up into composite systems or, conversely, we may divide a system into subsystems, sometimes referred to as cells.

Thermodynamic *processes* may be of various sorts, but what we consider primarily is the passage of some quantity X from one system to another. The systems are separated by *walls* that are *restrictive* of the passage of some of the quantities, say X_λ, and nonrestrictive of all the others denoted by X_j. Such a wall is symbolized by $W[X_j]$ on the assumption that the full set of X's has been given. The restrictions on a system imposed by its walls are referred to as *constraints*.

We consider now the set of all variables X', X'', $\cdots$ associated with the various subsystems. By choosing appropriate linear combinations, it is possible to classify these into *free* and *fixed* variables: the former are free to change while the latter are held constant by the existing constraints. We shall distinguish these types of variables by Roman and Greek subscripts respectively. The distinction between free and fixed variables X_k and X_λ is fundamental for STE, and the details of physical interpretation seem sometimes only secondary. Thus, there is a far-reaching analogy between processes of energy and mass transfer; nevertheless, the differences between such processes are by no means negligible. The universal role of the energy manifests itself thermodynamically in the fact that one can realize in practice a transfer of energy without a transfer of any other X, a true one-variable process, whereas if any of these other X's are transferred, the process is accompanied by a transfer also of energy and we have a process described by at least two variables.

We shall emphasize the close connection of our results with quantum mechanics, particularly with the fact that the energy may have a mixed discrete-continuous spectrum. However, we shall ignore those subtler effects for whose description the density matrix is essential. Accordingly we assume the following postulates in addition to those listed in I, A (MTE):

P a5. The values of the variables X are precisely measurable except, possibly, for an arbitrary constant.

Also connected with quantum mechanics is the following postulate through which the concept of the ground-state of a system is introduced.

P a6. For simple systems contained in a finite volume, the range of variation of the X's has a lower bound. In particular, the lowest energy of a system is a function of its fixed variables. In a context in which the latter are held constant, the lowest energy can be set equal to zero.

We shall not discuss in this paper the thermodynamic processes involving *work*, the discussion of which in STE adds little to the results available already in MTE. To be sure, the concept of work gives rise to significant statistical problems (*10*); but the discussion of these involves the "memory" or "persistence" of the previously realized states and belongs to a more subtle theory that deals with the interplay of randomness and correlations.

B. Definitions and Postulates on Statistical Equilibrium

We develop now the statistical concept of equilibrium characteristic of STE. As we have seen, the coupling of thermodynamic systems is effected by the exchange of additive invariants. We now postulate that this exchange be governed by statistical laws, the properties of which are expressed in merely *qualitative* postulates *P b1–4*. Together with those of Section A, these postulates suffice for the derivation of the *quantitative* laws of thermodynamic equilibrium, the effects of fluctuations included. These postulates taken together, then, implicitly define thermodynamic equilibrium. Contrary to what one might expect from a superficially critical analysis, this procedure is not circular because its conclusions are subject to a very exacting experimental verification, as we shall see in Section IV,D.

D b1. Similar systems [symbol $(x_k \mid x_\lambda)$]: A set of systems describable in terms of the same selection of variables X, and such that each fixed quantity X_λ has the same value x_λ in each system, while the values of the free variables X_k take on random values x_k in the various systems. We denote a particular, instantaneous value of a variable X by x.[4]

At least one of the X's has to be fixed, or else the system would be undefined. We shall usually consider the volume as a fixed variable; often also, the total mass, for one-component systems, preferably expressed in terms of the mole numbers.

D b2. Ensemble [symbol $e(x_k \mid x_\lambda)$, or $e(x_\lambda)$]: An unlimited set of similar systems.

Two different interpretations of this abstract definition are useful. In the first, the similar systems are distinct and independent replicas of some particular system, all subject to the same constraints as this latter. In the second, called a *temporal ensemble*, the similar systems represent simply the results of measurements on the free variables of a single system at discrete instants of time t_n $(n = 1, 2, \cdots)$.

A quantitative description of the values of the free variables can be given only in terms of statistical distribution functions (d.f.) the nature of which depends on the type of physical situation considered, and exhibits, in general, an interplay of randomness with correlations arising as a consequence of molecular dynamics.

The physical situation considered in STE is thermodynamic equilibrium.

[4] Throughout this paper we shall use the vertical bar to separate free and fixed variables in the symbols of ensembles and in expressions of probability. Since the specification of the system includes the proper selection of the set of all the variables X, strictly speaking, similar systems should be designated as $(x_k \mid X, x_\lambda)$.

The corresponding statistical theory is that of independent random variables. This means that correlation effects do not explicitly affect the statistical distribution of the free variables, although molecular correlations are in various ways implicit in the theory, in particular, in the selection of the fixed variables which determine the chemical identity of the system. While in (MTE) we assumed that the fixed variables determine the asymptotic equilibrium values of the free variables, we now make the weaker assumption that the fixed variables enter into the specification only of the d.f. of the free variables. In STE, then, equilibrium is characterized by the existence of a unique d.f., which we shall postulate in two typical situations: first, for the equilibrium of one finite system with another; and second, for the equilibrium of a finite system with an infinite environment called a reservoir.

P b1. Statistical equilibrium: Finite systems. Consider an ensemble $e(x_\lambda', x_\lambda'')$ of isolated systems, each of which is composed of a subsystem (x_λ') in contact with a subsystem (x_λ'') through a wall $W[X_k]$. For each member of e, let $(x_k \leqq X_k < x_k + dx_k)$. After a sufficiently long time ("aging"), the ensemble e is said to be in equilibrium with respect to the distribution of the X_k, and is then decomposed into $e'(x_\lambda')$ and $e''(x_\lambda'')$. The probability that, under these conditions, a member of e' should possess values of the X' less than x' is given by a unique conditional (cumulative) d.f.

$$\begin{aligned}\mathfrak{F}(x_k' \mid x_k\,;x_\lambda', x_\lambda'') &\equiv \mathfrak{F}(x_k' \mid x_k) \\ &= \Pr\{X_k' < x_k' \mid x_k \leqq X_k < x_k + dx_k\,; x_\lambda', x_\lambda''\}\end{aligned} \tag{1.1}$$

where Pr means "probability" and $x_k = x_k' + x_k''$.

In spite of its qualitative character, this postulate is very strong, since it states that $\mathfrak{F}$ is uniquely determined by the conditioning parameters to the right of the vertical bar in the argument. In order to appreciate the meaning of this requirement, we have to list some of the factors that *could* be relevant for the determining of the d.f. but actually are not. First, $\mathfrak{F}$ is determined by the present values of the fixed variables and does not depend on their past history. Thus there is implicitly assumed the existence of a memory-loss mechanism within systems approaching equilibrium.

Secondly, either of the interpretations of an ensemble given in *D b2* may be used; so far as $\mathfrak{F}$ is concerned, they are wholly equivalent. This equivalence of the two modes of ensemble-representation of a system is known as the *ergodic property*. We shall discuss this point in more detail in Section VI.

We shall also use the differential d.f.

$$d\mathfrak{F}(x_k' \mid x_k) = \mathfrak{F}(x_k' + dx_k' \mid x_k) - \mathfrak{F}(x_k' \mid x_k). \tag{1.2}$$

For continuous distributions one has

$$d\mathfrak{F}(x_k' \mid x_k) = \mathfrak{f}(x_k' \mid x_k)\, dx_k', \tag{1.3}$$

where f is a probability density. For discrete nonsingular distributions:

$$d\mathfrak{F}(x_k' \mid x_k) = f(x_{kn}' \mid x_k) = f_n(x_k' \mid x_k), \tag{1.4}$$

where n is an integer and the intervals of the X_k' have been chosen small enough so that f_n is the probability of a single "level." The distribution function, $d\mathfrak{F}$, as also the corresponding ensemble. is referred to as *microcanonical*.

In *P b1* the systems (x_λ') and (x_λ'') are on an equal footing. We now let the size of (x_λ'') tend to infinity. In this case *P b1* no longer ensures the existence of a d.f. for the X_k'; and if such a distribution exists, it is not conditioned by $x_k' + x_k''$, since this is infinite.

Although at first sight the transition to infinite systems or reservoirs seems to lead to difficulties, in reality this limiting process brings about important simplifications. There are many reservoirs that are equivalent to each other as environments of finite systems although they differ considerably in their intrinsic structure. Granted the possibility that the detailed structure of the reservoir has an effect on an individual observation of the state of a coupled finite system, still this effect is averaged out over the ensemble, and this is our only concern in the present theory. The precise definition of the equivalence of reservoirs constitutes the first element in a generalization of the zeroth law of thermodynamics.

D b3. X_k-Reservoir or, briefly, reservoir [symbol: R]: An infinite thermodynamic system that can be coupled by X_k-exchange to finite systems in such a way as to make $e'(x_k' \mid x_\lambda')$ and $e''(x_k'' \mid x_\lambda'')$ statistically independent of each other. Within STE the specification of this exchange will provide an adequate description of the interaction of the system and reservoir. Two reservoirs are equivalent if the time-average of X_k transfer between them is zero when they are coupled to each other by the intermediary of a finite system.

P b2. The equivalence relationship explained in *D b3* is independent of the nature of the coupling system, provided that it does not inhibit the X_k-exchange. Moreover, the relation is transitive, that is, if a reservoir R' is equivalent to R'' and R'' is equivalent to R''', then R' is equivalent to R'''.

As an immediate consequence of this equivalence relationship, we state the corollary:

C b1. The class of equivalent X_k reservoirs can be labeled by a set of parameters $\theta_1, \theta_2, \cdots$, briefly θ_s, in such a way that equivalent reservoirs are labeled by the same set θ_s, while nonequivalent ones differ in at least one of the θ. Reservoirs will be denoted also as $R(X_k, \theta_s)$ or $R(\theta_s)$.

Evidently the labels θ_s are closely related to the intensities of MTE, but they constitute much more rudimentary qualitative concepts. They lack a definite scale; they are not necessarily paired with the X_k; and they are even unspecified as to their number. Thus, for example, in the case of thermal coupling we do not postulate the existence of a *single* parameter, the temperature. The intensities, in their quantitative thermodynamic meaning are to be derived from the theory.

Other immediate consequences of *D b3* and *P b2* are the following: Equivalent reservoirs can be united to form another reservoir of the same class; the class of a reservoir remains unaffected if any finite system is joined to it. Two nonequivalent reservoirs are to be joined only by the intermediary of a finite system. However, this situation does not lead to equilibrium and is outside the scope of the present theory.

P b3. Statistical Equilibrium of a Finite System and Reservoir. Consider an ensemble $e(x_k \mid x_\lambda)$ that has been coupled for a sufficiently long time to a reservoir $R(X_k, \theta_s)$. Then the probability that the X_k of a particular member of e be less than some value x_k is given by a unique, (cumulative) d.f. $F(x_k \mid x_\lambda, \theta_s)$ which is completely determined by the nature of the system, indicated by its fixed variables x_λ, and by the class of the reservoir, specified by the θ_s. Or in symbols[5]:

$$F(x_k \mid x_\lambda, \theta_s) = \Pr\{X_k < x_k \mid x_\lambda, \theta_s\}. \tag{1.5}$$

Expressions analogous to (1.2), (1.3), and (1.4) can be written down in obvious fashion. The function F is called the (*generalized*) *canonical* d.f. More precisely, the d.f., or the corresponding ensemble, is canonical in X_k and microcanonical in X_λ. The d.f. that is canonical only in U and microcanonical in all others is called *canonical* in the narrower sense of the word. The d.f. that is microcanonical in the volume and canonical in all other X's is called the *grand canonical* d.f. A single system whose behavior is represented by a canonical ensemble is called a canonical system, or, from the experimental point of view, a system in thermodynamic *equilibrium* with $R(\theta_s)$.

Finally, we formulate a stability property for the canonical d.f.

P b4. Consider two ensembles e' and e'' with the canonical distributions $F'(x_k' \mid x_\lambda', \theta_s)$ and $F''(x_k'' \mid x_\lambda'', \theta_s)$, respectively, although they need not be in contact with a reservoir. If the two ensembles are coupled member by member through walls $W[X_k]$ and then separated again into e' and e'', the reconstituted ensembles have the same canonical d.f.'s as the original ones.

This postulate might be regarded as a statistical "zeroth law"[6]: two ensembles that have been in statistical equilibrium with a reservoir are in statistical equilibrium with each other, regardless of the presence or absence of the reservoir.

[5] In the literature, the vertical bar is used in the arguments of conditional probabilities. To the right of the bar are those random variables that assume fixed values for the events considered. Our use is somewhat more general, since θ_s are not random variables, but fixed properties of the reservoir. See also footnote 4.

[6] Cf. Thesis, pp. 37–38. More accurately: as the final element in the statistical generalization of the zeroth law begun in *D b3*. It can be shown that a similar law holds for any two objects ("object" being a canonical system, ensemble, or reservoir).

II. THE CANONICAL FORMALISM: SYSTEMS WITH A SINGLE RANDOM VARIABLE

A. Derivation of the Distribution Functions

The postulates of Section I allow one to derive the functional form of the generalized canonical d.f.: $dF(x_k \mid x_\lambda, \theta_s)$ and of the microcanonical d.f.: $d\mathfrak{F}(x_k' \mid x_k, x_\lambda', x_\lambda'')$. Although the discussion of the general case offers no formal complications, the conceptual interpretation of the formalism is greatly facilitated if the discussion is confined at first to the special case of the ordinary canonical d.f. in which the internal energy $X_1 = U$ is the only random variable of the system. We shall return in Section III to the general case of many variables.

Apart from a few refinements, we follow Szilard and analyze a simple sequence of thermodynamic operations on two ensembles and a heat reservoir, establish a functional equation involving dF and $d\mathfrak{F}$, and derive those solutions of this equation which satisfy the physical conditions of the problem.

We consider two ensembles $e'(x')$ and $e''(x'')$, and bring each member of e' into thermal contact with a member of e'' to form an ensemble of pairs of systems $e(x', x'')$. This ensemble of composite systems is then brought into thermal equilibrium with the reservoir $R(\theta_s)$. According to *P b3* and *D b3* the joint probability that the subsystems of a member of the ensemble e have the energies u' and u'' is

$$\Pr\{u' \leqq U' < u' + du', u'' \leqq U'' < u'' + du'' \mid x', x'', \theta_s\} = dF'(u' \mid x', \theta_s)\, dF''(u'' \mid x'', \theta_s).$$

This d.f. persists, regardless of whether the contacts between the ensemble and R, and the contacts between the subensembles are maintained, (*P b4*). If, then, the ensemble e is decoupled from R, we can describe this same d.f. from a different aspect: The above joint probability appears as the probability $dF(u \mid x', x'', \theta_s)$ of finding the energy $u = u' + u''$ in a member of the composite ensemble, multiplied by the conditional probability $d\mathfrak{F}(u' \mid u, x', x'')$ that this energy splits in the desired fashion among the subsystems (*P b1* and *P b3*). The comparison of the two expressions yields the basic functional equation:

$$dF'(u' \mid x', \theta_s)\, dF''(u'' \mid x'', \theta_s) = dF(u \mid x', x'', \theta_s)\, d\mathfrak{F}(u' \mid u, x', x''). \quad (2.1)$$

This equation reduces to the well-known Boltzmann relation if the microcanonical d.f. is a δ-function, i.e., if

$$d\mathfrak{F}(u' \mid u) = \delta(u' - u_0) \quad (2.2)$$

where u_0 is the amount of energy that would be trapped in the system (x') if the principle of thermostatistic determinism of MTE were rigorously valid.

To solve (2.1) for the analytic forms of dF and $d\mathfrak{F}$ we shall need the logarithm of these functions. For this reason, we must first examine the nature of the zeros of the distribution functions. Suppose that the d.f. of e' vanishes for a particular set of values of the energy and of the parameters: $dF'(u_0' \mid x', \theta_{s0}) = 0$. There is, in general, no reason why $dF(u \mid x', x'', \theta_{s0})$ should be simultaneously zero. Therefore, from (2.1), $d\mathfrak{F}(u_0' \mid u, x', x'') = 0$. Since this relation is θ_s-independent, (2.1) requires that $dF'(u_0' \mid x', \theta_s) = 0$ hold independently of the values of θ_s.

Consequently, there must be a component in $dF'(u_0' \mid x', \theta_s)$ which is θ-independent and which is zero for the entire set of forbidden values of U'. We may, therefore, exclude this forbidden set from consideration for the present.[7] In similar fashion, we exclude the forbidden sets of U'' and U and, so, all zero-values of dF and $d\mathfrak{F}$.

Taking the logarithm of (2.1) and writing $u' + u''$ for u, we have:

$$\ln dF'(u' \mid x', \theta_s) + \ln dF''(u'' \mid x'', \theta_s) = \ln dF(u' + u'' \mid x', x'', \theta_s) + \ln d\mathfrak{F}(u' \mid u + u'', x', x''). \tag{2.3}$$

Let c_s be an arbitrary but fixed set of values of the parameters θ_s, labeling some heat reservoir. We take, now, the difference between (2.3) for a general θ_s and (2.3) when evaluated at c_s. Introducing the notation

$$\eta(u, x, \theta_s, c_s) = \ln dF(u \mid x, \theta_s) - \ln dF(u \mid x, c_s), \tag{2.4}$$

we obtain

$$\eta'(u', x', \theta_s, c_s) + \eta''(u'', x'', \theta_s, c_s) = \eta(u' + u'', x', x'', \theta_s, c_s). \tag{2.5}$$

We shall now solve this functional equation under the simplifying assumption that η is a differentiable function of u.[8] Differentiating (2.5) with respect to u' at constant u'', and vice versa, we obtain

$$\frac{\partial\eta'(u', x', \theta_s, c_s)}{\partial u'} = \frac{\partial\eta(u' + u'', x', x'', \theta_s, c_s)}{\partial u'} = \frac{\partial\eta(u, x', x'', \theta_s, c_s)}{\partial u} \tag{2.6a}$$

$$\frac{\partial\eta''(u'', x'', \theta_s, c_s)}{\partial u''} = \frac{\partial\eta(u' + u'', x', x'', \theta_s, c_s)}{\partial u''} = \frac{\partial\eta(u, x', x'', \theta_s, c_s)}{\partial u} \tag{2.6b}$$

Expression (2.6a) is evidently independent of x'' and (2.6b) of x'. Since the

[7] That such forbidden sets exist is evident from the discrete energy spectra of quantum mechanics.

[8] This assumption may seem incongruous with the fact that the energy is, in general, defined only on a discontinuous set. However, the continuous solutions of (2.5) can be subsequently restricted to the allowed energy spectrum and, therefore, our procedure yields satisfactory solutions of (2.1). On purely mathematical grounds, Eq. (2.5) also has more general solutions but these can be excluded by physical considerations. At any rate, we shall see in Section II, B that the analytical character of the function $\eta(u)$ is consistent with the discontinuous aspects of the problem.

two expressions are equal, they cannot depend on the nature of the systems (x') and (x''), but only on θ_s and c_s. Denoting this universal function of the reservoir properties as $-B(\theta_s, c_s)$, we obtain after integration

$$\eta(u, x; \theta_s, c_s) = -B(\theta_s, c_s)u - A(x, \theta_s, c_s) \qquad (2.7)$$

with similar relations for η' and η''. Inserting these expressions into (2.5) we see that the integration constants A are additive

$$A'(x') + A''(x'') = A(x). \qquad (2.8)$$

From (2.7) and (2.4) we obtain

$$\ln dF(u \mid x, \theta_s) = \ln dF(u \mid x, c_s) - A(x, \theta_s, c_s) - B(\theta_s, c_s)u.$$

Since the left-hand side is independent of c_s, each term on the right must be individually separable into a term containing c_s and one not containing it, in such a way that the c_s-dependent terms cancel.[9] Designating the portions of dF, A, and B that are independent of c_s as dG, Φ, and β, respectively, we have the general solution of (2.1)

$$dF(u \mid x, \theta_s) = dG(u, x)e^{-\Phi(x,\theta_s)-\beta(\theta_s)u} \qquad (2.9)$$

with

$$\Phi'(x', \theta_s) + \Phi''(x'', \theta_s) = \Phi(x, \theta_s). \qquad (2.10)$$

Our derivation holds only for those values of u that have a nonvanishing probability. But we may now bring back the forbidden values of u, defining $dG(u_0) \equiv 0$.

The condition on normalization of the probability dF leads to

$$e^{\Phi(x,\theta_s)} = \int_u e^{-\beta(\theta_s)u}\, dG(u, x), \qquad (2.11)$$

where the integration extends over the entire range of u. Evidently this integral, and hence $\Phi(x, \theta_s)$, depends on the set of θ_s only through the intermediary of the single parameter β, and the canonical d.f. can be written in the form[10]

$$dF(u \mid \beta) = dG(u, x)e^{-\Phi-\beta u} = dG(u, x)e^{-\beta u}Z^{-1}(\beta, x), \qquad (2.12)$$

where

$$\ln Z(\beta, x) = \Phi(\beta, x). \qquad (2.13)$$

We see that the canonical distribution function dF is the product of three factors. One factor depends only upon the type of system and its energy; a

[9] A formal proof of this plausible statement is straightforward but tedious.

[10] We shall soon see that β is a satisfactory parameter for labeling reservoirs as stipulated in *C b1*. Therefore we shall henceforth use the parameter β without further regard to the θ_s.

second depends upon the type of system and the reservoir parameter β; the third is an exponential function that provides the coupling between u and β in a manner the same for every system.

With these facts established, we may, as a rule, omit explicit reference to the fixed variables x. In particular, we rewrite the normalization condition (2.11) as

$$Z(\beta) = \int_u e^{-\beta u}\, dG(u). \tag{2.14}$$

Moreover, from (2.10) and (2.13),

$$Z'(\beta)Z''(\beta) = Z(\beta) \tag{2.15}$$

The function $Z(\beta)$ is seen to be the Laplace-Stieltjes transform of $G(u)$. The requirement that the integral (2.14) exist imposes limitations on the ranges of β and u which we shall consider later. The use of Stieltjes integrals is convenient for handling the mixed, discrete-continuous energy spectrum of quantum mechanics. But if $G(u)$ is differentiable or varies in value only discontinuously, one has for sufficiently small energy intervals,

$$dG(u) = g(u)du, \tag{2.16a}$$

or

$$dG(u) = g(u_n) = g_n \tag{2.16b}$$

and (2.14) reduces to the more usual form of the Laplace (-Riemann) transform or to a sum respectively. Corresponding changes are brought about in the d.f. (2.12).

We shall call $G(u)$ the *structure function* of the system. This use is close to that of Khinchin *(11)* who calls $g(u)$ the structure function. $Z(\beta)$ will be referred to as the *generating function* or *partition function.*

Inserting (2.12) into (2.1) and using (2.15), we obtain the microcanonical d.f.

$$d\mathfrak{F}(u' \mid u) = dG'(u')dG''(u - u')/dG(u) \tag{2.17}$$

which is independent of β as it should be. The normalization of this probability yields

$$dG(u) = \int_{u'} dG'(u')\, dG''(u - u'). \tag{2.18}$$

Thus the composition of the structure functions of independent systems is obtained by convolution or "faltung." It is well known from the theory of the Laplace transform that this rule is equivalent to the product rule of composition (2.15) for the generating function.

B. Formal Properties of the Canonical Distribution

Before turning to the physical interpretation of the functions and parameters appearing in the canonical d.f., it is convenient to summarize some of the mathematical implications of the formalism.

From the theory of the Laplace transform, we know that if $Z(\beta)$ converges for one value of β, it will converge for all larger values of β (*12*, p. 37). The limit of convergence depends on the nature of the function $G(u)$. We shall return to this point below. Within the range of convergence of (2.14) $Z(\beta)$ is analytic (*12*, p. 57); thus it possesses finite derivatives of all orders. Therefore we obtain the canonical averages

$$\frac{d^n Z}{d\beta^n} = \int (-u)^n e^{-\beta u}\, dG(u) = (-)^n \overline{u^n} Z. \tag{2.19}$$

We shall need the important expressions for the average and the variance of the energy:

$$\bar{u} = -\partial\Phi/\partial\beta \tag{2.20}$$

$$\operatorname{Var} u = \overline{(u - \bar{u})^2} = \overline{(\Delta u)^2} = \partial^2\Phi/\partial\beta^2 > 0. \tag{2.21}$$

Strictly speaking, the inequality sign in this important convexity relation should be $\geqq 0$. It is, however, a simple matter to show that the variance of the canonical d.f. can vanish only in the limiting case in which all members of the ensemble are in the ground state (*P a6*).

For a given value of $\bar{u}$, Eq. (2.20) can be considered as an equation for the unknown β. For systems with an unbounded energy spectrum $\bar{u}$ is itself unbounded. It follows from the convexity of $\Phi = \Phi(\beta)$ that Eq. (2.20) has a single positive solution, $\beta = \beta_0$, for any positive $\bar{u} = u_0$ (*11*, p. 77). Thus we have established a one-to-one relation between the average ensemble energy $\bar{u}$ and the reservoir parameter β. Moreover, from (2.20) and (2.21), we have

$$d\bar{u}/d\beta < 0. \tag{2.22}$$

Thus if an ensemble $e(X)$ is brought in contact at first with the reservoir $R(\beta_1)$ and then with $R(\beta_2)$ where, say, $\beta_1 < \beta_2$, then the canonical averages satisfy the relation

$$\bar{u}(\beta_1) > \bar{u}(\beta_2). \tag{2.23}$$

Therefore the contacts mentioned will bring about an energy flow from $R(\beta_1)$ to $R(\beta_2)$.

Thus we see that β is a satisfactory parameter for labeling reservoirs in the sense of *C b1*. Of course, in addition, β has more specific properties and provides a numerical measure for the intuitive concept of the temperature, higher

β's corresponding to lower temperatures. This is not trivial, since our definition of the reservoir parameters did not specify a direction of the flow for the X_k or their averages between different reservoirs. Moreover, β has an "absolute" character in the sense that it is uniquely determined once the unit of energy is fixed and does not depend on the nature of the system chosen as the basis for the canonical ensemble or on any other arbitrary convention. Nevertheless, we are still not in a position to establish the connection between β and the thermodynamic temperature. This is possible only through the concept of entropy which has not yet been defined.

The canonical formalism exhibits an interesting duality. The properties of a particular ensemble $e(x)$ can be described either in terms of the generating function $Z(\beta, x)$ or in terms of the structure function $G(u, x)$ with the reservoir parameter β specified in addition. Either of these descriptions allows us to arrive at the other by means of the Laplace transformation or its inverse. The exponential kernel of this transformation corresponds physically to the coupling of the system and the reservoir.

Although we have two equivalent representations for canonical ensembles, these descriptions are far from symmetric, and the functions $Z = Z(\beta, x)$ and $G = G(u, x)$ play different roles in the conceptual structure of the theory. As we shall see in Section II,C, the elaboration of the properties of the former enables us to establish contact with MTE, whereas the latter is a connecting link with quantum mechanics. Moreover, the two representations exhibit also a formal mathematical asymmetry.

The "moment-generating function" $Z(\beta)$ is extremely smooth, with continuous derivatives of all orders in its interval of convergence. These derivatives provide the successive moments of the canonical d.f. as shown in (2.19). In contrast, G is usually discontinuous, corresponding to the quantum-mechanical discrete spectrum of the system. This discontinuous character of the structure function was taken into account in our solution of the functional equation (2.1). As a result, the assumption that $\eta(u, \beta)$ should be differentiable yielded the form

$$\eta(u, \beta) = -u\beta - \ln Z(\beta) \tag{2.26}$$

which is consistent with the discontinuous character of $G(u)$. We note that Szilard's method of solution required the differentiability of $G(u)$.

C. Structure Function and Entropy

The discussion of Section B is of a purely formal nature. In order to develop the physical interpretation of the theory, first, we connect the present formalism with that of quantum mechanics and, thus, pave the way for the microscopic calculation of the thermodynamic functions; second, we shall derive the formalism of MTE from the present basis and, thus, establish the connection of STE with experiment.

One simple postulate will do much toward joining STE and quantum mechanics.

P c1. The structure function $G(u, x)$ is to be interpreted as the number of linearly independent solutions of the Schrödinger equation which have an energy eigenvalue $E \leqq u$ under the condition that x be constant. These solutions are the possible microstates of the system and have equal *a priori* probabilities.

Accordingly, the canonical d.f. (2.12) can be written in the alternative forms:

$$f_n = g_n(x)e^{-\Phi-\beta u_n}, \tag{2.27a}$$

$$f_i = e^{-\Phi-\beta u_i}, \tag{2.27b}$$

where f_n is the probability of a g_n-fold energy level u_n and f_i is the probability of a single microstate of energy u_i.

It is easy to see that *P c1* is consistent with the properties of the structure function, in particular with the composition rule (2.18). Otherwise, this postulate is justified only by its success, which is considerable since *P c1* allows us to integrate with the present theory most quantum-statistical calculations of the structure functions of particular systems.

We are in a position now to specify the physical criteria for the numerical range of β. The various types of degrees of freedom give rise to two kinds of structure function, distinguished by their asymptotic behavior for high energies. For translational, rotational, and vibrational degrees of freedom, $G(u) \sim u^{\mathfrak{N}/2}$ where $\mathfrak{N}$ is the number of degrees of freedom. For internal angular momentum, $G(u)$ is bounded. Whence it is evident that $Z(\beta)$ converges for all positive values of β for all physical systems. Moreover, since all physical systems have at least one kind of degree of freedom with an unbounded spectrum, $Z(\beta)$ diverges for $\beta \leqq 0$.

There are, however, certain physical situations for which it is convenient to consider a particular degree of freedom as a thermodynamic system in its own right. For example, one may treat a spin system as a thermodynamic system, in spite of the fact that it occupies the same space as the crystal lattice, contrary to the requirement that thermodynamic systems be disjoint regions in space (*D a1*). For such "systems" $G(u)$ is bounded and β can be negative. We shall return to the interpretation of this situation in Section IV,B.

We turn now to the problem of establishing the connection of STE with MTE. The key to this problem is contained in the dualism of the canonical formalism pointed out in Section II,B. It is evident that the analytic function $Z = Z(\beta, x)$ and, hence, $\log Z \equiv \Phi(\beta, x)$, contains all statistical information about the system. As seen in (2.15), moreover, Φ is additive for independent systems and obeys the convexity relation (2.21). This fact allows us to establish a very smooth junction of the present theory with MTE. The latter theory is centered around the so-called fundamental equations, which are certain additive functions each of which contains all thermostatic information about the system. The various fundamental equations are associated with different physical situations. The one associated with a system in energy-exchange with a heat reservoir (the static equivalent of the canonical situation) is described in terms of the Massieu function[11]

$$\Psi(1/T, x) = S - U/T \tag{2.28}$$

where S, U, T are the thermostatic entropy, energy, and temperature. Hence we have to require that the $\Phi(\beta, x)$ of the present theory be proportional to the statistical counterpart of the Massieu function

$$\Psi = k\Phi(\beta, x). \tag{2.29}$$

The convexity relation (2.21) ensures the validity of the proper extremal properties of Ψ; and the postulates of MTE are satisfied for our construction of the fundamental equation. Thus the junction of MTE and STE is, in principle, completed for the limited choice of independent variables, and we could build up MTE from the basis of STE.

We seek now the formulation of the concept of entropy appropriate to the present canonical theory. In MTE the entropy is defined as that fundamental equation which is appropriate for isolated systems. Such a definition will be introduced in Section IV along with a complete and physically meaningful formalism for isolated systems; here we must seek another approach.[12] That different concepts of entropy are required for canonical and isolated systems is to be expected in a theoretical structure where such systems are no longer regarded as equivalent.

Guided by the thermostatic limit, given in (2.28), we define the canonical entropy as

$$\bar{s} = k\Phi - k\beta\frac{\partial\Phi}{\partial\beta} = k(\Phi + \beta\bar{u}) = -k\beta^2\frac{\partial(\Phi/\beta)}{\partial\beta}. \tag{2.30}$$

Since Eq. (2.20) enables us to express $\bar{u}$ and β in terms of one another, we have the option of considering $\bar{s}$ as a function of $\bar{u}$, corresponding to a fundamental equation of MTE, or as a function of β, as is more natural in the canonical formalism. The former alternative leads at once, through the Legendre transformation of $\bar{s}$, to a number of important thermostatic relations.

[11] Systems coupled with heat reservoirs can be described in terms of the Helmholtz function or the Massieu function. In the scheme based on the former, the virtual processes are (reversible) entropy transfers; in the latter, energy transfers.

[12] Thus, our problem here is the inverse of that encountered in statistical mechanics; there one starts with an isolated system, and the problem is that of transition to coupled and open systems.

Differentiating (2.30) at constant x, we obtain

$$d\bar{s} = k\,d\Phi + k\bar{u}\,d\beta + k\beta\,d\bar{u} = k\beta\,d\bar{u} \equiv d\bar{u}/T. \tag{2.31}$$

Hence,

$$\beta = \frac{1}{k}\left(\frac{\partial \bar{s}}{\partial \bar{u}}\right)_x = \frac{1}{kT} \tag{2.32}$$

where T is defined by (2.31). It is evident from our previous discussion of β and from (2.31) that T has all the properties of the absolute thermodynamic temperature.

Moreover,

$$k\Phi = \bar{s} - \bar{u}/T = -A/T, \tag{2.33}$$

where $A = A(T, x)$ is the Helmholtz free energy.

The convexity relation (2.21) for Φ, together with (2.20) and (2.32), yields the convexity relation for $\bar{s}$:

$$\partial^2\bar{s}/\partial\bar{u}^2 \leqq 0. \tag{2.34}$$

Expression (2.30) suggests the definition of a random function

$$s(u, \beta, x) \equiv k[\Phi(\beta, x) + \beta u]. \tag{2.35}$$

This function figures prominently in Gibbs' discussion of the extremal properties of the canonical ensemble and was called by him the index of probability; we shall call it the entropy function. Of course, the entropy proper is not a random function, but a property of the ensemble. With this fact in mind, there is no objection to this terminology, the advantages of which will be apparent in the next section. In case of ambiguity we shall refer to (random) entropy function and to average entropy, respectively.

From (2.35) and (2.12) we obtain

$$\bar{s} = k \int dF(u) \ln\,[dG(u)/dF(u)] \tag{2.36}$$

or, in the case of a discrete spectrum, by using (2.27), we have

$$\bar{s} = -k \textstyle\sum_i f_i \ln f_i\,. \tag{2.37}$$

Equations (2.36) and (2.37) are well-known and important relations which suggest a generalization of the entropy concept.

Let us consider a system the nature of which is specified, either in terms of the structure function $G(u, x)$ or in terms of its set of microstates. If the system is not in equilibrium, its state might be describable in terms of ensembles other than canonical. Suppose in the corresponding d.f. the probability of the micro-state i is ϕ_i with

$$\sum_i \phi_i = 1. \tag{2.38}$$

It is usual to define the functional

$$s^{(\phi)} = -k \sum_i \phi_i \ln \phi_i \tag{2.39}$$

as the entropy of the d.f. ϕ.

It is easy to show by the method of Lagrange multipliers that, among all normalized distributions having the same average energy, the average entropy takes on its maximum value for the canonical d.f.

$$s^{(\phi)} \leqq \bar{s} = k(\Phi + \beta\bar{u}) \tag{2.40}$$

or from (2.33)

$$\bar{u} - s^{(\phi)}T - A \geqq 0, \tag{2.41}$$

where the equality sign is valid only if $s^{(\phi)}$ is the canonical average entropy.

Relation (2.41) is formally similar to the extremum principle (5.20) in (MTE). Yet the two principles differ essentially because they refer to different comparison states. The noncanonical d.f.'s ϕ are not definable in MTE even as virtual states. We shall return to this matter in the next section. The fact that the canonical d.f. maximizes the entropy functional of a class of d.f.'s has also been used as a particularly direct method of derivation of this d.f. (*5*). We shall discuss the relation of this procedure to ours in Section VI.

III. THE CANONICAL FORMALISM FOR SEVERAL RANDOM VARIABLES

A. The Canonical Distribution Function

We proceed now to generalize the developments of Section II to several random variables.

We consider two ensembles $e'(x_k' \mid x_\lambda')$, $e''(x_k'' \mid x_\lambda'')$ coupled to a reservoir by means of the simultaneous exchange of the several quantities X_k. We stipulate that the volume be fixed, and the fluctuation confined to that of energy and mole numbers. The d.f. to be derived is thus the grand canonical d.f.[13]

[13] Although the theory can be applied without difficulty to the case of simultaneous random exchanges of volume and energy through diathermic pistons, this case is of slight importance in comparison with the diffusive coupling leading to the grand canonical d.f. On the other hand, an adiabatic piston produces mechanical accelerations and, hence, dynamic correlations, which cannot be accounted for on the present postulational basis, since the final equilibrium state depends upon such mechanical detail. The sweeping exclusion of processes involving random volume-changes saves us the trouble of hedging the general statements with complicated qualifications.

The simultaneous exchange of several quantities X_i between two systems (x') and (x'') can be represented as a vector $\{\delta x_1', \delta x_2', \cdots\}$ that is equal and opposite to $\{\delta x_1'', \delta x_2'', \cdots\}$. In this vector space we can choose a set of linearly independent vectors or, rather, directed lines (rays):

$$\delta x_i' \neq 0, \qquad \delta x_k' = 0, \qquad k \neq i \tag{3.1}$$
$$i = 1, 2, \cdots$$

the linear combinations of which represent all exchange processes. The basic processes (3.1) can be physically realized, although not necessarily by a simple application of walls.[14]

The derivation of the canonical d.f. can be carried out in such a step-by-step analogy to Eqs. (2.1)–(2.12) that we confine ourselves to giving the final answer with a few explanatory remarks. The crucial point in the argument is that the many-variable analog of Eq. (2.5) can be independently differentiated with respect to the x_i of Eq. (3.1). The term "independent" has many meanings and it may be worth while to point out that our basic processes are not dynamically independent, since we made no special assumptions on the system; they are not even statistically independent as will be evident from (3.6).[15] What we mean is that the virtual-exchange processes form a linear manifold because the X_i are additive invariants (*P a1*, *P a2*). Such a manifold can be spanned by linearly independent basis vectors; this is all that we need for the derivation of the final result:

$$dF(x_i \mid x_\lambda, \pi_i) = dG(x_i, x_\lambda) \exp\left[-\Phi_{x_\lambda}(\pi_i) - \sum \pi_i x_i\right], \tag{3.2}$$

where $G(x_i, x_\lambda)$ is the structure function, Φ and the π_i are parameters of the

[14] Thus, the process $\delta u' = 0$, $\delta n' \neq 0$ cannot be automatically ensured by walls. Mass transfer is always coupled to energy transfer, and $\delta u' = 0$ can be achieved only by carefully compensating for the energy transferred with the mass by producing heat flow in the opposite direction.

[15] In the microscopic sense the transfer processes exhibit a definite coupling pattern. Thus an individual transfer of a particle transfers also a fixed amount of energy. This coupling of the fluxes plays a central role in irreversible thermodynamics (*14*). However, the d.f. (3.2), that ignores the temporal sequence in the variation of the x_i, contains this coupling effect only in a strongly averaged form through the dependence of the function Φ_{x_λ} on all the $\pi_1, \pi_2, \cdots$. There is one striking exception in which the coupling between the δx_i essentially affects the equilibrium situation. In liquid helium II, a very thin capillary acts as an entropy filter; it is restrictive of entropy and of excitation energy, but unrestrictive of mass flow. The energy carried along with the latter is that of the ground state. Thus in this case we have a perfect coupling between the exchange of volume, mass, and energy. Two containers of helium coupled by such a capillary may have different temperatures, provided that they have correspondingly different pressures. The situation can be handled by thermostatic methods as shown (*MTE*, p. 38).

d.f., and the π_i are functions of the original θ_s which they completely replace. The canonical d.f. (2.12) is a special case of (3.2); hence, assuming $x_1 = u$, we have $\pi_1 = \beta$.

The implications of the d.f. (3.2) are so similar to those obtained in the one-variable case that we shall merely list the most important formulas. The normalization of the probability (3.2) yields

$$\int_{x_i} dG(x_i, x_\lambda) \exp [-\Phi_{x_\lambda}(\pi_i) - \sum \pi_i x_i] = 1. \tag{3.3}$$

The partition function is

$$Z_{x_\lambda}(\pi_i) \equiv \exp [\Phi_{x_\lambda}(\pi_i)] = \int_{x_i} dG(x_i, x_\lambda) \exp (-\sum \pi_i x_i). \tag{3.4}$$

Differentiating (3.3) with respect to π_i, we have

$$\bar{x}_i = -\partial\Phi/\partial\pi_i . \tag{3.5}$$

A second differentiation yields at once the covariances

$$\overline{\Delta x_i \Delta x_j} = \frac{\partial^2\Phi}{\partial\pi_i \partial\pi_j} \equiv \Phi_{ij} \qquad i, j = 1, 2, \cdots, \tag{3.6}$$

where $\Delta x_i = x_i - \bar{x}_i$.

The diagonal elements of the matrix $\| \Phi_{ij} \|$ of the covariances are obviously nonnegative:

$$\overline{(\Delta x_i)^2} = \partial^2\Phi/\partial\pi_i^2 = -\partial\bar{x}_i/\partial\pi_i \geqq 0, \tag{3.7}$$

where the limiting case of zero variance holds only when the random variable X_i is actually fixed, and there is no statistical problem—a situation that arises in the limiting case of absolute zero. The statement embodied in (3.7) can be immediately sharpened: the matrix $\| \Phi_{ij} \|$ must be positive semidefinite, since otherwise a transformation of the variables would lead to negative diagonal elements.

Equation (3.7) implies, as in the one-variable case, that, for reservoirs of different π_i but with all other π_j equal, $\pi_i' < \pi_i''$ requires $\bar{x}_i' > \bar{x}_i''$. Hence X_i will flow, on the average, from $R'(\pi_1, \cdots \pi_{i-1}, \pi_i', \pi_{i+1}, \cdots)$ to $R''(\pi_1, \cdots \pi_{i-1}, \pi_i'', \pi_{i+1} \cdots)$ when both are coupled to the same finite system. But no general statements hold for the direction of fluxes between reservoirs differing in two or more of the π_i.

We turn to the identification of the thermostatic functions in terms of the general canonical d.f. The situation continues to parallel the one-variable case. In analogy to (2.35) we define the random entropy function

$$s(x_i, \pi_i, x_\lambda) = k[\Phi_{x_\lambda}(\pi_i) + \sum \pi_i x_i]. \tag{3.8}$$

The canonical average entropy is

$$k(\Phi + \sum \pi_i \bar{x}_i) = k(\Phi - \sum \pi_i \partial\Phi/\partial\pi_i) = s(\bar{x}_i, \pi_i, x_\lambda) = \bar{s}. \tag{3.9}$$

By using Eq. (3.5), we can express the π_i by the $\bar{x}_i$ or vice versa. The condition of solubility is

$$\det |\Phi_{ik}| = \frac{\partial(\bar{x}_1, \bar{x}_2, \cdots)}{\partial(\pi_1, \pi_2, \cdots)} \neq 0. \tag{3.10}$$

As we have pointed out above, this relation is satisfied except at absolute zero. Another singularity of the theory arises at critical states, for which the determinant (3.10) becomes infinite. For a further discussion of these singularities we refer to (MTE). Equation (3.6) contributes the additional information that at critical states the fluctuations become infinite. While critical fluctuations are indeed very large, they cannot, of course, be infinite in a finite system. We return to this point in Section IV,E.

We define now new intensity parameters

$$P_i = -kT\pi_i \qquad i = 2, 3, \cdots. \tag{3.11}$$

Comparison with the canonical d.f. yields:

$$P_1 = T = (k\,\pi_1)^{-1} \tag{3.12}$$

By introducing a notation used in (*MTE*) we have

$$-kT\Phi_{x_\lambda}(\pi_i) = \bar{u} - T\bar{s} - \sum{}' P_i\bar{x}_i \equiv \bar{u}[T, P_i]. \tag{3.13}$$

In particular, we obtain for the important grand canonical ensemble,

$$\Omega(V, T, \mu_j) \equiv -kT\Phi_v(T, \mu_j) = \bar{u} - T\bar{s} + \sum \mu_j n_j \tag{3.14}$$

where μ_j is the chemical potential of the jth component. If the volume is a random variable, the corresponding intensity, is, of course, the negative of the pressure.

The formulas (2.36)–(2.39) remain unchanged in the many-variable case, provided only that expression (3.2) is used for the d.f. The maximum principle (2.40) is generalized to

$$\bar{u} - \langle s\rangle_\varphi T - \sum{}' P_i\bar{x}_i - \bar{u}[T, P_i] \geqq 0. \tag{3.15}$$

The equality sign is valid if and only if φ is the canonical d.f. (3.2).

We conclude this section with a few remarks concerning the range of the intensities π_i, inferred from the condition that the integral (3.4) be convergent. Since the structure function increases, at most, algebraically (power law) for large x_i, all positive values are admissible for the π_i. Zero and negative values are admissible as well if the structure function is bounded. Thus, the existence of a bounded $G(v)$, u = constant, for condensed phases indicates at once the possibility of negative pressures. If, as discussed above with reference to energy and mass transfer, δx_i cannot be kept zero when $\delta x_j \neq 0$, this fact must be considered in evaluating the ranges of the π_i. Thus, in the case of Fermi statistics, positive values of the chemical potential are possible because the high-density states have a sufficiently high energy to ensure the convergence of the integral (3.4).

B. Perfect Gases

The foregoing developments allow us to derive all of the quantitative information about a system that falls within the scope of STE and MTE, provided only that the structure function is obtained by solution of the Schrödinger equation. However, for systems of many degrees of freedom, this problem is usually insoluble. An effective way of overcoming this difficulty is to consider, at least as a point of departure, particularly simple systems called perfect gases. In statistical thermodynamics a perfect gas is defined as a system consisting of statistically independent constituents. As we have seen (Section II) the partition sum of such a system is the product of the partition sums of its parts. If the latter are sufficiently simple, the problem is easily manageable.

From the mechanistic point of view *statistical independence* is traced to *dynamic independence*: the perfect gas is an assembly of noninteracting particles. This simple connection was upset by quantum statistics. It became apparent soon enough that the dynamic independence of particles is no longer *sufficient* for statistical independence if the wave function of the system is subject to symmetrization (or antisymmetrization). However, it was recognized only gradually, as perfect-gas methods proved effective beyond all expectations in describing fluids of strongly coupled particles that dynamic independence is not *necessary* for statistical independence.

As we have pointed out, the requirement calling for the symmetrization of their wave functions renders *particles* unsuited to play the role of statistically independent elements for the building up of a gas. However, this role may be taken over by the single-particle *states*. In a fixed volume surrounded by impermeable walls these states will be spatially localized standing waves with characteristic energies ϵ_ν.

The total energy of the system is

$$U = \sum n_\nu \epsilon_\nu \tag{3.16}$$

where the occupation numbers n_ν are random variables. We have here, obviously, a transition from particles to fields. There is also a similarity to the definition of the composite system, only here the states are not spatially disjoint, but occupy the same volume. That they are, nonetheless, statistically independent is precisely what distinguishes a perfect gas from other systems. Thus the partition sum of the system will be

$$Z = \prod_{\nu} z_{\nu} \tag{3.17}$$

with

$$z_{\nu} = \sum_{n_{\nu}} x_{\nu}^{n_{\nu}} \tag{3.18}$$

and

$$x_{\nu} = e^{-\pi_1 \epsilon_{\nu} - \pi_2} = e^{(\mu - \epsilon_{\nu})/kT} \tag{3.19}$$

The summation is over 0 and 1 for Fermi-Dirac (F–D) statistics and over all nonnegative values of n_{ν} for Bose-Einstein (B–E) statistics.

Hence, we obtain

$$\varphi_{\nu} = \ln z_{\nu} = (1/\gamma) \ln (1 + \gamma x_{\nu}), \quad \text{with } \gamma = \begin{cases} +1 & \text{F–D.} \\ -1 & \text{B–E} \end{cases} \tag{3.20}$$

The average occupation number is

$$\bar{n}_{\nu} = -\frac{\partial \varphi_{\nu}}{\partial \pi_2} = \frac{1}{e^{(\epsilon_{\nu} - \mu)/kT} + \gamma} = \frac{1}{(1/x_{\nu}) + \gamma}. \tag{3.21}$$

An interesting limiting case is obtained if

$$e^{-\mu/kT} \gg 1, \tag{3.22}$$

where the lowest energy was chosen to be zero. Now the two distributions degenerate into a common distribution

$$\begin{aligned} \varphi_{\nu} &= \ln z_{\nu} \rightarrow x_{\nu} \\ \bar{n}_{\nu} &= x_{\nu}. \end{aligned} \tag{3.23}$$

Equation (3.23) can be rewritten as

$$\varphi_{\nu} = x_{\nu} = \ln e^{x_{\nu}} = \ln \left(1 + x_{\nu} + \frac{x_{\nu}^{2}}{2!} + \cdots \right) \tag{3.24}$$

and given a verbal formulation which is the well-known correction procedure for introducing the identity of particles into Boltzmann statistics.

The reason that we survey these well-known results is to show that they appear in a perspective that is different from the one of historical evolution. In the present context, Boltzmann statistics is the common degenerate limit of F–D and B–E statistics. Traditionally, however, the appearance of quantum effects is designated as "degeneracy," a usage which conflicts with every other accepted meaning of this term.

Finally, we note that the chemical potential in a B–E gas is negative (under the assumption that the lowest energy is zero). No sign restriction exists for F–D statistics. This is in agreement with the general rule stated at the end of Section III,A.

IV. THERMODYNAMIC THEORY OF MEASUREMENT

A. The Classification of Thermodynamic Experiments

The canonical d.f.'s contain three types of variables and, accordingly, enable us to consider three types of experimental questions.

(i) The response x_i of a known system (X_λ) is studied when in equilibrium with a reservoir $R(\pi_i)$ of known intensities $(X_\lambda, \pi_i \rightarrow x_i)$.

(ii) Information on the properties of an unknown environment is sought through measurements performed on a system with which it is in equilibrium $(X_\lambda, x_i \rightarrow \pi_i)$.

(iii) The investigation seeks the unknown aspects of the system from its response when it is in equilibrium with a known environment $(x_i, \pi_i \rightarrow X_\lambda)$.

First let us consider briefly the macroscopic approach of (MTE) to the aforementioned experimental questions. Question type (i) corresponds to the measurement of, say, energy U and volume V, in environments of given temperature and pressure. Measurements of this sort lead to the determination of the equations of state and of the fundamental equation, except for the fact that only differences rather than absolute values of entropy are obtained.

In experiments of type (ii), one measures, say, the energy of a finite system and infers the temperature of the environment with which it has been in equilibrium. The finite system can be considered as a thermometer. The "indicator property," such as the length of a fluid column, can, of course, be calibrated in terms of energy differences. Whereas the discussion of diverse indicator properties is very important from the practical point of view, only the energy measurement is significant in the present context. Similar types of measurement, involving other variables, are frequent and often of more than routine interest. Thus the system might be a fossil object, the isotopic composition of which is used to infer the temperature and the composition of the archaic ocean from which it was deposited (*16*).

Finally, type (iii) experiments belong to chemistry, in which the responses of an unknown system to known chemical environments are used for the chemical analysis of the system.

Our problem, now is, to reconsider the three types of experimental questions from the point of view of STE. An experiment of type (i) can be easily interpreted in probabilistic terms. The intensities π_i and the X_λ determine the d.f. and, hence, the averages and the higher moments of the random variables x_i.

Within the context of MTE, the principle of thermostatic determinism (*MTE*, p. 62) asserts that the intensities of the environment uniquely determine the extensive quantities within the finite system and vice versa, provided that the system is in equilibrium with the environment. Considering systems at constant volume, we may say that the intensities of the environment and the densities of the system are uniquely *matched* with each other in equilibrium. Of course, within STE this deterministic connection is weakened into a statistical one. Whence, in questions of type (ii), we attempt to infer the parameters of the d.f. from the measured values of the random variables. This inverse of the usual probabilistic problem is called the problem of parameter estimation, and is dealt with in mathematical statistics. Mandelbrot (*4*) seems to have been the first to point out that parameter estimation is the proper counterpart to thermometry. He called the procedure *retrodiction*,[17] and identified the method of estimation that is implicit in the existing techniques of statistical mechanics as the "method of maximum likelihood" (*18*).

The suggestion that the methods of mathematical statistics be used within the context of basic theory might produce, at first blush, some apprehension. The inversion of probability calculus has by no means the uniqueness of, say, the inversion of differentiation. Hence the problem of retrodiction is beset by ambiguities and disputes. However, our use of estimation theory is not affected by these difficulties. We shall show in Section IV,B that the application of the method of maximum likelihood to the canonical d.f. is very simple and can be justified also on the basis of physical plausibility.[18] We refer to the statistical literature for its additional justification in terms of its desirable mathematical properties (*18*).

The results obtained will be used in Section IV,C to develop a statistical thermodynamics that applies to isolated systems and is the counterpart of the probabilistic interpretation of the canonical formalism that refers to open systems.

Questions of type (iii) are outside the scope of classical statistical physics, but they can be handled successfully within quantum mechanics (*6*). We shall show

[17] This terminology may not be the best; it is to be understood only as a designation of the procedure to be described, and a number of possible connotations are to be avoided. Thus, the term was used by Watanabe (*17*) in the sense of inferring past observations from present observations, which would make retrodiction the symmetric inverse of prediction. The symmetry between prediction and retrodiction in analytic dynamics is so complete that it did not seem worth while even to discuss retrodiction. This may well be the origin of the widespread but unwarranted belief that science deals only with predictions. In order to emphasize the asymmetry, one might speak of *reconstruction* instead of retrodiction. Thus one reconstructs a sequence of events from a photographic plate taken in a bubble chamber. Jaynes (*5*) calls this procedure *interpretation*.

[18] One may note, however, that estimation-theoretical formalisms built on the microcanonical d.f. *are* beset by difficulties (cf. Thesis, p. 110) and, consequently, are of little use.

in Section IV,D that Nernst's law allows us to draw inferences with respect to experiments of type (iii). In Section IV,E we shall use the microcanonical ensemble to describe situations in which the finite size of the environment is significant.

B. The Measurement of Intensities as Parameter Estimation

We consider a single finite system that is to represent a thermometer. This system is brought into equilibrium with a reservoir $R(\beta)$, the temperature β^{-1} of which is unknown. The measurement consists in separating the system from $R(\beta)$ and ascertaining its energy u. Since the system is finite, it represents an ideal measuring device insofar as its coupling with the infinite reservoir does not affect the temperature of the latter.

The next step in our procedure is to estimate the value of β from the knowledge of u. The available data are clearly insufficient for certitude in this respect, since we have no way of knowing whether u is equal to $\bar{u}$. In particular, *P b3* precludes the system from retaining any more memory of the reservoir temperature than that contained in the energy measurement.

Let us consider a measurement of the set of instantaneous values x_i for a system in contact with a reservoir $R(\pi_i)$ as the specification of a member of a canonical ensemble. Then the probability for finding this set of values is the d.f. (3.2). We now propose to consider dF as a function of the π_i for these fixed values of the X_i. Of course, the x_λ are fixed as well. In this interpretation $dF(\pi_i)$ is called the *likelihood function*. This term expresses the fact that this function is akin to probability, although we have no basis for speaking of the probability of a reservoir β, nor is dF normalized; indeed, $\int dF(\pi_i)\, d\pi_i$ need not exist.[19]

We proceed to maximize dF with respect to the π_i and define the estimated, or most likely, intensities as the particular values $\hat{\pi}_i$ for which the likelihood function becomes a maximum, given the set of measured values x_i. As we have seen, $dF(\pi_i)$ is an analytic function of the π_i; thus, this maximization is achieved

[19] Of course, a likelihood function can be defined for any d.f. Our problem is simplified because we have established the fact that the interaction between system and reservoir is governed by the canonical d.f. We note that this d.f. may have a direct physical meaning, for example, a system serving as a thermometer in prolonged contact with $R(\beta)$. If numerous energy measurements are performed, their average is the canonical average. Another possibility is, however, that the system acquired its energy by some other means than through contact with a reservoir $R(\beta)$. Thus macroscopic work might be dissipated in the system after thermal isolation had been established. In this case we proceed with the estimation of β exactly as described above, although the canonical ensemble will have only fictitious, "virtual" existence. An interesting case is a spin system exhibiting "negative temperatures" (*19*), which has been brought into this state by the reversal of a magnetic field. The negative temperature is that of a fictitious reservoir that could have brought the system into the same state by contact.

by differentiation. Since dF is positive for the set of x_i that are at all observable, we may maximize instead of dF its logarithm. We have

$$\ln dF(x_i, \pi_i) = \ln dG(x_i) - s(x_i, \pi_i)/k, \tag{4.1}$$

where

$$s(x_i, \pi_i)/k = \Phi(\pi_i) + \sum \pi_i x_i \tag{4.2}$$

is the random entropy function, or index of probability. For the sake of brevity, we suppress the fixed variables x_λ which, in fact, are present in the arguments of each term of (4.1). Evidently, maximizing $\ln dF$ with respect to the π_i at constant x_i is identical to minimizing $s(x_i, \pi_i)$ under the same conditions. Hence we have

$$\frac{1}{k}\frac{\partial}{\partial \pi_i}[s(x_i, \pi_i)]_{\pi_i=\hat{\pi}_i} = \left(\frac{\partial \Phi}{\partial \pi_i}\right)_{\pi_i=\hat{\pi}_i} + x_i = 0 \tag{4.3}$$

with the stipulation that the matrix with the elements

$$\frac{1}{k}\frac{\partial^2 s}{\partial \pi_i\, \partial \pi_k} = \frac{\partial^2 \Phi}{\partial \pi_i\, \partial \pi_k} = \Phi_{ik} \tag{4.4}$$

be positive semidefinite.

Equation (4.3) can be written

$$\bar{x}_i(\hat{\pi}_i) = x_i\,. \tag{4.5}$$

The most likely values of the intensities π_i are defined by the requirement that the empirically given values x_i be equal to the canonical averages conditioned by the π_i.[20] Condition (4.3) is of a form identical to (3.5) and always possesses a single set of positive solutions. The matrix (4.4) is identical to the matrix of covariances (3.6) and, accordingly, it is positive semidefinite. However, in the limiting case of absolute zero we have $\det|\Phi_{ik}| = 0$. This means that the averages $\bar{x}_i(\pi_i)$ become insensitive to the variation of $\pi_1 = 1/kT$; and while the estimation of the $\bar{x}_i$ by x_i remains good, the inference with respect to π_i becomes, unreliable. This is the expression in our formalism of the well known difficulties of thermometry near absolute zero, which result from the flattening out of the temperature dependences of the thermometric indicators.

C. The Statistical Thermostatics of Isolated Systems

We shall now develop a thermostatic formalism for isolated systems that is formally similar to the canonical theory, but, instead of being based on the

[20] Although this result is plausible, it is not trivial. Without the above analysis we might have required that x be equal to the most probable value of the d.f. conditioned by π_i. No useful formalism results from such a postulate.

conjugate pairs of fixed intensities π_i and average extensive quantities $\bar{x}_i$, is expressed in terms of fixed extensive quantities x_i and most likely intensities $\hat{\pi}_i$.

In this formalism the role of entropy is played by a function that we call the estimated entropy and define as

$$\begin{aligned} \hat{s}(x_i) &= \min_{\pi_i} \{s(x_i, \pi_i)\} \\ &= \min_{\pi_i} k\{\ln Z(\pi_i) + \sum \pi_i x_i\}. \end{aligned} \tag{4.6}$$

Comparison of (4.6) with (4.3) and (4.4) indicates that this minimum problem has indeed a unique solution for $\pi_i = \hat{\pi}_i$.

Comparing Eq. (4.6) with (3.9) we see that, while the canonical entropy $\bar{s}(\pi_i)$ is obtained from the random entropy $s(x_i, \pi_i)$ by averaging over the x_i at fixed π_i, the estimated entropy is obtained from the same function by minimization with respect to the π_i at constant x_i. Since the minimum is reached for $\pi_i = \hat{\pi}_i$, we can also write

$$(1/k)\hat{s}(x_i) = \Phi(\hat{\pi}_i) + \sum \hat{\pi}_i x_i. \tag{4.7}$$

From (4.7) and (3.9), we have

$$\bar{s}(\hat{\pi}_i) = \hat{s}(x_i). \tag{4.8}$$

We see from (4.8) that the variational principle (4.6) can be expressed formally for the canonical entropy as follows:

$$\bar{s}(\hat{\pi}_i) = \min_{\rho_i} k\{\Phi(\rho_i) + \sum \rho_i x_i\}, \tag{4.9}$$

where the minimum is taken with respect to the dummy variables ρ_i at constant x_i. The minimum is reached for $\rho_i = \hat{\pi}_i$, and the x_i are eliminated by means of (4.5). This useful variational principle is valid independently of the method of the maximum likelihood estimate and was, indeed, obtained by Fowler (*20*), and used extensively by Khinchin (*11*). In their interpretation, however, the extremum-principle lacks the intuitive meaning provided by the maximum-likelihood estimate in the formulation given in Eq. (4.6).

We show now that

$$\hat{s} = \hat{s}(x_i) \tag{4.10}$$

can be considered as the fundamental equation of the system (X) with all the properties postulated for such equations in MTE. We have from (4.7) and (4.3)

$$\frac{1}{k}\frac{\partial \hat{s}}{\partial x_i} = \hat{\pi}_i + \sum_k \frac{\partial \hat{\pi}_k}{\partial x_i}\left[x_k + \frac{\partial \Phi(\hat{\pi}_k)}{\partial \hat{\pi}_k}\right] = \hat{\pi}_i. \tag{4.11}$$

In particular, and defining $k\hat{T} = 1/\hat{\pi}_1$, we have for $x_1 = u$, $\pi_1 = 1/kT$,

$$dx_i = 0 \ (i \neq 1)$$
$$d\hat{s} = du/\hat{T}. \tag{4.12}$$

Since du is a heat quantity, we see that $\hat{s}$, $\hat{T}$ are indeed satisfactory conjugate variables with some of the main connotations of entropy and temperature.

We shall show now that the relation (4.5) implies the same maximum property for the entropy $\hat{s}$ that has been postulated in (*MTE*) (*P b1*) for the macroscopic entropy.[21]

Suppose that we have two isolated systems (x') and (x''). The estimated entropy of the composite system is

$$\begin{aligned} \frac{1}{k}(\hat{s}' + \hat{s}'') &= \min\, \{\Phi + \sum \pi_i x_i\} \\ &= \min\, \{[\Phi' + \sum \pi_i' x_i'] + [\Phi'' + \sum \pi_i'' x_i'']\}, \end{aligned} \tag{4.13}$$

where $x_i' + x_i'' = x_i$. If the two systems are isolated and united only in a formal fashion, the two square brackets are to be minimized independently, and they reach their minima at $\hat{\pi}_i'$ and $\hat{\pi}_i''$, respectively. On the other hand, if the systems are coupled, they must have a common $\hat{\pi}_i$, with the x_i constant rather than the x_i' and x_i''. The restriction of the set of comparison states cannot decrease the minimum, hence

$$\hat{s} \geqq \hat{s}' + \hat{s}''. \tag{4.14}$$

The equality sign is valid only if $\hat{\pi}_i' = \hat{\pi}_i''$.

Relation (4.14) is valid also for the subdividing of a simple system by an adiabatic wall, an operation leading to an entropy that is less than or equal to the original value. A decrease of the entropy occurs if the subsystems are found with $\hat{\pi}_i' \neq \hat{\pi}_i''$. Hence we have to ask ourselves how these results are reconciled with the irreversible increase of entropy. The answer lies in the interplay of thermodynamic operations and processes. The former are manipulations of constraints, the latter are molecular processes that take place under the given set of constraints. The importance of this interplay for the description of composite temporal processes has been discussed elsewhere (*6*). Briefly, it is perfectly feasible, through a sequence of thermodynamic operations involving an expenditure of work, to prepare an initial state for which $\hat{\pi}_i'$ and $\hat{\pi}_i''$ of the isolated systems are considerably different from each other. In contrast, the probability of trapping the system by subdivision in a final state with values $\hat{\pi}_i'$ and $\hat{\pi}_i''$ is given by the microcanonical d.f. It is intuitively evident, and can be confirmed by the methods of Section V, that this probability decreases to negligible values if $\hat{\pi}_i'$ and $\hat{\pi}_i''$ are significantly different from each other.

[21] This proof is a simple generalization of the one given by Khinchin (*11*, p. 138) based on relation (4.9).

Relation (4.14) is identical to the entropy maximum principle *P b1* of (MTE) and can serve as a point of departure for developing the entire formalism of MTE. This connection can be brought out also by rewriting relation (4.6) as

$$\hat{s}(x_i) \leqq k(\Phi + \sum \pi_i x_i), \tag{4.15}$$

where the equality sign is valid if and only if the π_i are the estimated intensities corresponding to the given x_i. We shall say that such pairs of x_i, $\hat{\pi}_i$ are *matched.* By using (3.13) we rewrite (4.15) in the energy scheme as

$$U - \hat{s}T - \sum{}' x_i P_i - U[T, P_i] \geqq 0. \tag{4.16}$$

Relation (4.16) is formally similar to (3.15) in the canonical formalism but, in contrast with the latter, the present relation can be identified with the extremum principle (5.20) of (MTE). The concept of "matching" system quantities and reservoir intensities has a meaning in MTE: the variables are matched in equilibrium. Therefore the comparison states in the variational principle (4.15) or (4.16) are the same as in MTE. In contrast, in the canonical formalism the comparison states are noncanonical distributions which cannot be interpreted in MTE.

D. Equilibrium, Objective Entropy, and Nernst's Law

Taking stock of the connection between thermodynamic theory and experiment, we find that, at the present stage of the discussion, the situation is still beset by a serious limitation. Among the central concepts of thermodynamics is that of *equilibrium.* Without this concept we could not even express the first law of thermodynamics in a meaningful fashion, inasmuch as we could not construct an energy function in terms of macroscopic variables only. Yet there is no purely operational method for distinguishing a state of thermodynamic equilibrium from a quiescent nonequilibrium state. Experiment allows us to determine the entropy difference between any two equilibrium states, provided that they can be connected by a quasi-static path. The requirement of consistency for different paths provides a *necessary* criterion that the actual paths are reasonably quasi-static. However, this criterion is not *sufficient.* This is particularly significant if the system has internal degrees of freedom, such as the orientation of molecular groups, or nuclear or electronic spins. A consistent value for the entropy difference may mean either that the hidden degrees of freedom are in equilibrium throughout, or that they are completely frozen. In order to decide which of these alternatives is relevant one must be able to integrate empirically between two states of known entropy. It is precisely the knowledge of the entropy in two reference states that is provided by a sufficiently sharp formulation of Nernst's law (*21*). The two reference states are those for which either the temperature or the density tends to zero. The first limiting case can be handled in STE only by the same postulational method as that in MTE.

Pc 2. The entropy of every thermodynamic system in equilibrium approaches, in the limit of vanishing temperature, $k \ln g_0$ where g_0, the degeneracy of the ground-state, is a small integer whose order of magnitude is independent of the size of the system.

We may add that this limit is reached sufficiently smoothly so that the entropy practically vanishes at attainable temperatures.

At the end of this section we shall make a few remarks about the microscopic meaning of this postulate. Meanwhile, we turn to the case of vanishing density. Practically every system subject to thermostatic study can be transformed into a nearly perfect-gas state. Under these conditions, the calculation of the entropy of the system is reduced to that of a molecule (see Section III, B), and the latter calculation can, in many cases, be performed from spectroscopic data.

The formalism of MTE and of STE is meaningful only when it is considered *relative* to the set of chemical reactions used to define the independent components (*C a3*) of (*MTE*).

In view of this situation, we have to compute the entropy in relation to various plausible assumptions concerning the relevant independent components. (Decisions have to be made concerning the inclusion of isotopic constitution, nuclear spin, and the like.) Once this choice has been made, the entropy has an absolute meaning in the sense that it depends only on the counting of quantum states without any arbitrary or conventional constants.[22] By integrating the experimental caloric data from this gaseous state to the vicinity of absolute zero, it is possible to decide whether the low-temperature system is in equilibrium with respect to the various intrinsic variables. The method involves more than a simple routine, and for details we have to refer to the literature. In the hands of experienced workers it has led to significant detailed insights concerning equilibrium with a measure of assurance which would have been inconceivable without the careful dovetailing of experimental and theoretical procedures (*21*).

It is apparent that the procedure described deals with a type (iii) experimental situation (Section IV, *A*). *P c1* is a *qualified prediction* that can be used to perform retrodictions concerning the extent of the equilibrium prevailing in the system. Most of the discussions of Nernst's law within the context of the principles of thermodynamics are obscured by the fact that they do not evaluate this law in terms of its actual use, but are concerned with the fact that it does not conform to the supposed norm of the laws of physics, not being expressed as an *unqualified* prediction. At any rate, the literature on the subject is full of

[22] The fact that the entropy is in one sense *relative* and yet in another sense *absolute* offers the possibility of semantic confusion, which beclouds much of the discussion of the subject. Another misconception is the assumption that the *relativity* of the entropy concept makes it also *subjective*. While there is, of course, a subjective element in the use of entropy, inasmuch as we may choose to ignore detail, we certainly cannot increase the detail beyond counting quantum-mechanical pure states.

reformulations that propose to correct this alleged shortcoming. One of the best known versions claims that "it is impossible to reach absolute zero." We do not question the interest inherent in this statement, when considered as a corollary of *P c2*. However, this formulation cannot be taken as an adequate replacement of this postulate since unqualified predictions have no retrodictive power.

Finally, we make a short remark concerning the microscopic interpretation of *P c2*. Let us consider a composite system consisting of N cell divisions and make an attempt to describe this system in terms of the present theory, that is, by assuming statistical independence of the cells even for small cells and large values of N. Under the assumption that the lowest state of each cell is g-fold degenerate, the lowest possible value for the entropy is of the order of $Nk \ln g$. This is in contradiction with *P c2*, unless $g = 1$. It is indeed widely believed that such an assumption is the proper explanation for Nernst's law (*22*). However, this assumption is unwarranted. For instance, a system containing an odd number of electrons has at least a double spin degeneracy. Yet there is no breakdown of Nernst's law and the system becomes, say, ferromagnetic or antiferromagnetic. The responsibility for the difficulty is undoubtedly the breakdown of the statistical independence of the cells. If the cells are strongly correlated, the degree of degeneracy of the system of N cells need not be larger than that of a single cell. The entropy per unit cell is then $s_0 = (k \ln g)/N$, and this quantity tends to zero as N goes to infinity. Hence Nernst's law is satisfied even in the presence of degeneracy of the microcells.

E. The Microcanonical Formalism

Up to this point our whole discussion has centered around the coupling of a finite system with an infinite reservoir. Although the assumption of infinite environments is almost always a reasonable assumption, it is of obvious interest to consider the coupling of two or more finite systems with each other. The formalism governing this situation is already implicit in the theory as it has been developed thus far. The so-called microcanonical d.f. describing the distribution of the energy of a system carved out of a finite system is given by (2.17). We can easily generalize that expression for the distribution of any set of x_i :

$$d\mathfrak{F}(x_i' \mid x_i) = \frac{dG'(x_i')\, dG''(x_i'')}{dG(x_i)}. \tag{4.17}$$

The generalization to more than two subsystems is obvious. The rigorous computation of this function is cumbersome, since the structure function of composite systems is obtained from that of its parts by convolution, (2.18). This contrasts with the much more manageable product rule for the generating function $Z(\pi_i)$.

Evidently, the complementary system X'' can be made into a reservoir by means of the limiting process $X'' \to \infty$, $X' =$ constant. Accordingly, the microcanonical d.f. tends to the canonical d.f. $d\mathfrak{F}(x_i' \mid x_i) \to dF(x_i', \pi_i)$. Since the latter is easier to handle, the microcanonical d.f. should be used in the present theory only if there are special reasons for considering this formalism more adequate for the problem at hand.

Two problems of this sort have been treated in the literature. The first is the emission of neutrons by heavy nuclei, considered as an evaporation process (*23*). Emission of neutrons by an infinite extension of nuclear matter would correspond exactly to the thermodynamic problem to be treated by the grand canonical method (Section III). The "reservoir" of nuclear matter would remain unaffected by the loss of one or of any finite number of neutrons. In reality, the original nucleus is substantially different from the product nucleus, a circumstance which is, of course, allowed for in the microcanonical separation problem. A second problem in which the finite size of the reservoir is significant is the calculation of critical fluctuations. Here the finiteness of the reservoir is essential to ensure finite fluctuation (*24*).

In the literature the use of the microcanonical ensemble has a much greater prominence than the foregoing discussion would suggest. The reason for this is that in traditional statistical mechanics it is imperative to use isolated systems as a point of departure, since the use of Liouville's theorem depends on such a procedure. The aforementioned difficulties of the microcanonical method are solved or circumvented by means of simplifying assumptions that are not dictated by the nature of the problem. We shall discuss these asymptotic assumptions in the next section.

V. ASYMPTOTIC THEORIES

It is an important feature of STE that the canonical d.f. is derived for systems in equilibrium with infinite reservoirs. While this same point of view is taken in refs. *8*, *4*, and *5*, the most widespread approach in the literature is different: the point of departure is a large but finite isolated system. In this microcanonical theory, the canonical d.f. is established approximately for a small subsystem of the isolated system. This subsystem is the system of interest; the complementary system plays the role of the reservoir. According to our results, of course, the canonical d.f. is not rigorously correct in this situation but only in the asymptotic case for which the complementary system tends to infinity.

Our first problem in this section is to establish asymptotic relations that are valid for large, but still finite, complementary systems. This is easily achieved in the present context, since we have the entire canonical formalism at our disposal, this formalism having been established without the asymptotic assumptions that are under scrutiny. The situation was much more complicated in the historical context in which the asymptotic methods were used to establish the canonical d.f. The complications produced by the finite size of the complementary system were solved either by means of conceptual oversimplifications or at the expense of formal complications and unduly restrictive conditions of validity. We shall attempt to clarify the nature of these conceptual oversimplifications, which have been severely criticized (see refs. *20*, *11*, *11a*, *4*) but are still quite widely used.

We start by establishing an approximate expression for the structure function that plays a central role in the microcanonical theory.[23] For the sake of simplicity, we confine ourselves to canonical d.f.'s with a single random variable, the energy. By using (2.12) and (2.35) we obtain

$$\ln dG(u) = s(u, \beta)/k + \ln dF(u \mid \beta), \tag{5.1}$$

where β is considered as a fixed numerical parameter. Since the probability dF is smaller than unity, we have

$$k \ln dG(u) < s(u, \beta). \tag{5.2}$$

The interesting aspect of these relations is that by choosing appropriate values for u and du one can guarantee that $dF(u \mid \beta) \approx 1$. If this condition is satisfied, the last term in (5.1) becomes negligible and the inequality (5.2) turns into an approximate equality. To render this idea more quantitative, we approximate the canonical d.f. by a Gaussian d.f., the parameters of which are such that the energy and its variance have their correct canonical values:

$$dF(u \mid \beta) \approx \frac{du}{\sqrt{2\pi}\sigma} e^{-(u-\bar{u})^2/2\sigma^2}, \tag{5.3}$$

where

$$\sigma^2 = \mathrm{Var}\,(u) = \overline{(\Delta u)^2}. \tag{5.4}$$

Since moments higher than the second are hardly ever important, the approximation (5.3) is expected to be satisfactory, at least in the neighborhood of the peak of the d.f.[24]

Inserting (5.3) into (5.1), we obtain, for $u = \bar{u}$,

$$\ln dG(\bar{u}) \approx s(\bar{u})/k + \ln du/\sqrt{2\pi}\sigma, \tag{5.5}$$

where $dG(\bar{u})$ is the number of quantum states between $\bar{u}$ and $\bar{u} + du$. (As usual, β in $s(u, \beta)$ can be expressed in terms of $\bar{u}$.) We define the function

$$\tilde{s}(u) \equiv k \ln dG(u), \tag{5.6}$$

[23] The idea of this approximation is found in Khinchin (*11*), although our interpretations differ, as we shall see at the end of this section.

[24] The Gaussian d.f. is justified also by the central limit theorem (*11*) whenever the system consists of many statistically independent parts.

and obtain in excellent approximation

$$\bar{s}(\bar{u}) = s(\bar{u}). \tag{5.7}$$

Equation (5.7) holds either if we put

$$du = \sqrt{2\pi}\,\sigma \tag{5.8}$$

or if we go to the limit of many degrees of freedom, $\mathfrak{N} \to \infty$ while keeping du constant. Since $\bar{u}$, $s(\bar{u})$ and σ are all proportional to $\mathfrak{N}$, we have

$$\bar{s}(\bar{u}) = s(\bar{u})[1 + O(\ln \mathfrak{N}/\mathfrak{N})]. \tag{5.9}$$

The function $\bar{s}(u)$, defined in Eq. (5.6), is the well known Boltzmann entropy.[25]

Following Boltzmann, the identification of $\bar{s}$ with the entropy of the system is usually justified in terms of the relation

$$dG'(u')\, dG''(u'') = dG(u) \tag{5.10}$$

where $u = u' + u''$ is the energy of the composite system. However, (5.10) is correct only for isolated systems and, in the case of coupling, we have to use either (2.17) or (2.18) instead. Thus, for coupled systems, $\bar{s}(u)$ does not have the additivity properties of the entropy. To be sure, the effects of this discrepancy are numerically negligible whenever relation (5.7) holds. Nevertheless, this discrepancy is a symptom of a conceptually unsatisfactory situation.

The definition of the canonical entropy given in Section II involves a sequence of well structured mathematical operations which we can represent symbolically as follows:

$$dG \to \Phi \to dF \to \bar{s}(\beta) \to \bar{s}(\bar{u}). \tag{5.11}$$

Or, formally, by using (2.30), we have

$$\bar{s}(\beta) = -k\beta^2 \frac{\partial}{\partial\beta}\left[\frac{1}{\beta} \ln \int_u e^{-\beta u}\, dG(u)\right] \tag{5.12}$$

and, finally, β is expressed in terms of $\bar{u}$ by inverting the relation

$$\bar{u} = -\frac{\partial}{\partial\beta}\left[\ln \int e^{-\beta u}\, dG(u)\right]. \tag{5.13}$$

These expressions mean that the Legendre transform of the logarithm of the Laplace transform leads us back, in virtue of (5.6) and (5.9), very nearly to the logarithm of the original starting point $dG(u)$. In spite of this fact, the telescoping

[25] The entropy definition (5.6) has been used by both Boltzmann and Gibbs. Yet it has a place in the "neo-Gibbsian" theory of Sections I–IV only in the peripheral microcanonical formalism (IV,E), though it plays a central role in the theories of Boltzmann (*2*) and Einstein (*1*).

of the sequence of operations symbolized in (5.11) leads to an unwarranted fusion of the concepts of "structure function," "probability," and the exponential of the "entropy." Thus the structure function $dG(u)$ is usually called the "thermodynamic probability" of the system; and the entropy, the logarithm of this probability.

Actually, dG is not a probability, nor can it be normalized in any obvious way. In principle, the function $G(u)$ is obtained from the time-independent Schrödinger equation and does not have a statistical character, although in practice it is often computed by approximate statistical methods. Likewise, it is preferable not to call $e^{s/k}$ a probability but, rather, to associate microstates with probabilities and macrostates with d.f.'s; the entropy is a functional of these d.f.'s. A theory built upon the aforementioned identifications of concepts appears from the point of view of STE as a conceptually degenerate system, although, from the point of view of historical evolution, one should rather speak of a conceptually undifferentiated theory.

We may avail ourselves of relations (5.5)–(5.9) without any conceptual ambiguity wherever the situation warrants the use of such asymptotic relations. Thus we obtain a connection between entropy and probability by inserting (5.6) into (2.17):

$$d\mathfrak{F} = \exp\{[\tilde{s}'(u') + \tilde{s}''(u'') - \tilde{s}(u)]/k\}. \tag{5.14}$$

If we insert into this relation the quantity $s(\bar{u})$ from (5.5), we have the point of departure of Einstein's fluctuation theory (*1*).[26]

It is instructive to rewrite the canonical d.f. by using (5.1), (5.6), and (2.33):

$$\begin{aligned} dF(u \mid \beta) &= \exp[\{\tilde{s}(u) - s(u,\beta)\}/k] = \exp[\tilde{s}(u)/k - \Phi(\beta) - \beta u] \\ &= \exp[\beta(\bar{u} - u) - \{\bar{s}(\bar{u}) - \tilde{s}(u) - \}/k] \\ &= \exp[\{A(T) - (u - T\tilde{s})\}/kT]. \end{aligned} \tag{5.15}$$

Each of these expressions implies an inequality based upon the relation $dF \leqq 1$. Thus,

$$\tilde{s}(u)/k - \beta u - \Phi(\beta) \leqq 0 \tag{5.16a}$$

$$u(\tilde{s}) - T\tilde{s} - A(T) \geqq 0 \tag{5.16b}$$

where the equality sign is valid only if $dF = 1$. This relation, of course, cannot

[26] It has been argued by Greene and Callen (*25*) that Einstein's fluctuation theory is rigorously equivalent to the canonical formalism. Their argument is inconclusive, since it hinges on the statement that "there is no distinct isothermal and adiabatic thermodynamics." This is true in MTE, but in STE it does not have even a precise meaning. The same functional form for the entropy function describes an isolated system in terms of $\hat{s}(u)$ and a diathermally coupled system in terms of $\bar{s}(\bar{u})$, provided that we use the formalisms of Sections IV and II of the present paper. However, the functional form of $\tilde{s}(u)$ is only asymptotically equal to the functions just considered. This is not good enough for the argument of Greene and Callen, who do not neglect the higher powers of Boltzmann's constant. This criticism does not affect their elegant formalism for the computation of fluctuation moments.

be exact. However, it is valid to an excellent approximation if the conjugate variables u and β(or $\bar{s}$ and T) are matched with each other, that is, if they are associated with a system and a reservoir which would be in equilibrium with each other according to MTE.

It is worthwhile to add a remark concerning the width du associated with the d.f. dF. The latter approaches unity for matched variables only if du is at least of the order of the standard deviation of dF. If the width of the d.f. is set to be much sharper, dF may become much smaller than unity. Yet, for macroscopic systems, the equality sign in (5.16) is still reached to a very good approximation. The excess over zero of the left-hand side is of the order of $k \ln \mathfrak{N}$, which is to be compared with the order of $k\mathfrak{N}$ in the case of nonmatching variables. This is again an instance of the insensitivity of the functions of MTE to statistical assumptions.

Consequently for matching variables, relation (5.7) is valid and $\bar{s} = s(\bar{u})$ becomes identical to the entropy of MTE. In contrast to the superficially similar inequality (2.41), the relation (5.16b) is identical with the extremum principle (5.20) of (*MTE*). The statistical interpretation is particularly simple for (5.16a), that is, in the entropy scheme rather than in the energy scheme: for a reservoir of given β, the matching energy is that of maximum probability (most probable energy). For given energy, the matching β is that obtained from the maximization of likelihood.

We shall discuss now somewhat more closely the approximate handling of the microcanonical distributions. Let us consider the partition of the energy u of an isolated system between two subsystems. The probability $d\mathfrak{F}(u' \mid u)$ of a particular division $u' + u'' = u$ is given by Eq. (2.17). The approximate procedure is to select those values u_0', u_0'' for which the probability is maximum.[27] It is apparent from the above discussion and, in particular, from (5.9) and (5.14) that the probability $d\mathfrak{F}$ can be made practically unity for the correct value of the energy and of the width of the distribution. The correct choice of the energy is the main part of the procedure and is sufficiently obvious. However, the proper choice of the width du' of the distribution, i.e., according to (5.8), is usually achieved in a rather hidden form by omitting the last term in the Stirling approximation

$$\ln N! \approx N \ln (N/e) + \tfrac{1}{2} \ln (2\pi N). \tag{5.17}$$

[27] This is usually expressed by saying that the distribution is "the most probable" one. There are several difficulties with this expression. It was pointed out by T. A. Ehrenfest (*26*) that there is no reason to expect that only the most probable values of a random variable should be observed. Moreover, we are justified in speaking of the probability of microstates, but the probability of a d.f. has no obvious meaning. However, as Mandelbrot points out (*4*), we may rationalize the traditional procedure by speaking of the maximization of $d\mathfrak{F}$ considered as "likelihood" (see Section IV). The maximization is carried out by means of Lagrange multipliers. These are the parameters estimated by the procedure.

The effect of this term is negligible if N is very large. This term is neglected, however, also under conditions in which the distribution is over many microstates and N is quite small. If this term were kept, the method of the most probable distributions would lead to wrong results; the probability of the most probable state, taken in an exact sense, would be much smaller than one. Thus, the use of the most probable distributions and the truncation of the Stirling formula are compensating mistakes. That the compensation is as perfect as it is seems fortuitous and, also, unfortunate, inasmuch as it has encouraged conceptually unsound practices.

We conclude this section with a short discussion of the work of Khinchin (*11*, *11a*), which occupies a special position among the theories that derive the canonical d.f. from the microcanonical d.f. This theory is conceptually correct; it was Khinchin, in fact, who introduced the concept of the structure function and contributed considerably to the clarification of the situation brought about by the degeneracy of the asymptotic theories. However, Khinchin's idea is that the mathematical basis of statistical mechanics is the central limit theorem (*11*); in the case of quantum statistics sharper limit theorems are used (*11a*). Physically, this means that Khinchin assumes that even the "small" subsystem consists of a large number of statistically independent parts. The use of this assumption, necessary in a rigorously mechanical theory, leads to considerable formal complications. On the other hand, in STE there is no need to use either this assumption or the corresponding limit theorems. One should compare, consequently, the simple and rigorous derivation of the quantum distribution laws in Section III, B with Khinchin's complicated approximate derivation (*11a*). Moreover, the door is left open in STE for further microscopic extensions of the theory.

VI. DISCUSSION

The canonical formalism developed in STE seems not unlike the one used in statistical mechanics. Nevertheless, the thermodynamic interpretation has aspects without a mechanical analog. A case in point is the theory of measurement of Section IV. In this theory the distinction between free and fixed variables plays an important role. The same distinction has also some bearing on the ergodic problem[28] and on the relation of entropy and information.

We have noted in Section I that our statistical postulates imply an ergodic property. We state now somewhat more precisely that this property is identical to that of the mathematical ergodic theorem (*11*). Curiously, in the literature this theorem is considered to be either insufficient or irrelevant for the purposes

[28] Unfortunately, the term "ergodic" is used to mean many things. It is not within the scope of this paper to deal with this question either with great precision or from many different angles. We shall outline only the particular point to the clarification of which we hope to contribute.

of statistical physics. In the texts of statistical mechanics the ergodic hypothesis is usually formulated in essence as follows: the time average of a phase function of a dynamical system is equal to the average of the function taken over the *entire* energy surface. This is attributed to the underlying dynamics according to which the phase point representing the system covers the energy surface sufficiently densely. This statement implies that, except for the volume and the number of invariant particles, the only fixed variable of the system is the energy. We shall call this the "strong" form of the ergodic hypothesis.

The mathematical ergodic theorem is a weaker statement, inasmuch as the proof is subject to the condition that the system be metrically transitive or, equivalently, that its phase space be metrically indecomposable. This restriction guards against the possibility that the system might be trapped in one of several domains of the energy surface which are invariant under the transformation representing the dynamical law. If such domains exist, the averaging is to be performed only over the domain in which the system happens to be trapped. This state of affairs is quite in keeping with the phenomenological situation described in STE. Since we have not committed ourselves in advance concerning the full listing of the fixed variables, we can introduce an appropriate set to be associated with each domain.

The diversity of situations that may arise can be illustrated by two typical examples. There are realizable experimental conditions in which the number of ortho- and para-hydrogen molecules are separately conserved. The methods of STE are still applicable provided that additional fixed variables are admitted. Another case is furnished by a system such as glass. Here we have a huge number of fixed variables distributed at random which cannot be explicitly managed; and the methods of STE do not strictly apply. We have a frozen-in nonequilibrium, also called a false equilibrium.

Since the tendency to form stable configurations is a basic property of matter, we cannot hope to provide a proof of the strong version of the ergodic hypothesis, because that would be contrary to fact. Of course, it is legitimate to explore the problem of the validity of the strong ergodic hypothesis for special systems consisting of inert particles. However, this problem is of no basic importance for STE.

The distinction between free and fixed variables allows us to throw some light on the interpretation of expression (2.37). This expression for the entropy conveys, as well, the notion of statistical uncertainty; it arises also as a quantitative measure of information (*5a*). Accordingly, there have been many suggestions that entropy be considered as a measure of subjective ignorance and that information be related to negative entropy. Although there is undoubtedly some value in these suggestions, the conceptual unification they tend towards is bought at the price of losing important shades of meaning.

To avoid this difficulty, we propose the introduction of a generic term, designating (2.36) or (2.37), with $k = 1$, as the *dispersal of the distribution*. Of course, this term is to be distinguished from the *dispersion of a random variable*, a synonym for its variance or relative quadratic moment. Thus in a uniform distribution over g degenerate states the dispersion of the energy vanishes whereas the dispersal of the distribution is $\ln g$.

We consider now three typical situations in which the dispersal of the d.f. assumes the character of entropy, information, and uncertainty, respectively. However, these cases are by no means exhaustive. In particular, the concept of information has many ramifications which are outside the scope of the present discussion.

(i) The first case is that of an ergodic system in thermodynamic equilibrium, identified by invariant fixed variables. The system makes incessant transitions among states specified by free variables. The entropy is the dispersal of the d.f. over these transient states. The fact that we calculate the properties of the system by averaging over the free variables is not because of our *ignorance*, but because we *know* that the system actually exhibits a behavior that is the average of that of the transient states. The microscopic dynamics responsible for these transitions and, hence, for ergodicity, brings about the adjustment of the system to new external conditions, a prerequisite for the possibility of extracting the maximum amount of work out of it. Thus ergodicity is a prerequisite for the entropy's having the standard thermodynamic meaning.

(ii) We consider now a distribution of molecules that belong to a finite number of classes that are isomers or other well defined variants of each other. In polymer chemistry one speaks of poly-dispersivity. The situation is described in terms of d.f.'s which, in our terminology, are taken over the fixed variables that distinguish the above mentioned classes. These d.f.'s are nonergodic and their dispersal provides a numerical measure of poly-dispersivity.

In the case of genetic material the "poly-dispersivity" of the DNA molecule stores genetic information. If a subensemble of molecules is selected, that are all in identical configurations described by the same set of fixed variables (a "pure strain"), the dispersal of the new d.f. is zero. The difference between the dispersals of the original d.f. and the new one is taken as the measure of information stored up by the selection of the subensemble.

Of course, the macromolecules partake also in thermal agitation (internal vibration and rotation). The dispersal of the d.f. over these states is their entropy, an entirely different quantity. The present point of view also disposes of the difficulties that arise if negative entropy is taken as a measure of the degree of organization. The paradoxical character of this alleged connection has been demonstrated in terms of a simple numerical example by Klein (*27*).

(iii) Systems in false equilibrium, such as a glass, are characterized by a huge number of quasi-fixed variables distributed at random. These variables are practically constant compared, say, with the free variables specifying the phonon states. Nevertheless they are subject to a slow drift.

The dispersal of the free-variable distribution could be used as an entropy. However the usefulness of this quantity is impaired by the fact that we cannot precisely identify the system to which it belongs. It is possible, in principle, to consider a virtual ensemble of glasses described by a d.f. over the quasi-fixed variables distributed at random over a continuum. The dispersal of this d.f. would be, indeed, a measure of our ignorance. The usefulness of this quantity has not yet been demonstrated, however, and we bring up the case only as a contrast with cases (i) and (ii).

We conclude with a few remarks concerning the relation of STE to MTE. Evidently, the former reduces to the latter as Boltzmann's constant tends to zero. Using a terminology introduced before *(6)*, STE is "dominant" to MTE. An interesting aspect of this relation is that the fundamental equation of MTE remains valid in STE and has even the same functional form. The difference is in the conceptual interpretation of the conjugate couples of variables, X and π, of MTE. In STE there are two alternative interpretations: we have either $\bar{x}$ and π, that is, the actual intensities of reservoirs and averages of the additive invariants of systems, or x and $\hat{\pi}$, fixed values of the invariants coupled with estimated reservoir intensities. The transition to MTE means that these two interpretations coalesce.

In view of this situation the following possibility arises for the presentation of MTE. Instead of introducing the entropy postulationally as an undefined concept, as was done in Section I, B of *(MTE)*, we could start, with some adjustments, from the present postulational basis and develop in this fashion the entire material contained in *(MTE)*. We do not propose actually to do this, however, since we believe that STE is no more than a stepping stone toward another dominant theory that should be an even more satisfactory starting point for the structuring of thermodynamics. In such a future theory, correlations and time-dependent processes and operations should be handled explicitly. There is, of course, an extensive literature devoted to these problems. It seems however, that these always involve procedures that are, in one way or another, analogous to the asymptotic procedures discussed in Section V. The development of a theory that can be considered as a generalization of the formalism contained in Sections I–IV still has major difficulties to overcome.

Acknowledgment

The authors are glad to acknowledge stimulating discussions with Dr. Benoit Mandelbrot.

Received: January 21, 1963

REFERENCES

1. A. Einstein, *Ann. Physik* **33,** 1275 (1910).
2. L. Boltzmann, "Vorlesungen zur Gastheorie." Barth, Leipzig, 1896.
3. J. W. Gibbs, "Collected Works," Vol. I. Thermodynamics, Vol. II, Elementary Principles in Statistical Mechanics. Scribner, New York, 1902. (Reprinted by Yale Univ. Press, New Haven, 1948.)
4. B. Mandelbrot, *Symp. Inform. Theory, M. I. T., September 1956; IRE Trans. Inform. Theory* **IT-2** (1956); *Ann. Math. Statist.* **33,** 1021 (1962).
5. E. T. Jaynes, *Phys. Rev.* **106,** 620 (1957); *ibid* **108,** 171 (1957); R. S. Imgarden and K. Urbanik, *Acta Phys. Polon.* **21,** 281 (1962).
5a. C. E. Shannon and W. Weaver, "The Mathematical Theory of Communication." Univ. of Illinois Press, Urbana, Ill., 1949.
6. L. Tisza, *Rev. Mod. Phys.* **35,** 151 (1963).
7. L. Tisza, *Ann. Phys.* (*N. Y.*) **13,** 1 (1961).
8. L. Szilard, *Z. Physik* **32,** 753 (1925).
9. M. R. Schafroth, *Helv. Phys. Acta.* **32,** 349 (1959).
10. L. Rosenfeld, *Physica* **27,** 67 (1961).
11. A. I. Khinchin, "Mathematical Foundations of Statistical Mechanics." Translated by G. Gamow. Dover, New York, 1949.
11a. A. I. Khinchin, "Mathematical Foundations of Quantum Statistics." Translated and edited by Irwin Shapiro. Graylock Press, Albany, N. Y., 1960.
12. D. V. Widder, "The Laplace Transform." Princeton Univ. Press, Princeton, N. J., 1941.
13. H. Cramer, "Mathematical Methods of Statistics." Princeton Univ. Press, Princeton, N. J., 1946.
14. S. R. deGroot, "Thermodynamics of Irreversible Processes." North-Holland, Amsterdam, 1952.
15. K. R. Atkins, "Liquid Helium." Cambridge Univ. Press, 1959.
16. H. C. Urey, H. A. Lowenstam, S. Epstein, and C. R. McKinney, *Bull. Geol. Soc. Am.* **62,** 399 (1951); S. Epstein, R. Buschbaum, H. A. Lowenstam, and H. C. Urey, *Bull. Geol. Soc. Am.* **64,** 1315 (1953).
17. S Watanabe, *Rev. Mod. Phys.* **27,** 179 (1955).
18. R. A. Fisher, *Trans. Roy. Soc.* (*London*) **222,** 309 (1921); "Contributions to Mathematical Statistics." Wiley, New York, 1950.
19. N. F. Ramsey, *Phys. Rev.* **103,** 20 (1956).
20. R. H. Fowler, "Statistical Mechanics," 2nd ed. Cambridge Univ. Press, 1936.
21. F. Simon, *Ergeb. Exakt. Naturw.* **9,** 222 (1930); L. Pauling, *J. Am. Chem. Soc.* **57,** 2680 (1935); K. Clusius, L. Popp, and A. Frank, *Physica* **4,** 1105 (1937); J. O. Clayton and W. F. Giauque, *J. Am. Chem. Soc.* **54,** 2610 (1932); F. E. Simon, 40th Guthrie Lecture, *in* "Year Book of the Physical Society." London, 1956.
22. L. Landau and E. M. Lifshitz, "Statistical Physics," p. 66. Translated by E. and R. F. Peierls. Addison-Wesley, Reading, Mass., 1958.
23. V. F. Weisskopf, *Phys. Rev.* **52,** 295 (1937).
24. M. J. Klein and L. Tisza, *Phys. Rev.* **76,** 1861 (1949).
25. R. F. Greene and H. B. Callen, *Phys. Rev.* **83,** 1231 (1951).
26. T. Ehrenfest-Afanassjewa, *Am. J. Phys.* **26,** 388 (1958).
27. M. J. Klein, *Brit. J. Phil. Sci.* **4,** 158 (1953).

Reprinted from THE PHYSICAL REVIEW, Vol. 76, No. 12, 1861–1868, December 15, 1949
Printed in U. S. A.

Theory of Critical Fluctuations

M. J. KLEIN* AND L. TISZA
*Department of Physics and Research Laboratory of Electronics, Massachusetts Institute of Technology, Cambridge, Massachusetts***

(Received August 24, 1949)

A macroscopic system is subdivided into cells of identical size and shape arranged in a regular spatial array. The method of canonical ensembles would consider one of the cells, schematizing the rest into a "reservoir." The present "cellular method" treats the cells on an equal footing and is appropriate to deal with the fluctuations near the critical point for which the standard theory yields infinite results. Earlier theories dealing with the same problem appear as special cases of the present treatment. In particular, the critical points are defined generally enough to include the so-called λ-points in solids. The macroscopic system is invariant under the group of translations which displaces one cell into another. The macroscopic quantities (e.g., the thermodynamic parameters) are invariants of this group.

I. INTRODUCTION

THE role of fluctuations in statistical thermodynamics is primarily a negative one. Indeed, the very existence of thermodynamics can be said to be based on the fact that the fluctuations in the macroscopic thermodynamic variables are negligible compared to the mean values of these variables. The proof of this fact is essential in the statistical foundation of thermodynamics. The agreement in the results obtained from different statistical representations of a physical system is also a consequence of the negligible character of the fluctuations. Thus, whether one uses a microcanonical ensemble which corresponds to a completely closed system, or a canonical ensemble which corresponds to a system in contact with a reservoir, the results are the same since fluctuations can be neglected.

It is therefore of fundamental interest that fluctuations are not negligible at the so-called critical points of physical systems, of which the gas-liquid critical point is the best known case. As a matter of fact, if the fluctuations are calculated according to standard procedures, infinite results are obtained. Experimental results indicate that critical-point fluctuations, although finite, are orders of magnitude larger than usual and give rise to macroscopic phenomena, e.g., critical opalescence.

Several theories[1–4] have been proposed for treating the problem of critical-point fluctuations, each based on a different analysis of the physical situation. There are several reasons for our reconsidering the question. In the first place, the relation between the above-mentioned theories is rather obscure and none of them can claim general acceptance. In the second place, all of them appear to be too specialized. (In fact, one of the authors has recently shown[5] that critical points are of rather common occurrence, since the well-known λ-point phenomena observed in solids and some liquids may be considered as critical points.) The above-mentioned theories refer only to the gas-liquid critical points and to the critical mixing points in liquids. Finally, these theories are often based, in part, on rather special molecular models and do not really treat the basic statistical problem.

In Section II we sum up the results of the conventional theory of fluctuations in a form general enough for our purposes. The difficulties arising at the critical point are discussed in detail.

In Section III we introduce the "cellular method" which can be considered a refinement of the ordinary treatment of generalized canonical ensembles. This method allows one to treat the correlation of fluctuations in different volume elements, an effect which is of great importance at critical points and which is ignored by the standard method. The importance of these correlations was first pointed out by Ornstein and Zernike,[2] and our method is a development of their idea. This development, on its formal side, leads to a general criterion for the definition of macroscopic variables, i.e., those variables which can be used in discussing macroscopic phenomena.

In Section IV we present a calculation of fluctuations and their correlations.

Section V is devoted to a discussion of the relationships between our theory and those previously mentioned, as well as to some clarification of the relationships of these theories to each other. In particular, the question of critical opalescence is discussed.

II. FLUCTUATIONS AND CRITICAL POINTS

The thermodynamic properties of a closed system can be adequately described in terms of a so-called

* Present address: Department of Physics, Case Institute of Technology, Cleveland, Ohio.

** This work has been supported in part by the Signal Corps, the Air Materiel Command, and the ONR.

[1] M. V. Smoluchowski, Ann. d. Physik. **IV** (Folge) **25**, 205 (1908).

[2] L. Ornstein and F. Zernike, Proc. Amst. Akad. Sci. **17**, 793 (1914); **18**, 1520 (1916); **19**, 1321 (1917); Physik. Zeits. **19**, 134 (1918); **27**, 761 (1926).

[3] Y. Rocard, J. de phys. et rad. **4**, 165 (1933).

[4] J. Yvon, Actualités Scientifiques et Industrielles, Nos. 542, 543 (Paris, 1937).

[5] L. Tisza, "General theory of phase transitions," National Research Council Conference on Phase Transformations in Solids, 1948 (to appear in book form).

fundamental equation

$$U = U(X_1, X_2, \cdots, X_{r+1}), \tag{1}$$

where U is the internal energy and $X_1, X_2, \cdots, X_{r+1}$ are the extensive variables characterizing the system. (The X_i include volume, entropy, mole numbers, etc.) We shall assume our system to be spatially homogeneous on a macroscopic scale so that U is a first-order homogeneous function of its arguments. It is then most convenient to let $X_{r+1} = V$, the volume, and to define the "densities":

$$x_i = \frac{X_i}{X_{r+1}}; \quad u = \frac{U}{X_{r+1}}. \tag{2}$$

The fundamental equation can then be written

$$u = u(x_1, x_2, \cdots, x_r). \tag{3}$$

The generalized forces conjugate to X_k or x_k are

$$P_k = \partial U/\partial X_k = \partial u/\partial x_k. \tag{4}$$

In order to obtain the fluctuations of the quantities x_k about their mean values, we shall consider our system as a part of a large closed system, e.g., one cm³ of a gas in a large container. We now have the possibility of processes in which the quantities X_k flow from the sub-system to the remainder of the system, called a "reservoir," and vice versa; these processes are subject only to the condition that there is no change in the value of X_k for the whole closed system. The assumed difference in size between the sub-system and the reservoir implies that these processes may lead to a noticeable change in the state of the sub-system while the quantities x_k and P_k are essentially unchanged in the reservoir.

It is well known[6] that the probability of finding the sub-system in a state $x_{1\alpha}, x_{2\alpha}, \cdots, x_{r\alpha}$, which may differ from the equilibrium state $x_1, \cdots, x_r$, is given by a generalized canonical distribution function:

$$p_\alpha = w_\alpha \exp 1/kT\left[\psi - E_\alpha + \sum_{k=2}^{r} P_k x_{k\alpha}\right]. \tag{5}$$

Here p_α is the probability of finding a state of the system described by the index α, w_α is the statistical weight (degeneracy) of this state, E_α its energy, and $x_{k\alpha}$ the value of the variable x_k in this state. The index k runs from 2 to r omitting x_1 which is the entropy density. The function ψ is the thermodynamic potential

$$\psi = U - \sum_{k=1}^{r} P_k x_k. \tag{6}$$

We can put the expression for p_α into a more symmetrical form as follows. Let us define:

$$S_\alpha = k \ln w_\alpha.\text{[a]} \tag{7}$$

Then:

$$p_\alpha = \exp \frac{\left[U - E_\alpha - \sum_{k=1}^{r} P_k (x_k - x_{k\alpha})\right]}{kT}, \tag{8}$$

where the summation now goes from 1 to r (including the entropy term). We can also write:

$$p_\alpha = \exp \frac{[\psi - \psi_\alpha]}{kT} \tag{9}$$

where:

$$\psi_\alpha = E_\alpha - \sum_{k=1}^{r} P_k x_{k\alpha}. \tag{10}$$

This "microscopic free energy" ψ_α will be of interest to us subsequently.

The physical meaning of $\psi - \psi_\alpha$ is the minimum work required to bring about the state α. Hence Eq. (9) expresses the well-known principle of Boltzmann. Let us introduce the notation:

$$\delta x_k = x_{k\alpha} - x_k. \tag{11}$$

If we expand E_α as a function of the δx_k, we obtain

$$E_\alpha = \bar{E}_\alpha + \sum_{k=1}^{r} \left(\frac{\partial E_\alpha}{\partial x_k}\right)_0 \delta x_k + \frac{1}{2} \sum_{k,l=1}^{r} \left(\frac{\partial^2 E_\alpha}{\partial x_k \partial x_l}\right)_0 \delta x_k \delta x_l, \tag{12}$$

keeping second-order terms. Now $\bar{E}_\alpha = U$, $(\partial E_\alpha/\partial x_k)_0 = P_k$; and we can define

$$\left(\frac{\partial^2 E_\alpha}{\partial x_k \partial x_l}\right)_0 = u_{kl} = \frac{\partial^2 U}{\partial x_k \partial x_l}. \tag{13}$$

Then to this approximation we find

$$p_\alpha = c \exp\left\{\frac{-\frac{1}{2} \sum_{k,l=1}^{r} u_{kl} \delta x_k \delta x_l}{kT}\right\}, \tag{14}$$

where c is a normalization constant which is needed because Eq. (14) is only an approximation to Eq. (8).

In terms of this Gaussian distribution, one can immediately write down the fluctuations in which we are interested:

$$\langle \delta x_k \delta x_l \rangle_{\text{Av}} = \left\{\frac{\int (\delta x_k)(\delta x_l) p_\alpha d(\delta x_1) \cdots, d(\delta x_r)}{\int p_\alpha d(\delta x_1) \cdots, d(\delta x_r)}\right\} = kT v_{kl}, \tag{15}$$

[6] E. A. Guggenheim, J. Chem. Phys. 7, 103 (1939).

[a] S_α is the entropy one assigns to a system known to be in a state α of statistical weight w_α.

where the matrix $\|v_{kl}\|$ is the inverse of the matrix $\|u_{kl}\|$. These matrices have been called stiffness $\|u_{kl}\|$ and compliance $\|v_{kl}\|$ matrices.[5]

The smallness of kT compared to the internal energy of macroscopic systems assures us that the relative fluctuations are small, hence the microcanonical and canonical methods are essentially equivalent. Speaking more physically, the behavior of the sub-system is the same whether it is isolated from the rest by a real wall or isolated only in imagination.

This situation is different, however, at the so-called critical point where $\|u_{ik}\|$ is positive semidefinite ($\text{Det}\|u_{ik}\|=0$) and where the matrix elements v_{ik} and hence the fluctuations become infinite.[5] Actually, the experimentally observed fluctuations become extremely high at the critical point, as manifested in the phenomenon of critical opalescence, for example. The fluctuations are, however, finite and our formalism must be improved, so as to provide a finite result.

The weak point in the above discussion, which is essentially the same as the standard discussion of fluctuations, is in the use of the canonical ensemble to represent the behavior of a sub-system of our whole system while idealizing the remainder by treating it as a reservoir. When the fluctuations are large, as in the vicinity of a critical point, the microcanonical and canonical ensembles no longer give the same results, and it is essential to treat the system as a whole in a proper manner. This will involve a modification of the usual canonical ensemble which will be carried out in the next section. At this stage, we merely need to point out that such a modification is necessitated by the large critical-point fluctuations which indicate a breakdown of the usual approach discussed above.

Before going on to examine the statistical problem in more detail, it seems convenient to simplify the definition of the critical point. As presented above, it appears as a result of the "co-operation" of all extensive variables $x_1, x_2, \cdots, x_r$. Instead, we can choose $P_1, P_2, \cdots, P_{r-1}, x_r$ as independent variables, and introduce the corresponding free energy

$$A(x_r)=u-\sum_{k=1}^{r-1}P_kx_k. \tag{16}$$

(A is a function of $P_1, P_2, \cdots, P_{r-1}$ as well as x_r, but it is the latter dependence which will interest us.) It can be shown[5] that a critical point may also be defined as follows:

$$\frac{\partial^2 A}{\partial x_r^2}=0,\quad \frac{\partial^3 A}{\partial x_r^3}=0,\quad \frac{\partial^4 A}{\partial x_r^4}>0. \tag{17}$$

The advantage of this formulation is that it refers explicitly to a single independent variable. Of course, this can be chosen to be any of the $x_1, x_2, \cdots, x_r$. The various formalisms arising are not identical, but lead to the same conclusions.

III. THE CELLULAR METHOD

We generalize the concept of a canonical ensemble by introducing what we call the cellular method. Instead of concentrating our attention on a sub-system and schematizing the rest into a "reservoir," we divide the whole system into cells which we treat on an equal footing. We choose these cells as identical in size and shape and arranged in a regular spatial array. If we are dealing with a fluid system, we can choose a simple cubic array of cubical cells. If we deal with a crystalline system, it will be natural to let our array have the symmetry of the crystal; the individual cells can then consist either of single unit cells of the crystal or of groups of contiguous unit cells. To avoid the necessity of introducing the reciprocal lattice, we shall work with a cubic array, but our considerations can easily be transcribed so as to apply to the general case.

The position of any cell is then specified by a vector $\mathbf{k}$:

$$\mathbf{k}=k_1\mathbf{j}_1+k_2\mathbf{j}_2+k_3\mathbf{j}_3, \tag{18}$$

where k_1, k_2, k_3 are integers between 0 and $M-1$ if there are M^3 cells, and $\mathbf{j}_1, \mathbf{j}_2, \mathbf{j}_3$ are the unit vectors defining the primitive translations of our array. (In the rest of the paper the symbols $\mathbf{k}$, $\mathbf{l}$, $\mathbf{m}$, $\mathbf{j}_1$, $\mathbf{j}_2$, and $\mathbf{j}_3$ designate vectors, whereas k_1, k_2, k_3, l_1, l_2, l_3, m_1, m_2, and m_3 are integers.)

In the previous section, we pointed out that if one works with the appropriate thermodynamic potential $A(x_r)$, the conditions for a critical point can be expressed in terms of the one variable x_r. We are now interested in considering states of our large system specified by the set of values $x_{r\mathbf{k}}$, where $\mathbf{k}$ varies over all the M^3 cells; $x_{r\mathbf{k}}$ is the value of x_r in the cell $\mathbf{k}$. (We take the variables $P_1, P_2, \cdots, P_{r-1}$ as fixed for our system.)

By applying an argument similar to that used in the preceding section, one can easily show that the probability of finding a state $\{x_{r\mathbf{k}}\}$, is given by:

$$p=C\exp\left(-\frac{1}{2}\sum_{\mathbf{k}\mathbf{l}}a_{\mathbf{k}\mathbf{l}}z_{\mathbf{k}}z_{\mathbf{l}}\right), \tag{19}$$

where[b]

$$a_{\mathbf{k}\mathbf{l}}=\frac{1}{kT}\left(\frac{\partial^2 A}{\partial x_{r\mathbf{k}}\partial x_{r\mathbf{l}}}\right)_0=\frac{1}{kT}\left(\frac{\partial^2 A}{\partial z_{\mathbf{k}}\partial z_{\mathbf{l}}}\right); \tag{20}$$

$$z_{\mathbf{k}}=x_{r\mathbf{k}}-\bar{x}_r. \tag{21}$$

In spite of the great formal similarity between Eq. (14) and Eq. (19), one must be very careful to distinguish between the two quadratic forms appearing in the exponential in these two equations. The first, Eq. (14), refers to a set of different thermodynamic variables $x_1\cdots, x_r$; the second, Eq. (19), refers to the set of values $x_{r\mathbf{k}}$ of one thermodynamic variable x_r in the different cells $\mathbf{k}$ of the whole system. Both quadratic

[b] It is necessary to average over all of the "internal coordinates" of the system, e.g., thermal vibration coordinates; in other words, we preserve only the dependence on the set of variables $z_{\mathbf{k}}$.

forms express the minimum work necessary to carry the system from its equilibrium state to one deviating from equilibrium.

We should point out an important restriction on the values of the z_k, the deviations of x_{rk} from its average value x_r; namely, we must have:

$$\sum_k z_k = 0. \tag{22}$$

This expresses the fact that our whole system is closed, so there can be no net fluctuation in x_r if we sum over all cells.

The eigenvalues and eigenvectors of this form are of considerable interest, so we shall now determine them. This is simple, if we notice that the set of coefficients a_{kl} for fixed k, l varying over all cells, is just a permutation of the set $a_{k'l}$ for any other fixed k'. This expresses the fact that the interactions (as measured by a_{kl}) of any cell with its neighbors of all orders are independent of the particular cell considered. (We impose periodic boundary conditions on our whole system, so that the cells near the boundary "surfaces" need not be considered separately.)

The eigenvalues λ_m and eigenvectors ξ_{lm} must satisfy

$$\sum_l a_{kl}\xi_{lm} = \lambda_m \xi_{km}; \tag{23}$$

(k, m varying over all cells, i.e., $m = m_1 j_1 + m_2 j_2 + m_3 j_3$, $m_i = 0, 1, \cdots, M-1$). The solutions to Eq. (23) are:

$$\lambda_m = \sum_l a_{kl} \exp\left[\frac{2\pi i}{M} m \cdot (l-k)\right], \tag{24}$$

$$\xi_{lm} = \frac{1}{M^{\frac{3}{2}}} \exp\left[\frac{2\pi i}{M} l \cdot m\right], \tag{25}$$

where

$$\sum_k \xi_{km}\xi_{km'}^* = \delta(m-m'). \tag{26}$$

Since a_{kl} depends only on $l-k$, λ_m is actually independent of k as it must be.

The eigenvalues can be written in a simpler form if we note that it is sufficient to take only two distinct values of the a_{kl}:

$$\begin{aligned} &a_{kk} = a_0; \quad a_0 > 0 && \text{(all } k\text{)}, \\ &a_{kl} = a_1 && (k, l \text{ nearest neighbors}), \\ &a_{kl} = 0 && \text{(otherwise)}. \end{aligned} \tag{27}$$

The first of these conditions states that all cells are equivalent. The second condition, nearest-neighbor "interactions" only, is not necessary for most of the following work but simplifies the form of the equations to some extent.

The eigenvalues are now:

$$\lambda_m = a_0 + 2a_1 \sum_{i=1}^{3} \cos\frac{2\pi}{M} m_i; \quad (m = m_1 j_1 + m_2 j_2 + m_3 j_3). \tag{28}$$

The eigenvector corresponding to λ_0 ($\lambda_0 = \lambda_{M(j_1+j_2+j_3)} = a_0 + 6a_1$) is of especial significance. It is

$$\xi_0 = \frac{1}{M^{\frac{3}{2}}}(1, 1, \cdots, 1).$$

This is the only eigenvector which does not involve phase differences between the different cells. Now, while the eigenvalues λ_m depend on the properties of the quadratic form of Eq. (19), the eigenvectors really furnish an appropriate and flexible coordinate system, as it were, for describing our physical system.

We may use the unique character of the eigenvector ξ_0 to suggest a characterization of those variables of our system which can be handled in a macroscopic discussion; namely, only variables which transform like ξ_0, i.e., which are the invariants of the translation group, can qualify as macrovariables, since all others involve unobservable phase differences as one goes from cell to cell.

The variables which satisfy this criterion are of two rather different kinds. In the first class, we have thermodynamic variables such as magnetic moment per unit volume, average particle density, etc. The additional characteristic feature of these variables is that their conjugate "forces" exist and, consequently, the values of these variables are subject to the control of the experimenter. In the second class are variables which might be called quasi-thermodynamic ones, such as the atomic coordinates in a crystal and the long-range order in alloys, etc. Here there are no conjugate forces, so the variables in question cannot be controlled freely. Nevertheless, they are of a macronature in the sense described above and, correspondingly, are subject to measurement—by their coherent x-ray scattering, for example.

The variables which do not behave like ξ_0 are perhaps typified by the thermal vibrations or phonons which essentially involve phase differences between cells and must always be averaged over in macroscopic treatments.

Let us finally apply our study of the eigenvalues to the case of the critical point. This is now redefined as the point where:

$$\frac{1}{2}\sum_{k,l} a_{kl} z_k z_l,$$

becomes positive semidefinite, i.e., one of its eigenvalues must vanish. It is not difficult to see that the eigenvalue which vanishes (the others remaining positive) is λ_0. Thus, at the critical point,

$$\lambda_0 = a_0 + 6a_1 = 0. \tag{29}$$

This result is, of course, to be expected in the light of the above discussion, but it can be proved directly and, therefore, is an example in support of the arguments advanced.

In fact, we notice that if $a_0+6a_1=0$, all other λ_m are positive; but if $\lambda_m=0$, $m\neq 0$, then some $\lambda_m<0$, contradicting our requirement that the quadratic form be positive semidefinite. In the absence of correlations $a_1=0$ and Eq. (29) reduces to the previous definition of critical points.

IV. FLUCTUATIONS WITH CORRELATIONS

We can now turn to the calculations of the mean-square fluctuation in the value of x_r in cell k and the correlations of the fluctuations in different cells. We shall use the approach of the preceding section.

We have already written in Eq. (19) the expression for the probability of finding a state of the system characterized by a given set of values of the x_{rk}. It has also been pointed out that there is a restriction, Eq. (22), on the values of the set of variables z_k, because our system is closed. We now want to calculate

$$\langle z_k z_l\rangle_{\text{Av}}=\frac{\int z_k z_l \exp\left[-\frac{1}{2}\sum_{k,l}a_{kl}z_k z_l\right]d\tau'}{\int \exp\left[-\frac{1}{2}\sum_{k,l}a_{kl}z_k z_l\right]d\tau'}, \tag{30}$$

where $d\tau'$ means that we integrate over the M^3-1 dimensional space of the z_k, subject to the restriction of Eq. (22).

In order to take into account this restriction in the simplest way, we go over to the coordinate system determined by the eigenvectors of the matrix $\|a_{kl}\|$. The calculation is given in the Appendix; we quote only the result here:

$$\langle z_k z_l\rangle_{\text{Av}}=\frac{1}{M^3}\sum_p{}'\frac{\cos(2\pi/M)p\cdot(k-l)}{\lambda_p}. \tag{31}$$

The primed summation means that we sum over all p except $p=0$, i.e., that the eigenvalue λ_0 does not appear in the sum. This follows immediately from the restriction of Eq. (22), as is shown in the Appendix.

Let us first analyze this result for the case $k=l$. Introducing the explicit expressions for the eigenvalues, we have

$$\langle z_k^2\rangle_{\text{Av}}=\frac{1}{M^3}\sum_{p_1,p_2,p_3}{}'\times\frac{1}{a_0+2a_1[\cos(2\pi/M)p_1+\cos(2\pi/M)p_2\cos(2\pi/M)p_3]}. \tag{32}$$

Now, far from the critical point, where the interactions between cells should not be important, this reduces to the conventional result. For, in this case, it follows from Eq. (29) that

$$a_0\gg -6a_1, \tag{33}$$

and, therefore,

$$\lambda_p\simeq a_0, \tag{34}$$

$$\langle z_k^2\rangle_{\text{Av}}\simeq\frac{1}{a_0}=\frac{kT}{(\partial^2A/\partial x_r^2)_{P_1,\ldots,P_{r-1}}} \tag{35}$$

which is the result obtained by the method of Section II in this particular case (one variable).

Closer to the critical point, we can transform the sum in Eq. (32) into an integral, since M is as large as we please;

$$\langle z_k^2\rangle_{\text{Av}}=\frac{1}{a_0}\frac{1}{(2\pi)^3}\int_0^{2\pi}\int_0^{2\pi}\int_0^{2\pi}\times\frac{dq_1dq_2dq_3}{\{1+(2a_1/a_0)[\cos q_1+\cos q_2+\cos q_3]\}}. \tag{36a}$$

Equation (36a) shows that the fluctuations become larger as we approach the critical point. At the critical point, where $(2a_1/a_0)=-\frac{1}{3}$, the integral still converges[c] and can even be evaluated exactly. Thus

$$\langle z_k^2\rangle_{\text{Av}}=\frac{1}{a_0}\frac{3}{\pi^3}\int_0^{\pi}\int_0^{\pi}\int_0^{\pi}\frac{dq_1dq_2dq_3}{3-(\cos q_1+\cos q_2+\cos q_3)}$$
$$=\frac{3}{a_0}(18+12\sqrt{2}-10\sqrt{3}-7(6)^{\frac{1}{2}})\left(\frac{2K_2}{\pi}\right)^2, \tag{36b}$$

where K_2 is the complete elliptic integral of modulus $(2-\sqrt{3})(\sqrt{3}-\sqrt{2})$. Numerically:

$$\langle z_k^2\rangle_{\text{Av}}=1.5164/a_0, \tag{36c}$$

at the critical point.

Let us now analyze the physical meaning of our results. In the first place we know that at the critical point $a_0=-6a_1\neq 0$. Since thermodynamics assures us that $a_0>0$, it follows that $a_1<0$ at and near the critical point. This means that a fluctuation in any cell increases the probability of a fluctuation of like sign in the neighboring cells. This is most readily seen from Eq. (40) below.

Secondly, we see that the fluctuations are kept finite by the physical requirement that the entire system be closed with respect to the thermodynamic variable x_r whose fluctuations we are considering. This entered our formalism in the restriction of Eq. (22) which automatically led to the omission of the term in λ_0 from the sum of Eq. (32). Thus the convergence of the integral in Eq. (36b) and the closure of our whole system lead to finite critical fluctuations.

Considering now the case $k\neq l$, i.e., the correlation of the fluctuations in different cells, we may make several remarks. In the first place, it is clear that, in the limit of vanishing interaction between cells, all $\langle z_k z_l\rangle_{\text{Av}}$ are

[c] The proof of convergence and the more difficult evaluation of the integral are due to G. N. Watson, Quart. J. Math. (Oxford) **10**, 266 (1939).

We should like to express our sincere thanks to Dr. Elliott Montroll for informing us of this reference.

zero ($k\neq l$). For, in this case, $\lambda_p=a_0$ and the phase factors cancel out.[d] Second, as $|k-l|$ increases the cancellation between successive terms increases and $\langle z_kz_l\rangle_{\text{Av}}$ becomes small.

This point can be brought out more clearly by another method which we now discuss. This method is of interest in itself and it establishes the connection between our work and that of Ornstein and Zernike[2] (see Section V). The essential point is to derive an equation relating the correlation coefficient of the fluctuations in different cells to the basic coefficients a_{kl} which are derivatives of the free energy and thus express the forces within our system.

The correlation coefficient g_{kl} is defined[7] by

$$g_{kl}=\frac{\langle z_kz_l\rangle_{\text{Av}}}{(\langle z_k^2\rangle_{\text{Av}}\langle z_l^2\rangle_{\text{Av}})^{\frac{1}{2}}} \tag{37}$$

Instead of using the expressions given above for these quantities, it is more convenient to evaluate them in terms of the minors of the determinant of $\|a_{kl}\|$. Let A be the determinant of $\|a_{kl}\|$ and A_{kl} be the co-factor (with its appropriate sign) of a_{kl} in A. Then[8]

$$\langle z_kz_l\rangle_{\text{Av}}=\frac{\int z_kz_l\exp\left[-\frac{1}{2}\sum_{k,l}a_{kl}z_kz_l\right]d\tau}{\int\exp\left[-\frac{1}{2}\sum_{k,l}a_{kl}z_kz_l\right]d\tau}=\frac{A_{kl}}{A}. \tag{38}$$

Since $A_{kk}=A_{ll}$ for all k, l as a result of the properties of $\|a_{kl}\|$, we have

$$g_{kl}=A_{kl}/A_{ll}. \tag{39}$$

Let us define

$$f_{kl}=-a_{kl}/a_{kk}. \tag{40}$$

The quantity f_{kl} is the average of z_k when $z_l=1$ and all other variables are zero (linear regression coefficient).[7] We see that f_{kl} is a direct index of the "energetic" interaction between cells k and l, and vanishes under our assumptions unless k and l are nearest neighbors. Using Eqs. (39) and (40), one can easily show[e] that

$$g_{kl}=f_{kl}+\sum_{m\neq k,l}f_{km}g_{ml}.\quad (k\neq l) \tag{41}$$

[d] The cancellation is not complete and $\langle z_kz_l\rangle_{\text{Av}}=(-1/a_0)(1/M^3)$ because we do not sum over the eigenvalue λ_0. This term is, of course, negligible for large M.

[7] H. Cramér, *Mathematical Methods of Statistics* (Princeton University Press, Princeton, 1946).

[8] The restriction of Eq. (22) has not been imposed here. It can be shown [M. J. Klein, Ph.D. thesis in physics, M.I.T. (1948)] that this restriction plays no important role in these considerations.

[e] For:

$$\sum_{m\neq k,l}f_{km}g_{ml}=-\sum_{m\neq k,l}\frac{a_{km}A_{ml}}{a_{kk}A_{ll}}$$
$$=-\frac{1}{a_{kk}A_{ll}}\left[\sum_m a_{km}A_{ml}-a_{kk}A_{kl}-a_{kl}A_{ll}\right]$$
$$=-\frac{A\delta_{kl}}{a_{kk}A_{ll}}+g_{kl}-f_{kl}=g_{kl}-f_{kl}\quad\text{as}\quad k\neq l.$$

Equation (41) has a simple interpretation: it states that the correlation coefficient g_{kl} (which measures the statistical effect of cell l on cell k) is equal to the sum of two terms. The first is the direct interaction f_{kl}, present only for nearest neighbors; the second is the sum giving the effect g_{ml} of cell l on all other cells m which have a non-vanishing interaction f_{km} with cell k.

We need consider the set of Eqs. (41) only for some definite value of l, since the set g_{kl} for this l value is the same, except for permutation, as any other set g_{km}. Hence we shall fix l and, with this understanding, drop it from our notation. We then have, as our final result,

$$g_k=f_k+\sum_{m\neq k}f_{km}g_m. \tag{42}$$

V. DISCUSSION

In this section, we shall point out the connections between the theory presented in this paper and the earlier work done on this problem, particularly with respect to the theory of critical opalescence.

It was shown in Section II that the customary statistical treatment of fluctuations leads to infinite critical-point fluctuations. The physical reason for this was that the minimum work necessary to produce such a fluctuation vanishes at the critical point in the usual approximation. It is clear from the nature of the problem that any additional contribution to the work, usually neglected, will limit the fluctuations to finite values at the critical point. A number of theories have been advanced which have this result, and it is our purpose to point out the nature of the additional term in each case and to discuss the interrelations of these theories in the light of our own.

In the first place, it should be said that virtually all previous work on the subject has been concerned with the ordinary gas-liquid critical point of a pure substance or with the critical mixing point of a binary mixture of liquids. Our theory has been set up in such a manner as to be applicable to λ-points as well as critical points in the ordinary sense. For this reason, we shall be interested in discussing only those other theories which are capable, in principle, of such generalization. Hence we have not considered the mechanism proposed by Yvon,[4] who has pointed out that infinite fluctuations are avoided if one considers the effect of gravity. Briefly, near the ordinary critical point the compressibility is large and the gravitational field of the earth gives rise to a non-uniformity of density, "thickening" the gas-liquid interface into a layer. Although this effect undoubtedly exists, it seems to be special, insofar as it cannot be of importance in λ-points in solids, for example.

The original suggestion for obtaining finite critical-point fluctuations is due to Smoluchowski.[1] His proposal was to continue the expansion of the free energy to the next non-vanishing term (quartic term), at the critical point (see Eq. (12)). This insures a non-zero minimum work and gives finite results. This method cannot be considered satisfactory because it is based on the consideration of one volume element with the remainder treated as a reservoir, and when the fluctuations are large (though finite) the difference of the states of neighboring volume elements and their correlations must be taken into account.

The first major attempt to give a more adequate treatment of critical-point fluctuations considering correlations is due to Ornstein and Zernike.[2] We have already mentioned that their papers served as a stimulus to the present work. Their chief contributions were the introduction of the idea of correlations and the development of an integral equation for determining the correlation coefficient in terms of the intermolecular forces. This integral equation, which they derived by other methods, follows from our Eq. (42) if one redefines f_k and g_k by dividing by the volume of one cell. Going over to the continuum, one then obtains

$$g(\mathbf{r})=f(\mathbf{r})+\int f(\mathbf{r}'-\mathbf{r})g(\mathbf{r}')d\mathbf{r}'. \tag{43}$$

They have shown that if $f(r)$ is a short-range isotropic function, one can transform Eq. (43) into a differential equation whose asymptotic solution is

$$g(r)\sim e^{-\kappa r}/r, \tag{44}$$

where

$$\kappa^2=\frac{2(1-F)}{E};\quad F=\int f(r)d\tau;\quad E=\frac{1}{3}\int r^2 f(r)d\tau. \tag{45}$$

This solution is instructive, since it shows that for $\kappa>0$ (above the critical point) $g(r)$, the correlation coefficient of the fluctuations, drops to zero exponentially as the distance increases. At the critical point where $\kappa=0$,[f] $g(r)\sim 1/r$, a long-range function, so that fluctuations make their influence felt over long distances.

Having indicated how the equations of Ornstein and Zernike follow from our work, under suitable approximations, we now turn to another important development, the theory of Rocard.[3] Rocard's starting point is the observation that the thermodynamic variable in question, our x_r (the density in his case), is actually a function of position in the system, due to the fact that fluctuations occur locally rather than over the system as a whole. Although this remark is in agreement with the theories given here and by Ornstein and Zernike, Rocard's method of working out its consequences is different. Since Rocard himself has stressed this difference, it is of interest that his equations also can be made to follow from our theory if certain approximations are made.

Rocard's method is to consider the modification in the effective "force," P_r (pressure in his case), due to the spatial variation of x_r. P_r is modified by an additive term proportional to the Laplacian of x_r, thus giving rise to a corresponding term in the minimum work. This latter term does not vanish at the critical point and, thus, finite fluctuations are obtained.

We shall now show how our theory can be made to give results of this form. The effective force P_{rk} (or simply P_k) in the cell k is given by

$$P_k=\frac{\partial}{\partial z_k}\left[\frac{1}{2}\sum_{k,\,l} a_{kl}z_kz_l\right]=\sum_l a_{kl}z_l, \tag{46}$$

since the quadratic form expresses the minimum work. Applying the nearest-neighbor approximation of Eq. (27) we obtain:

$$P_k=a_0z_k+a_1(z_{k+j_1}+z_{k-j_1}+z_{k+j_2}+z_{k-j_2}+z_{k+j_3}+z_{k-j_3})\simeq(a_0+6a_1)z(k)+a_1\nabla^2z(k), \tag{47}$$

where we have gone over to a continuous position vector k and we have approximated

$$z(k+j_1)+z(k-j_1)-2z(k) \text{ by } \frac{\partial^2 z(k)}{\partial x^2}, \text{ etc.}$$

It is seen that Eq. (47) is essentially equivalent to Rocard's assumptions, since the first term gives rise to a vanishing work at the critical point where $a_0+6a_1=0$, and the second term is Rocard's additional term. It is worth pointing out that Rocard's approach does not take full account of the correlation effects since, after setting up an equivalent to Eq. (47), he considers only one cell and not its interactions with its neighbors.

The last theory to be considered is that of Yvon.[4] His fundamental approach is essentially similar to ours, but his method of treating the system as a whole is based on a rather different viewpoint. Our theory is based on treating the system as an assembly of interacting cells whose basic properties we do not compute but in terms of which we work. Yvon attempts a completely molecular description of the system (liquid in his case); he works in terms of an Ursell development but, due to the complexity of his method, he is able to handle only the terms corresponding to "binary clusters" of molecules when it comes to making explicit calculations. Since there is no reason to believe that this limitation is plausible in the neighborhood of the critical point, we do not think this method is fruitful in practice although it is probably correct in principle.

Let us conclude by comparing the predictions of the various theories discussed so far as critical opalescence[9–11] is concerned. We have not made here an explicit calculation of our own, but we have indicated that both the Ornstein-Zernike and the Rocard theories are, in a sense, special cases of ours.

[f] $F=\sum_l f_{kl}=-6a_1/a_0$; hence $F=1$ at the critical point and $\kappa=0$.

[9] H. A. Stuart and H. G. Trieschmann, "Lichtzerstreuung," Hand und Jahrbuch d. Chem. Phys. Bd. 8, Abschn. II (Leipzig, 1936).

[10] J. Cabannes and Y. Rocard, *La Diffusion Moléculaire de la Lumière* (Les Presses Universitaires de France, Paris, 1929).

[11] G. Oster, Chem. Rev. 43, 319 (1948).

Our interest is only in the dependence on wave-length and the angular distribution.

The Smoluchowski theory predicts

$$I/I_0 \propto 1/\lambda^4(1+\cos^2\theta)\beta, \tag{48a}$$

$$I/I_0 \propto 1/\lambda^4(1+\cos^2\theta)(\gamma)^{\frac{1}{2}}. \tag{48b}$$

Here I/I_0 is the ratio of scattered to incident intensity of light of wave-length λ observed in the direction θ; β is the compressibility,

$$\beta = -1/V(\partial V/\partial P)_T,$$

and $\gamma = \partial^3 V/\partial P^3$. The expression (48a) is valid away from the critical point, and (48b) at the critical point, in this theory.

Ornstein and Zernike's theory predicts

$$\frac{I}{I_0} \propto \frac{1}{\lambda^4}(1+\cos^2\theta)\frac{1}{\{1/\beta + d^2[\sin(\theta/2)/\lambda]^2\}}. \tag{49}$$

The constant d^2 is a measure of the radius of the intermolecular forces and is proportional to E defined in Eq. (45). The additional term in the denominator arises from the correlation of fluctuations in different volume elements giving rise to a $1/\lambda^2$ dependence and an enhancement of the forward scattering at the critical point. The result is not quite complete as it stands, since it makes I/I_0 infinite for $\theta = 0$ at the critical point. Placzek[12] has shown that taking the finite volume of the scattering medium into account introduces a factor preserving the finiteness without modifying the result for experimentally realizable situations.

Rocard's theory predicts

$$\frac{I}{I_0} \propto \frac{1}{\lambda^4}(1+\cos^2\theta)\frac{1}{[(1/\beta)+c]}, \tag{50}$$

where c is a constant arising from the additional term in the pressure ($\propto \nabla^2\rho$).

The predictions of the Ornstein-Zernike and Rocard theories differ in two ways: the wave-length dependence, which becomes $1/\lambda^2$ at the critical point in the former theory, and the angular distribution, which is preferentially forward at the critical point in this theory. The two effects must go together here, as can be seen from the derivation of Eq. (49). (The theory of Yvon gives a wave-length dependence $\sim A\lambda^{-4} - B\lambda^{-6}$ where A, B are positive. We discuss this no further for reasons given above.)

Experimental work on critical opalescence has not settled the question in any clear fashion. The results of Andant[13] and Battacharya[14] show a change in wave-length dependence in the immediate vicinity of the critical point. Both find a dependence on $1/\lambda^2$ at the critical point and Andant has studied the change in exponent in (λ^{-n}) quite carefully. On the other hand, Rousset,[15] working with fluid mixtures (for which the theoretical predictions are essentially the same), finds no such definite results. For several types of fluids, the λ^{-4}-law and the angular symmetry (forward and backward) are preserved up to the critical point. In other cases, the exponent n decreases to a value of about 3 and, at the same time, an asymmetry between forward and backward scattering appears.

Summing up, it seems safe to say that there are experimental indications that the correlation effects are real and observable, although not in all cases.

[12] G. Placzek, Physik Zeits. 31, 1052 (1930).

[13] A. Andant, J. de phys. et rad. 5, 193 (1924).

[14] D. K. Battacharya, Proc. Ind. Assoc. Cultiv. Sci. 8, 277 (1923).

APPENDIX

We wish to evaluate

$$\langle z_k z_l \rangle_{\text{Av}} = \frac{\int z_k z_l \exp\left[-\frac{1}{2}\sum_{k,l} a_{kl} z_k z_l\right] d\tau'}{\int \exp\left[-\frac{1}{2}\sum_{k,l} a_{kl} z_k z_l\right] d\tau'}. \tag{A1}$$

Let us introduce the (real) eigenvectors of $\|a_{kl}\|$:

$$\eta_{kl} = \left(\frac{2}{M^3}\right)^{\frac{1}{2}} \cos\left(\frac{2\pi}{M} k\cdot l + \frac{\pi}{2}\frac{l_1+l_2+l_3}{3M}\right); \quad l \neq 0. \tag{A2}$$

$$\eta_{k0} = \left(\frac{1}{M^3}\right)^{\frac{1}{2}}.$$

The eigenvalues are given in Eq. (24). We use real eigenvectors to avoid complex integration.

Let

$$z_k = \sum_l \eta_{kl} y_l; \quad y_l = \sum_k \eta_{lk}^{-1} z_k = \sum_k \eta_{kl} z_k; \tag{A3}$$

(since $\|\eta_{kl}\|$ is orthogonal).

The restriction of Eq. (22) is now simple:

$$\sum_k z_k = \sum_{k,l} \eta_{kl} y_l = \sum_l y_l \sum_k \eta_{kl} = M^{3/2} y_0 = 0. \tag{A4}$$

Hence, all we need to do to observe the restriction is to set $y_0 = 0$ and drop this variable.

$$\sum_{k,l} a_{kl} z_k z_l = \sum_k \lambda_k y_k^2. \tag{A5}$$

$$\langle z_k z_l \rangle_{\text{Av}} = \frac{\sum_{m,n} \eta_{km}\eta_{ln} \int y_m y_n \exp\left[-\frac{1}{2}\sum_k{}' \lambda_k y_k^2\right] d\tau'}{\int \exp\left[-\frac{1}{2}\sum_k{}' \lambda_k y_k^2\right] d\tau'}, \tag{A6}$$

$$\langle z_k z_l \rangle_{\text{Av}} = \frac{\sum_m \eta_{km}\eta_{lm} \int y_m^2 \exp\left[-\frac{1}{2}\sum_k{}' \lambda_k y_k^2\right] d\tau'}{\int \exp\left[-\frac{1}{2}\sum_k{}' \lambda_k y_k^2\right] d\tau'}. \tag{A7}$$

The integrals are now simple and we obtain:[7]

$$\langle z_k z_l \rangle = \sum_m{}' \frac{\eta_{km}\eta_{lm}}{\lambda_m}, \tag{A8}$$

Let us put this expression into a simple explicit form by introducing the values of the η_{km} from Eq. (A2).

$$\langle z_k z_l \rangle_{\text{Av}} = \frac{2}{M^3}\sum_m{}' \frac{\cos[(2\pi/M)k\cdot m + (\pi/2)(m_1+m_2+m_3/3M)]}{\lambda_m} \times \cos[(2\pi/M)l\cdot m + (\pi/2)(m_1+m_2+m_3/3M)]. \tag{A9}$$

Expanding the cosines and making use of the fact that $\lambda_m = \lambda_{M(j_1+j_2+j_3)-m}$, we obtain

$$\langle z_k z_l \rangle_{\text{Av}} = \frac{1}{M^3}\sum_m{}' \frac{\cos(2\pi/M)(k-l)\cdot m}{\lambda_m}, \tag{A10}$$

which is the result given in the text, Eq. (31).

[15] A. Rousset, Ann. de physique 5, 5 (1936).

PAPER 9

THE FLUCTUATION OF PATHS

1. Introduction

In the traditional presentation there is a clean-cut distinction between thermodynamics and statistical mechanics. The former operates with more or less directly measurable variables such as energy, mass, entropy, and temperature; the second with "hidden" molecular coordinates and momenta. The application of statistical methods to the molecular parameters bridges the gap between the two points of view.

This sharp distinction is somewhat blurred in case of fluctuation phenomena that are intermediate between the macroscopic and the microscopic levels. Fluctuations originate in the molecular activity but manifest themselves in the random variations of the thermodynamic variables. The theory necessarily involves statistical methods, but these can be applied either to molecular or to thermodynamic parameters. I shall be concerned only with the latter type of theory and will refer to it as the *thermodynamic fluctuation theory*. Its aim is to establish connections between macroscopic observable quantities rather than to compute these from molecular data.

Thermodynamic fluctuation theory deals with two types of problems, and this division corresponds closely to the division of classical thermodynamics into a "thermostatics" of equilibrium and irreversible thermodynamics, or kinetics dealing with irreversible processes. I shall refer to these problems as *static* and *kinetic* fluctuation problems, respectively. In the former, one deals with the probability distribution of, say, the energy or mass density in a system in equilibrium. In the kinetic theory one is interested also in the temporal sequence, in the rhythm of the fluctuating values. The distribution of Brownian motion paths and the "thermal noise" in communication networks are instances where this temporal sequence of the fluctuations is important.

The basic paper of the static fluctuation theory is by Einstein [1], who computed the light scattering of the atmosphere. Kinetic fluctuation theory is summarized in the excellent review of Wang and Uhlenbeck [2]. The phenomena considered were Brownian motion and noise in electric networks. The connection with irreversible thermodynamics was established by Onsager and Machlup [3] and further developed among others by Tisza and Manning [4]. I shall present here the unified method of the last paper to which I shall refer as TM. This paper and also reference 3 contain additional details with bibliography.

2. Fluctuations and Probability

Let us consider a single fluctuating thermodynamic parameter α. The fluctuation is supposed to occur around $\alpha = 0$; the fluctuation path $\alpha(t)$ is qualitatively represented in Figure 1*a*. For concreteness sake we

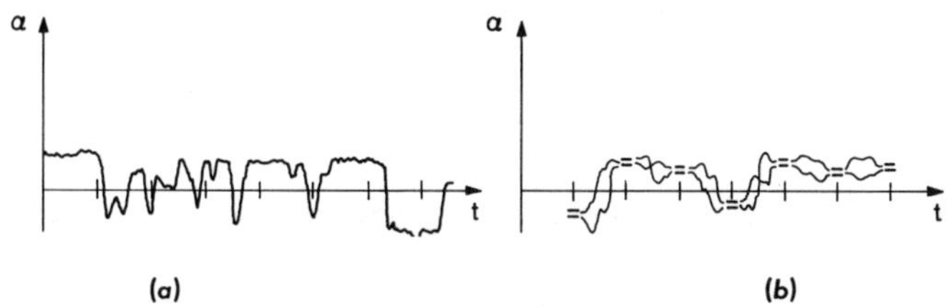

Fig. 1 (*a*) Typical fluctuation path, (*b*) specification of paths in terms of gates.

may think of α as the charge on a condenser connected through a resistance $\mathcal{R}$ at a temperature T; or else $\alpha = U$, the energy of a volume element in a closed system. The energy in equilibrium is taken to be the zero value.

For the purposes of this section the physical meaning of the random variable is irrelevant. We shall be concerned with the concepts of probability theory required for the discussion of fluctuations. When treated fully and rigorously, these concepts are of a rather sophisticated nature beyond the scope of elementary texts. Yet, the basic ideas can be explained in fairly simple terms.

Ideally one might try to obtain the probability distribution for the paths $\alpha(t)$. However, this problem has only a trivial solution: Every probability is zero. Just as we cannot ask for the probability of selecting a mathematical point in an area but have to bunch the points into finite domains, here we have to bunch the paths by what is sometimes called the gate method represented in Figure 1*b*. We put all the paths into the same bunch which pass at the discrete instances of time $t_1 < t_2 < \cdots < t_p$ within specified limits:

$$\alpha_k < \alpha(t_k) < \alpha_k + d\alpha_k, \qquad k = 1, 2, \cdots, p \tag{1}$$

The probability of finding a path within this bunch we denote by

$$W_p \begin{pmatrix} \alpha_1 & \cdots & \alpha_p \\ t_1 & \cdots & t_p \end{pmatrix} d\alpha_1 \cdots d\alpha_p \tag{2}$$

where W_p is a p-gate probability density function (p.d.f.).

The p.d.f. is nonnegative

$$W_p \geqq 0 \tag{3}$$

and normalized

$$\int \cdots \int W_p \, d\alpha_1 \cdots d\alpha_p = 1 \tag{4}$$

The further discussion is the simplest if the intervals of time

$$\tau_k = t_k - t_{k-1} \tag{5}$$

are long enough so that the values $\alpha(t_{k-1})$ and $\alpha(t_k)$ are statistically independent from each other. In this case

$$W_p \begin{pmatrix} \alpha_1 & \cdots & \alpha_p \\ t_1 & \cdots & t_p \end{pmatrix} = W_1 \begin{pmatrix} \alpha_1 \\ t_1 \end{pmatrix} \cdots W_1 \begin{pmatrix} \alpha_p \\ t_p \end{pmatrix} \tag{6}$$

Moreover, since the process is stationary in time,

$$W_1 \begin{pmatrix} \alpha_k \\ t_k \end{pmatrix} = W_1(\alpha_k) \tag{7}$$

This assumption leads to the static fluctuation theory and will be further developed in the next section.

If the intervals τ_k are short enough, the system will at t_k still "remember" the value of α at t_{k-1}. In order to describe this situation, we introduce the concept of conditional p.d.f. We define P by means of the relation

$$W_p \begin{pmatrix} \alpha_1 & \cdots & \alpha_p \\ t_1 & \cdots & t_p \end{pmatrix} = P \left(\begin{matrix} \alpha_p \\ t_p \end{matrix} \middle| \begin{matrix} \alpha_1 & \cdots & \alpha_{p-1} \\ t_1 & \cdots & t_{p-1} \end{matrix} \right) W_{p-1} \begin{pmatrix} \alpha_1 & \cdots & \alpha_{p-1} \\ t_1 & \cdots & t_{p-1} \end{pmatrix} \tag{8}$$

where P is the probability that $\alpha(t_p) = \alpha_p$, provided $\alpha_1(t_1) = \alpha_1, \cdots, \alpha(t_{p-1}) = \alpha_{p-1}$. In other words, everything to the right of the vertical bar is fixed.

Equation 8 is quite general; it is simply the definition of the conditional p.d.f. P. The various types of random processes can be classified in terms of assumptions with respect to P. Thus if the α are statistically independent from each other, we have

$$P \left(\begin{matrix} \alpha_p \\ t_p \end{matrix} \middle| \begin{matrix} \alpha_1 & \cdots & \alpha_{p-1} \\ t_1 & \cdots & t_{p-1} \end{matrix} \right) = W \begin{pmatrix} \alpha_p \\ t_p \end{pmatrix} = W(\alpha_p) \tag{9}$$

and we fall back to Equations 6 and 7.

The case which ranks next in complication is defined by the relation

$$P\begin{pmatrix} \alpha_p & \alpha_1 & \cdots & \alpha_{p-1} \\ t_p & t_1 & \cdots & t_{p-1} \end{pmatrix} = P\begin{pmatrix} \alpha_p & \alpha_{p-1} \\ t_p & t_{p-1} \end{pmatrix} \tag{10}$$

or, in other words: The probability of α_p at $t = t_p$ is determined if we know the value of α at $t = t_{p-1}$. Nothing is gained by specifying α also at $t_1 \cdots t_{p-2}$. Such processes are called Markov processes. They are the natural probabilistic generalizations of processes described by differential equations of first order in the time variable. For the latter, the initial state determines the trajectory, the present determines the future, and nothing is gained by specifying also the past. The same is true for Markov processes provided the *probability* of the future state is substituted for the exact knowledge of it.

Iterating Equation 8 and by using Equation 10, we obtain

$$W_p\begin{pmatrix} \alpha_1 & \cdots & \alpha_p \\ t_1 & \cdots & t_p \end{pmatrix} = P\begin{pmatrix} \alpha_p & \alpha_{p-1} \\ t_p & t_{p-1} \end{pmatrix} \cdots P\begin{pmatrix} \alpha_2 & \alpha_1 \\ t_2 & t_1 \end{pmatrix} W_1\begin{pmatrix} \alpha_1 \\ t_1 \end{pmatrix} \tag{11}$$

Since fluctuations around equilibrium are stationary in time, P depends only on the time lag $t_k = t_k - t_{k-1}$, and we can use the simplified notation

$$P(\alpha_k, \tau_k | \alpha_{k-1}) \equiv P\begin{pmatrix} \alpha_k & \alpha_{k-1} \\ t_k & t_{k-1} \end{pmatrix} \tag{12}$$

Clearly the entire process is known if the basic probabilities in Equations 7 and 12 are given.

Finally we define the p-gate conditional p.d.f. as

$$P_p(\alpha_p, \tau_p; \cdots; \alpha_2, \tau_2 | \alpha_1) \equiv \frac{W_p\begin{pmatrix} \alpha_1 & \cdots & \alpha_p \\ t_1 & \cdots & t_p \end{pmatrix}}{W_1(\alpha_1)} \tag{13}$$

From Equations 11, 12, and 13 we have

$$P_p(\alpha_p, \tau_p; \cdots; \alpha_2, \tau_2 | \alpha_1) = P(\alpha_p, \tau_p | \alpha_{p-1}) \cdots P(\alpha_2, \tau_2 | \alpha_1) \tag{14}$$

This is the basic chain relation of Markov processes and may be compared with the relation 6 for independent random variables. It is interesting to use the foregoing formalism to discuss the problem of microscopic reversibility and the irreversible trend toward equilibrium.

It is a matter of experimental fact and physical principles [3, 4] that the fluctuation paths of aged system which reached their equilibrium are symmetric with respect to future and past. On applying the transformation $t \rightarrow -t$, each path transforms into another possible path and the two are equally probable. In formulas:

$$W_2\begin{pmatrix} \alpha_1 & \alpha_2 \\ t_1 & t_2 \end{pmatrix} = W_2\begin{pmatrix} \alpha_2 & \alpha_1 \\ t_1 & t_2 \end{pmatrix} \tag{15}$$

From formulas 11, 12, and 7 we have

$$\frac{P(\alpha_2, \tau_2 | \alpha_1)}{P(\alpha_1, \tau_2 | \alpha_2)} = \frac{W(\alpha_2)}{W(\alpha_1)} \tag{16}$$

If $|\alpha_2| > |\alpha_1|$, then $W(\alpha_2) < W(\alpha_1)$, expressing the fact that larger deviations from equilibrium become less probable. Hence the ratio 16 is < 1 and

$$P(\alpha_2, \tau_2 | \alpha_1) < P(\alpha_1, \tau_2 | \alpha_2) \tag{17}$$

Thus reversibility is expressed by Equation 15 and the unidirectional trend to equilibrium by formula 17.

In order to obtain more specific results we have to know more about the basic probability 7 and 12. This will be achieved by thermodynamic methods.

3. Fluctuations and Thermodynamics

We proceed now to find thermodynamic expressions for the basic probabilities 7 and 12. The method is tentative; it is based on an intuitively recognized analogy between the probabilistic and thermodynamic quantities, and it is eventually justified by success.

We start with the case of $W_1(\alpha)$ which is well known in the literature. The argument runs somewhat as follows.

In thermodynamics, equilibrium is characterized by the maximum of the entropy; in other words, $S = S(\alpha)$ is to be maximized with respect to the variations of α. In statistics, on the other hand, it seems reasonable to expect that equilibrium corresponds to the maximum of the p.d.f. $W_1(\alpha)$. Therefore, one expects some relation among these quantities; i.e.,

$$W_1(\alpha) = f(S(\alpha)) \tag{18}$$

where f denotes some unknown functional relationship. Once a relation of this sort is assumed, the functional form of f is uniquely determined. In fact, let us consider two noninteracting systems. The composition rules for the entropy and the probability are

$$S(\alpha) = S'(\alpha') + S''(\alpha'') \tag{19}$$

$$W(\alpha) = W'(\alpha') \cdot W''(\alpha'') \tag{20}$$

where the single and double primes denote the two systems. The only function that satisfies the relations 18 through 20 is the exponential. Hence

$$S = k \ln W_1 + \text{const} \tag{21}$$

or

$$W_1(\alpha) = \text{const}\ e^{S(\alpha)/k} \tag{22}$$

where k is Boltzmann's constant, the value of which depends on the choice of units and is determined by experiment: $k = 1.37 \times 10^{-16}$ erg/degree, mole.

Equations 21 and 22 are, of course, equivalent, but they admit different interpretations.

Equation 21 is usually called Boltzmann's principle. It is used to compute the entropy from the statistical-mechanical evaluation of the probability W. On the other hand, Equation 22 was first used by Einstein [1] to compute the probability for such states where the entropy can be obtained by macroscopic methods.

We expand the entropy with respect to α:

$$S(\alpha) = S(0) + X\alpha - \tfrac{1}{2}s\alpha^2 + \cdots \tag{23}$$

where

$$X = \frac{dS}{d\alpha}, \qquad s = -\frac{d^2S}{d\alpha^2} \tag{24}$$

Since the entropy is maximum for $\alpha = 0$, we have

$$X(0) = 0, \qquad s > 0 \tag{25}$$

As an example let us interpret α as the energy of a system in equilibrium with a heat reservoir, the equilibrium energy being chosen as the zero value. In this case

$$X = \Delta\left(\frac{1}{T}\right), \qquad s = \frac{1}{TC_v} \tag{26}$$

where T is the absolute temperature, C_v the constant volume heat capacity, and the Δ refers to the difference between system and reservoir.

If α is the charge on a condenser, then

$$X = \Delta\frac{\varphi}{T}, \qquad s = \frac{1}{TC} \tag{27}$$

where $\Delta\varphi$ is electric potential difference and C is the capacitance.

From Equations 22, 23, and 25 we have

$$W_1(\alpha) = \text{const}\ e^{-(\Delta/2k)\alpha^2} \tag{28}$$

This is a Gaussian distribution, and the average is (cf. Appendix)

$$\langle\alpha^2\rangle = \frac{k}{s} \tag{29}$$

In case of the energy, $\alpha = \Delta U$, one has

$$\langle(\Delta U)^2\rangle = kC_vT^2 \tag{30}$$

It is noteworthy and characteristic of the thermodynamic method that the right-hand side of Equation 30 contains, apart from Boltzmann's constant, only macroscopic quantities.

We turn now to the more intricate problem of finding a thermodynamic analog to the conditional p.d.f. $P(\alpha_2, \tau|\alpha_1)$ [3, 4].

Let us assume at first that the nonequilibrium state α regresses to equilibrium according to the linear phenomenological relation

$$R\dot{\alpha} = X \tag{31}$$

with

$$X = \frac{dS}{d\alpha} = -s\alpha \tag{32}$$

hence

$$R\dot{\alpha} + s\alpha = 0 \tag{33}$$

Here R is a sort of resistance.

The solution of this equation is

$$\alpha = Ae^{-\gamma t} \tag{34}$$

with

$$\gamma = \frac{s}{R} \tag{35}$$

The foregoing macroscopic formalism is inadequate to describe fluctuations since it accounts only for the regression and not for the building up process.

We generalize Equation 34 formally by considering

$$\alpha = Ae^{-\gamma t} + Be^{\gamma t} \tag{36}$$

We shall refer to the two terms in Equation 36 as the forward and the reversed processes. The latter are "unnatural" processes precluded on the macroscopic level by the second law of thermodynamics. While Equation 36 exhibits the symmetry of forward and reverse processes necessary for describing fluctuations, we shall have to satisfy thermodynamics by assuring that the latter should have a smaller probability.

This can be achieved by means of a function

$$\mathcal{O}(\alpha) = \frac{1}{4R}(R\dot{\alpha} - X)^2 = \frac{R}{4}(\dot{\alpha} + \gamma\alpha)^2 \tag{37}$$

which is identical within a multiplying factor to that introduced by Onsager and Machlup [3]. We shall refer to it as the OM function.

This function has a number of remarkable properties: $\mathcal{O}(\alpha) \geqq 0$. The phenomenological path 34 is characterized by the fact that it is the only one to give $\mathcal{O}(\alpha)$ its minimum value, i.e., zero. Inserting instead path 36 we obtain

$$\mathcal{O}(\alpha) = \mathcal{O}(Be^{\gamma t}) = R\gamma^2B^2e^{2\gamma t} \tag{38}$$

which can be seen to be equal to the rate of entropy decrease associated with the reverse path. Consequently we have

$$\int_{t_1}^{t_2} \mathcal{O}(\alpha)\, dt = \tfrac{1}{2} s B^2 (e^{2\gamma t_2} - e^{2\gamma t_1})$$
$$= S_B(t_1) - S_B(t_2) = -\Delta S_B \tag{39}$$

where $-\Delta S_B$ is the entropy decrease associated with the reverse component of the path 36. Thus the OM function has a peculiar "filtering" property. The entropy decrease associated with the reverse component is registered while the forward component is "passed," leaving the corresponding entropy increase unrecorded. It is apparent that expression 39 has the property to express a bias against the reverse component in thermodynamic language. Thus we have a thermodynamic analog to the bias found in connection with the conditional p.d.f. in Equations 16 and 17. The precise formulation of this analogy takes some care, and for the details we have to refer to Section 4 of TM.

We confine ourselves to give the final result:

$$P(\alpha_2),\ \tau(\alpha_1) = \text{const} \exp\left[-\frac{1}{k} \int_{t_1}^{2} \mathcal{O}(\alpha_{\text{exp}})\, dt \right] \tag{40}$$

where α_{exp}, called the *exponentially smoothed* path is defined as follows:

$$\alpha_{\text{exp}}(t_1) = \alpha_1, \qquad \alpha_{\text{exp}}(t_2) = \alpha_2$$

For $t_1 < t < t_2$ α_{exp} has the form 36. The coefficients are found by simple algebra:

$$A = \frac{\alpha_1 e^{-\gamma\tau} - \alpha_2}{e^{\gamma\tau} - e^{-\gamma\tau}}, \qquad B = \frac{\alpha_2 - \alpha_1 e^{-\gamma\tau}}{e^{\gamma\tau} - e^{-\gamma\tau}} \tag{41}$$

where $t = t_2 = t_1$.

We obtain now from Equation 40 by using Equations 36, 41, 37, and 38

$$P(\alpha_2, \tau \mid \alpha_1) = \text{const} \exp\left[-\frac{s}{2k} \frac{(\alpha_2 - \alpha_1 e^{-\gamma\tau})^2}{1 - e^{-2\gamma\tau}} \right] \tag{42}$$

This relation was first derived in connection with Brownian motion theory by Uhlenbeck and Ornstein [5]. It served as the starting point for the derivation of Equation 40 by Onsager and Machlup [3].

Equation 42 has a simple intuitive meaning. Let a number of systems with $\alpha_1(t_1) = \alpha_1$ be released at the time t_1. Both the average and the most probable evolution follows the phenomenological relation $\alpha_2(\tau) = \alpha_1(0)e^{-\gamma\tau}$ where we have set $t_1 = 0$ for simplicity. The actual paths will be distributed around the most probable one according to a Gaussian distribution. The variance starts with zero (we have released the systems in sharply defined states) and for $\gamma\tau \gg 1$ becomes k/s, the normal variance of α.

4. The Noise Spectrum

Equation 40 has a number of applications [3, 4, 6, 7]. Perhaps the most important point is that the present method can be combined with the method of Rice [5] which uses the Fourier representation of the fluctuation path

$$\alpha(t) = \sum_{n=1}^{\infty} (a_n \cos \omega_n t + b_n \sin \omega_n t) \tag{43}$$

with

$$\omega_n = \frac{2\pi n}{\theta} \tag{44}$$

where n is an integer and θ is a long interval of time.

We introduce the probability

$$f(a_1, a_2, \cdots; b_1, b_2, \cdots)\, da_1\, da_2 \cdots db_1\, db_2 \cdots \tag{45}$$

that an observed $\alpha(t)$ has Fourier coefficients within the limits a_i and $a_i + da_i$; b_i and $b_i + db_i$. Expression 45 is equal to [4, 6]

$$f = \text{const} \exp \left[-\frac{1}{k} \int_0^{\theta} \mathcal{O}\{\alpha(t)\}\, dt \right] \tag{46}$$

where $\alpha(t)$ is given by Equation 43. This procedure results in a Gaussian distribution for the Fourier coefficients.

A straightforward calculation yields (cf. Section 6 of TM):

$$\langle \alpha^2 \rangle = \frac{2k}{\pi} \int_0^{\infty} \frac{R\, d\omega}{R^2\omega^2 + s^2} \tag{47}$$

Equation 47 is often expressed in the form that the fluctuation path $\alpha(t)$ has the spectral density

$$G_\alpha(\omega) = \frac{2k}{\pi} \frac{R}{R^2\omega^2 + s^2} \tag{48}$$

It is interesting to introduce

$$R\dot{\alpha} + s\alpha = \epsilon \tag{49}$$

In the absence of fluctuation $\epsilon = 0$. Hence ϵ is a measure of an apparent force which would produce the observed fluctuations.

Equation 49 has a particularly intuitive meaning in the electrical analogy: an RC network. We have for an RC network (see also Equation 27):

$$\mathcal{R}\dot{q} + \frac{q}{C} = V \tag{50}$$

where $\mathcal{R} = RT$, $q = \alpha$, $C = (1/Ts)$, $V = \epsilon T$.

Here q is the charge, $\mathcal{R}$ the electrical resistance, C the capacitance, and V the so-called noise voltage originating in a fictitious noise generator which has an effect equivalent to the thermal fluctuations.

We have from Equations 48, 49, and 50 for the spectral density of the noise generator

$$G_\alpha(\omega) = \frac{2}{\pi} kR \tag{51}$$

or

$$G_V(\nu) = 2\pi G_V(\nu) = 4kTR \tag{52}$$

In the last formula we have replaced ω by $2\pi\nu$ in order to obtain the well-known Nyquist formula [8] of thermal noise in its conventional form.

References

1. A. Einstein, *Ann. Physik* **33**, 1275 (1910).
2. M. C. Wang and G. E. Uhlenbeck, *Rev. Mod. Phys.* **17**, 323 (1945).
3. L. Onsager and S. Machlup, *Phys. Rev.* **91**, 1505 and 1512 (1953).
4. L. Tisza and I. Manning, *Phys. Rev.* **105**, 1695 (1957). (Paper 10 of this volume.)
5. G. E. Uhlenbeck and L. S. Ornstein, *Phys. Rev.* **36**, 823 (1930).
6. N. Hashitsume, *Prog. Theoret. Phys.* (*Japan*) **15**, 369 (1956).
7. A. Siegel, *Phys. Rev.* **106**, 609 (1957).
8. H. Nyquist, *Phys. Rev.* **32**, 110 (1928).

APPENDIX. Averages for Gaussian Distributions

Let

$$W(\alpha) = \text{const } e^{-1/2\, \lambda\alpha^2} \tag{A.1}$$

be a Gaussian p.d.f. We compute the variance defined as

$$\langle \alpha^2 \rangle = \frac{\int_{-\infty}^{+\infty} \alpha^2 e^{-1/2\, \lambda\alpha^2}\, d\alpha}{\int_{-\infty}^{+\infty} e^{-1/2\, \lambda\alpha^2}\, d\alpha} \tag{A.2}$$

We denote

$$I(\lambda) = \int_{-\infty}^{+\infty} e^{-1/2\, \lambda\alpha^2}\, d\alpha \tag{A.3}$$

If we introduce $\lambda^{1/2}\alpha = x$ as integration variable,

$$I(\lambda) = \int_{-\infty}^{+\infty} e^{-1/2\, x^2}\, dx \lambda^{-1/2} = \text{const } \lambda^{-1/2} \tag{A.4}$$

From Equations A.2, and A.3

$$\langle \alpha^2 \rangle = -2 \frac{d \ln I}{d\lambda}$$

By Equation A.4

$$\frac{d \ln I}{d\lambda} = \frac{1}{2\lambda}$$

Hence (A.5)

$$\langle \alpha^2 \rangle = \frac{1}{\lambda}$$

Reprinted from THE PHYSICAL REVIEW, Vol. 105, No. 6, 1695–1705, March 15, 1957
Printed in U. S. A.

Fluctuations and Irreversible Thermodynamics*

LASZLO TISZA AND IRWIN MANNING†
Department of Physics and the Research Laboratory of Electronics, Massachusetts Institute of Technology, Cambridge, Massachusetts
(Received August 16, 1956)

The time-dependent theory of fluctuations is based on a combined application of the phenomenological theory of dissipation and the stochastic theory of random processes. The traditional method of joining these theories into a uniform scheme proceeds by adding a random perturbation to the differential equation of the phenomenological kinetic theory (the Langevin equation in case of Brownian motion). In the present approach the problem is considered as an essentially statistical one. The role of the differential equation is to fix the form of the distribution function over the manifold of fluctuation paths in function space. The solutions of the equation constitute the most probable region in function space and the fluctuations appear with their appropriate probabilities. The connection between the phenomenological equation and the distribution function is stipulated by means of a postulate, the essential ingredient of which is the auxiliary function recently introduced by Onsager and Machlup. Heuristically this postulate was suggested by the kinetic analog of Boltzmann's principle established by these authors. In the present logical structure it is sufficient to join this postulate to the standard assumptions of the phenomenological and the stochastic theories for the derivation of the entire time-dependent fluctuation theory. A number of statements usually postulated appear as theorems in this presentation. The reversible and irreversible aspects of time play an essential part in the argument. The calculation of fluctuations is carried out in the temporal and the spectral descriptions. The relation of the two schemes is discussed along with the scope and limitations of the theory.

1. INTRODUCTION

THE thermodynamic theory of fluctuations deals with two types of problems. On a simpler level one considers the fluctuation of a quantity as registered at a single instant of time over an ensemble of systems.

In the problems of the second type one observes each member of the ensemble over a time interval rather than a single instant, yielding a fluctuation path $\alpha=\alpha(t)$ for each system; we are concerned with the distribution of these paths over an ensemble.

This classification of fluctuation problems corresponds closely with the division of phenomenological thermodynamics into *thermostatics* dealing with equilibrium and *irreversible thermodynamics;* or *kinetics* devoted to the study of time-dependent dissipative processes.

Whenever ambiguity might arise we shall qualify a fluctuation problem as "time-independent," "static" or "thermostatic," on the one hand and as "time-dependent" or "kinetic" on the other.

The time-independent theory is based on Einstein's[1] interpretation of Boltzmann's principle. The definition of the entropy function is extended to a class of non-equilibrium states and used to establish a probability density function (p.d.f.).

Recently Onsager and Machlup[2,3] proved a theorem which is the kinetic analog of the Boltzmann-Einstein principle. They established a probability distribution function (p.d.f.) for fluctuation paths in terms of an auxiliary function (henceforth: the OM function), which in turn was defined by means of the phenomenological kinetic equation.

The Boltzmann principle plays a central role in the time-independent theory. It allows one to compute fluctuation moments and, more generally, serves as a point of departure for the theory of the generalized canonical ensembles.

In this paper we embark on a program in which the Onsager-Machlup theorem is used as a similar organizing principle for the development of the time-dependent theory.

According to the central idea of this theory there is close connection between fluctuation and dissipation phenomena.[2–10] In the usual approach this connection is accounted for by the method of stochastic differential equations. The phenomenological equations are augmented by a random term producing the correct amount of fluctuation. The equation obtained in case of Brownian motion is called the Langevin equation.[2,6]

We shall show in this paper that the same problem can be attacked also from another point of view. Instead of focusing attention on differential equations, one considers the set of fluctuation paths and describes various physical situations in terms of appropriate distribution functions. The statistical point of view

* This paper is a development of the Ph.D. thesis submitted by Irwin Manning to the Department of Physics, Massachusetts Institute of Technology. The work was supported in part by the Army (Signal Corps), the Air Force (Office of Scientific Research, Air Research and Development Command), and the Navy (Office of Naval Research).

† Present address: Physics Department, Syracuse University, Syracuse, New York.

[1] A. Einstein, Ann. Physik 33, 1275 (1910).
[2] L. Onsager and S. Machlup, Phys. Rev. 91, 1505 (1953).
[3] S. Machlup and L. Onsager, Phys. Rev. 91, 1512 (1953).
[4] A. Einstein, Ann. Physik 17, 549 (1905).
[5] P. Langevin, Compt. rend. 146, 520 (1908).
[6] H. Nyquist, Phys. Rev. 32, 110 (1928).
[7] L. Onsager, Phys. Rev. 37, 405 (1931) and 38, 2265 (1931).
[8] H. B. Callen and R. F. Greene, Phys. Rev. 86, 702 (1952).
[9] M. C. Wang and G. E. Uhlenbeck, Revs. Modern Phys. 17, 323 (1945).
[10] N. Hashitsume, Progr. Theoret. Phys. (Japan) 8, 461 (1952) and *Proceedings of the International Conference of Theoretical Physics, Kyoto and Tokyo, September, 1953* (Science Council of Japan, Tokyo, 1954), p. 495.

inherent in this approach is similar to the one taken in the static fluctuation theory. However, the situation is complicated by the emergence of time as an independent variable. To clarify some of the problems connected with the role of time is the main purpose of this paper.

Sections 2 and 3 contain a short summary of the basic definitions and postulates of the phenomenological theory of dissipation and the stochastic theory of fluctuations. In Sec. 4 we formulate a postulate which allows us to combine these disparate elements into a unified theory of kinetic fluctuations. We shall arrive at a derivation of the OM p.d.f. which differs from the original proof of Onsager and Machlup. The behavior of the OM function on time reversal plays an essential part in the argument.

We then go on to deal with the calculation of averages. The theory exhibits a duality corresponding to two methods for the description of fluctuation paths. In the *temporal description* the values of the fluctuating quantity are specified within narrow limits at closely spaced time intervals. In the *spectral description* one deals with Fourier coefficients considered as random variables.

We shall show that the OM p.d.f. has two equivalent formulations, one in the temporal, the other in the spectral description. We demonstrate the application of the p.d.f. in both schemes.

The theory will be developed within certain restrictions to be specified in detail later. Some of these assumptions, such as the limitation to a single fluctuating variable, are made primarily in the interest of simplicity. We hope to make extensions to more general situations in a future paper. On the other hand, assumptions such as the limitations to Gaussian Markoff processes and to linear phenomenological equations represent more essential restrictions on the scope of the present method.

2. PHENOMENOLOGICAL RELATIONS

a. Thermostatistics

Let us consider a closed thermodynamic system which is specified in equilibrium by its fundamental equation

$$S=S(Q_1,Q_2,\cdots Q_r), \tag{2-1}$$

where S is the entropy and the Q_i are a set of extensive variables obeying conservation laws. Associated with each variable Q_i is a conjugate thermostatic force,

$$F_i=\partial S/\partial Q_i, \quad i=1,\,2,\,\cdots r. \tag{2-2}$$

The most important pairs of conjugate variables are U, $1/T$; N_γ, $-\mu_\gamma/T$; q, $-\varphi/T$ where U is the internal energy, T is the absolute temperature, N_γ the number of moles of the chemical species γ, μ_γ the chemical potential of this species, q the electric charge, and φ the electrostatic potential.

The thermostatic theory based on Eq. (2-1) applies only to systems which have reached their state of thermodynamic equilibrium. In order to extend the formalism to *processes*, we use the device of *composite systems*. The original system is partitioned into closed subsystems each of which is allowed to reach equilibrium. A process is initiated by the relaxation of an internal constraint; it consists of the redistribution of the quantities Q_i among the subsystems, a transition from a constrained to an unconstrained (or less constrained) equilibrium.

In this paper we shall consider only the one-variable case in which a single quantity Q is redistributed between two subsystems.

The entropy of the composite system is in this case

$$S(Q';Q'')=S'(Q')+S''(Q''), \tag{2-3}$$

where S' and S'' refer to the subsystems. For all processes in the composite system,

$$Q'+Q''=\text{constant}.$$

The deviation from unconstrained equilibrium is described in terms of the variable

$$\alpha=Q'-Q_e'=-(Q''-Q_e''), \tag{2-4}$$

where Q_e, Q_e' are the equilibrium values.

We define the thermodynamic force as

$$X=\frac{\partial S}{\partial\alpha}=\frac{\partial S'}{\partial Q'}-\frac{\partial S''}{\partial Q''}=F'-F''. \tag{2-5}$$

In unconstrained equilibrium $X=0$.

We note that the process corresponding to the differentiation (2-2) leaves the system homogeneous, but not closed, while the process of increasing Q is performed. In contrast, (2-5) refers to an internal displacement within a closed, but no longer homogeneous system.

We shall need the expansion of the total entropy around equilibrium

$$S=S_0-\tfrac{1}{2}s\alpha^2+\cdots. \tag{2-6}$$

Here

$$s=-\frac{\partial^2 S}{\partial\alpha^2}=-\frac{\partial^2 S'}{\partial Q'^2}-\frac{\partial^2 S''}{\partial Q''^2}. \tag{2-7}$$

It is a consequence of the entropy maximum principle that $s>0$. If the primed system is a reservoir then

$$\frac{\partial^2 S'}{\partial Q'^2}=\frac{\partial F'}{\partial Q'}=0.$$

There are two obvious physical interpretations of the single-variable theory. First, we have $Q=U$, $F=1/T$, and $s=1/(T^2C_v)$, where C_v is the constant volume heat capacity; the process is heat conduction under conditions of negligible thermal expansion ($C_v \simeq C_p$).

Second, we have $Q=q$, $F=\varphi/T$, $s=1/(TC)$, where C is the electric capacity; the process is electric conduction.[11]

In all that follows, we shall assume that the composite system is close enough to unconstrained equilibrium (α small) so that the series (2-6) can be broken off after the second term. To within this approximation the thermodynamic force (2-5) becomes

$$X=-s\alpha. \tag{2-5a}$$

b. Kinetics

We now turn to the temporal aspects of the regression process. From the thermostatic point of view we can merely say that relaxation of a constraint leads eventually to a new equilibrium with an entropy which is not smaller than the initial entropy.

In irreversible thermodynamics, or kinetics, we describe the rate of the regression process in terms of a phenomenological equation connecting the forces and fluxes. We assume this relation to be linear. Hence, in the single variable theory we have

$$R\dot{\alpha}=X. \tag{2-8}$$

Here X is defined by (2-5), $\dot{\alpha}$ is the current or "flux" within the composite system and R is the resistance of the partition or of the connecting wire which establishes the coupling between the two systems. It is assumed that this resistor does not contribute significantly to the entropy of the composite system.

In view of (2-5a) the kinetic equation can be written also as

$$R\dot{\alpha}+s\alpha=0. \tag{2-8a}$$

The form (2-8) or (2-8a) of the kinetic equation involves a number of simplifying assumptions. First, the thermostatic formalism developed for constrained equilibria is assumed to hold even in the absence of constraints. One speaks in this case of *local equilibrium*. This assumption may fail far from equilibrium and in the presence of relaxation phenomena.[12]

The second simplification inherent in (2-8a) is the linearity of the relation. This may fail not only for large values of the current, but also for small ones. An extreme instance for such a behavior is a superconductor in which Ohm's law breaks down as the current falls below a critical value.

Finally, we have assumed also that the forces governing the variations of α are independent of the "velocities" $\dot{\alpha}$, i.e., there are no external magnetic fields or Coriolis forces present, nor is the entropy dependent on $\dot{\alpha}$.

The solutions of (2-8a) are of the form

$$\varphi(t)=Ae^{-\gamma t}, \tag{2-9}$$

with

$$\gamma=s/R, \tag{2-10}$$

and A a constant.

We consider now the function

$$\mathcal{O}=\frac{1}{4R}(R\dot{\alpha}-X)^2=\frac{R}{4}(\dot{\alpha}+\gamma\alpha)^2, \tag{2-11}$$

which is identical within a multiplying factor to that introduced by Onsager and Machlup.[2] We shall refer to it as the OM function.

Obviously $\mathcal{O}\geqq 0$ and $\mathcal{O}_{\min}=\mathcal{O}(\varphi)=0$, i.e., $\mathcal{O}$ vanishes if and only if its argument is a solution of Eq. (2-8a).

The OM function can be written also in the form

$$\mathcal{O}=\tfrac{1}{2}[\phi(\dot{\alpha})+\psi(X)-\dot{S}], \tag{2-12}$$

where

$$\phi\equiv\tfrac{1}{2}R\dot{\alpha}^2, \tag{2-13}$$

$$\psi\equiv\tfrac{1}{2}R^{-1}X^2, \tag{2-14}$$

and

$$\dot{S}=\frac{\partial S}{\partial\alpha}\dot{\alpha}=X\dot{\alpha}. \tag{2-15}$$

The quantities ϕ and ψ are the dissipation functions, and their sum is the rate of total entropy production.

For processes which satisfy the kinetic equation,

$$\phi(\varphi)=\psi(X)=\tfrac{1}{2}\dot{S}. \tag{2-16}$$

On the basis of this relation the quantities 2ϕ and 2ψ are sometimes used interchangably with the entropy production $\dot{S}$. Actually, however, (2-16) is no longer valid if fluctuation paths are taken into consideration.

An extension of the phenomenological theory to include fluctuations will be developed in Sec. 4. Meanwhile, in the next section we summarize the definitions and postulates involving probability functions needed for that purpose.

3. STOCHASTIC PROCESSES

In this section we confine ourselves to a brief formulation of the basic definitions and postulates, appropriate to our case, which will provide the mathematical equipment for dealing with stochastic processes. For an integration of this formalism into a wider context we refer to Onsager and Machlup.[2,3,13]

[11] This statement calls for some qualification. A displacement of electric charge is always associated with a transfer of energy. Hence, we have a coupled-flux problem. This point of view is the basis for the theory of thermoelectric effects. However, if all parts of the composite system are maintained at a constant temperature, the energy transfer need not be made explicit in the theory. This will be justified from the general thermodynamic formalism in a subsequent paper.

[12] It is interesting to note that relaxation effects become amenable to treatment in a limiting case which is in a way complementary to that of local equilibrium. Thus in the case of a spin-lattice system, there may be approximate equilibrium within the spin system and within the phonon system in the face of a local disequilibrium between the two.

[13] Further details concerning the formalism will be found also in Wang and Uhlenbeck's review on the theory of Brownian motion (reference 9). There is a considerable formal analogy between the kinetic fluctuation theory and the theory of Brownian motion paths.

We introduce the notation that

$$W_p\begin{pmatrix}\alpha^{(1)} & \cdots & \alpha^{(p)} \\ t_1 & \cdots & t_p\end{pmatrix} d\alpha^{(1)}\cdots d\alpha^{(p)} \tag{3-1}$$

be the probability that the fluctuation path $\alpha(t)$ satisfies the following specifications: at the times $t_1 < t_2 \cdots < t_p$

$$\alpha^{(k)} \leq \alpha(t_k) \leq \alpha^{(k)} + d\alpha^{(k)}, \quad k=1, 2, \cdots p, \tag{3-2}$$

where $\alpha^{(1)}, \cdots, \alpha^{(p)}; d\alpha^{(1)}, \cdots, d\alpha^{(p)}$ are given numbers. We say that $\alpha(t)$ is to pass through the "gates" of infinitesmal width specified by (3-2), and we will call the above equations a *finite gate specification.* W_p is called the joint, or p-gate, probability density function (p.d.f.) for the fluctuation path.

We proceed to formulate some basic assumptions, more or less standard in the theory of stochastic processes.

Postulate I: We assume that *the process is Markoffian;* that is, that W_p can be written in the form

$$W_p\begin{pmatrix}\alpha^{(1)} & \cdots & \alpha^{(p)} \\ t_1 & \cdots & t_p\end{pmatrix} = P_2\left(\begin{array}{c|c}\alpha^{(p)} & \alpha^{(p-1)} \\ t_p & t_{p-1}\end{array}\right) W_{p-1}\begin{pmatrix}\alpha^{(1)} & \cdots & \alpha^{(p-1)} \\ t_1 & \cdots & t_{p-1}\end{pmatrix}, \tag{3-3}$$

where

$$P_2\left(\begin{array}{c|c}\alpha^{(p-1)} & \alpha^{(p)} \\ t_{p-1} & t_p\end{array}\right) d\alpha^{(p)}$$

s the probability that if $\alpha(t_{p-1}) = \alpha^{(p-1)}$ then

$$\alpha^{(p)} \leq \alpha(t_p) \leq \alpha^{(p)} + d\alpha^{(p)}.$$

P_2 is called the conditional p.d.f.

Postulate II: The process is stationary, i.e., W_p and hence, P_2 depend only on the time lags

$$\tau_k = t_k - t_{k-1}, \quad k=2, \cdots, p. \tag{3-4}$$

The above postulate suggests the following simplified notation:

$$W_1(\alpha^{(1)}) \equiv W_1\begin{pmatrix}\alpha^{(1)} \\ t_1\end{pmatrix}. \tag{3-5}$$

$$P_2(\alpha^{(2)}, \tau_2 | \alpha^{(1)}) \equiv P_2\left(\begin{array}{c|c}\alpha^{(2)} & \alpha^{(1)} \\ t_2 & t_1\end{array}\right). \tag{3-6}$$

Postulate III: The fluctuations satisfy the principle of *microscopic reversibility:*

$$W_p\begin{pmatrix}\alpha^{(1)} & \cdots & \alpha^{(p)} \\ t_1 & \cdots & t_p\end{pmatrix} = W_p\begin{pmatrix}\alpha^{(p)} & \cdots & \alpha^{(1)} \\ t_1 & \cdots & t_p\end{pmatrix}. \tag{3-7}$$

Since the W's and P's are probability densities, they are normalized, non-negative, and satisfy the consistency relations

$$W_k\begin{pmatrix}\alpha^{(1)} & \cdots & \alpha^{(k)} \\ t_1 & \cdots & t_k\end{pmatrix} = \int \cdots \int W_p\begin{pmatrix}\alpha^{(1)} & \cdots & \alpha^{(p)} \\ t_1 & \cdots & t_p\end{pmatrix} d\alpha^{(k+1)} \cdots d\alpha^{(p)}, \tag{3-8}$$

where $k < p$.

It follows from (3-3) that

$$W_p\begin{pmatrix}\alpha^{(1)} & \cdots & \alpha^{(p)} \\ t_1 & \cdots & t_p\end{pmatrix} = P_2(\alpha^{(p)}, \tau_p | \alpha^{(p-1)}) \cdots P_2(\alpha^{(2)}, \tau_2 | \alpha^{(1)}) W_1(\alpha^{(1)}). \tag{3-9}$$

Thus the entire process is determined by the two basic probabilities W_1 and P_2. However, these are not independent of each other, but are related according to (3-7) and (3-9) as

$$P_2(\alpha^{(2)}, \tau_2 | \alpha^{(1)}) W_1(\alpha^{(1)}) = P_2(\alpha^{(1)}, \tau_2 | \alpha^{(2)}) W_1(\alpha^{(2)}). \tag{3-10}$$

We shall use also the p-gate conditional probability density defined as

$$P_p(\alpha^{(p)}, \tau_p; \cdots; \alpha^{(2)}, \tau_2 | \alpha^{(1)}) \equiv W_p\begin{pmatrix}\alpha^{(1)} & \cdots & \alpha^{(p)} \\ t_1 & \cdots & t_p\end{pmatrix} \Big/ W_1(\alpha^{(1)}). \tag{3-11}$$

Accordingly, we write the Markoff chain relation (3-9) in the form

$$P_p(\alpha^{(p)}, \tau_p; \cdots \alpha^{(2)}, \tau_2 | \alpha^{(1)}) = P_2(\alpha^{(p)}, \tau_p | \alpha^{(p-1)}) \cdots P_2(\alpha^{(2)}, \tau_2 | \alpha^{(1)}). \tag{3-12}$$

Using the terminology of statistical mechanics, we shall also refer to the probability density functions W_p and P_p as "distributions."

The foregoing postulates form a precise basis for further deductions. Before proceeding on this line of thought, however, it is natural to ask the question: Under what conditions can we expect that the distributions defined in terms of the above postulates correspond to the actual behavior of physical systems? We do not wish to discuss this delicate question in any detail at this point, but the following hints will convey an idea of the factors involved.

The strongest assumption is the Markoffian postulate I. It requires in the first place that the system has a short "memory", in the sense that a knowledge of the values of $\alpha(t)$ at the times t_{k-1}, t_k yields information of no greater value in predicting $\alpha(t_{k+1})$ than if we knew $\alpha(t_k)$ alone. This is apparently a good approximation if the time interval $t_k - t_{k-1}$ is long compared to the mean collision time. Systems exhibiting slow relaxation processes and frozen-in molecular disorder are omitted from consideration.

The Markoffian assumption also requires that the number of α variables chosen is adequate for the specification of the statistical properties of the system. This means that the application of the present one-variable theory is severely limited. The rules governing the choice of variables will be considered in another paper dealing with more general systems.

Finally, the fact that the formalism operates in terms of probabilities renders it inadequate for the description of effects in which the quantum mechanical interference of *probability amplitudes* manifests itself in macroscopic phenomena. Hence such phenomena as superconductivity and superfluidity are presumably beyond the scope of the present formalism.[14]

4. DERIVATION OF THE DISTRIBUTION FUNCTION

The previous section contains the formal framework for the description of a general class of random processes. In order to interpret the latter as thermodynamic fluctuations we shall express the basic probability densities W_1 and P_2 in terms of the phenomenological coefficients of Sec. 2.

This junction of the phenomenological and stochastic theories into a unified theory of fluctuation and dissipation calls for the formulation of two postulates. The first of these is well known.[1] It refers to W_1 and we summarize the argument only since it casts light on the subsequent discussion of the p.d.f. P_2.

Postulate IV: The p.d.f. $W_1(\alpha)$ can be expressed as a function of the entropy function:

$$W_1(\alpha)=f(S(\alpha)). \tag{4-1}$$

This postulate is motivated by the intuitively recognized parallel between the one-gate p.d.f. and the entropy function.[15] It is justified by the correctness of its implications.

Once the qualitative relation (4-1) is granted, one concludes in well-known fashion that the unknown function f is of the form

$$W_1(\alpha)\propto \exp[-S(\alpha)/k], \tag{4-2}$$

where k is Boltzmann's constant. On using the expansion (2-6), we obtain the distribution law in the Gaussian approximation. After normalization we have

$$W_1(\alpha)=(s/2\pi k)^{\frac{1}{2}}\exp(-s\alpha^2/2k). \tag{4-3}$$

From here the variance of α is

$$\langle\alpha^2\rangle=ks^{-1}. \tag{4-4}$$

The notation on the left-hand side indicates an ensemble average.

The discussions of this paper will be confined to Gaussian distributions. In particular, we shall use (4-3) rather than the exact distribution (4-2). The scope and limitations of this approximation will be considered in the final discussion.

We proceed now to formulate a postulate which is the kinetic analogue of (4-1) and which expresses a qualitative connection between the conditional p.d.f. P_p and the OM function. In order to render the analogous properties of these functions apparent and, in fact, even to state the postulate in a precise fashion, we need an auxiliary concept to be referred to as the *exponentially smoothed fluctuation path.* This is introduced in a natural fashion in the context of the problem of time reversal.

It is a generally accepted principle[7,16] that fluctuation paths around thermodynamic equilibrium are symmetric with respect to the transformation which reverses the sense of time: $t\rightarrow -t$. We shall denote the corresponding operator by K. This principle finds its expression in the symmetry of the *joint* probability densities W_p (postulate III). On the other hand, assuming that $\alpha^{(1)}>\alpha^{(2)}$, one obtains from (3-10) and (4-3)

$$P_2\begin{pmatrix}\alpha^{(2)} & \alpha^{(1)}\\ t_2 & t_1\end{pmatrix}>P_2\begin{pmatrix}\alpha^{(1)} & \alpha^{(2)}\\ t_2 & t_1\end{pmatrix}. \tag{4-5}$$

Hence, the irreversible trend toward equilibrium is expressed in terms of the *conditional* probabilities.

This reconciliation of the reversible aspect of fluctuation and the irreversible trend toward equilibrium within a discrete stochastic model has been known for a long time.[16] However, the situation is less clear within the continuum model of the phenomenological theory. This gap can be filled by utilizing a unique property of the OM function.

We start from the fact that the α variables, considered as averages over molecular parameters, are invariant under time reversal: $K\alpha=\alpha$ (reference 7). The thermodynamic force is defined by (2-5a) as $X\equiv -s\alpha$; hence $KX=X$. In contrast, the fluxes transform according to $K\dot{\alpha}=-\dot{\alpha}$. They are β variables in Casimir's terminology.[17]

We have seen in Sec. 2 that the OM function is minimized by the phenomenological curve $\varphi(t)$ of (2-9). The concept of minimization implies that we may insert into $\mathcal{O}$ virtual paths which do not satisfy the phenomenological equation. We shall construct such virtual paths by means of the time reversal operator. We obtain at once $K\varphi(t)=Ae^{\gamma t}$ and shall refer to $\varphi(t)$ and $K\varphi(t)$ as the forward and reversed paths, respectively.

The dissipation functions ϕ and ψ are positive definite and are invariant under time reversal. In contrast $K\dot{S}=-\dot{S}$; in other words, the reverse paths are associated with a decrease of total entropy. These are "unnatural" processes excluded by the second law

[14] F. London, *Superfluids* (John Wiley and Sons, Inc., New York, 1950), Vol. 1, p. 142.

[15] We note that the entropy function corresponding to the local equilibrium state α is a random function. This is in contrast to the thermodynamic entropy of the equilibrium state.

[16] P. Ehrenfest and T. Ehrenfest, Encykl. Math. Wiss. 4, 2, ii (1911), and Physik. Z. **8**, 311 (1907).

[17] H. B. G. Casimir, Revs. Modern Phys. **17**, 343 (1945).

of thermodynamics. Therefore we obtain from (2-13) and (2-17)

$$\mathcal{O}(\varphi)=0, \quad \mathcal{O}(K\varphi)=-\dot{S}>0; \tag{4-6}$$

or, expressed in words: *The OM function vanishes for forward paths and is equal to the rate of entropy decrease for the reverse paths.*[3]

This selective property of the OM function with respect to the sense of time remains valid even under more general conditions. Let us consider the virtual path

$$\alpha=Ae^{-\gamma t}+Be^{\gamma t}, \tag{4-7}$$

where A and B are constants, and γ is defined by (2-10). We have

$$\mathcal{O}(\alpha)=\mathcal{O}(Be^{\gamma t})=R\gamma^2B^2e^{2\gamma t}, \tag{4-8}$$

and

$$\int_{t_1}^{t_2}\mathcal{O}(\alpha)dt=\tfrac{1}{2}sB^2(e^{2\gamma t_2}-e^{2\gamma t_1})$$
$$=S_B(t_1)-S_B(t_2)=-\Delta S_B, \tag{4-9}$$

where $-\Delta S_B$ is the entropy decrease associated with the reverse component of the path (4-7).

Equations (4-6) to (4-9) show that the OM function has a peculiar "filtering" property. The entropy decrease associated with the reverse component is registered while the forward component is "passed," leaving the corresponding entropy increase unrecorded.

Motivated by the above, we associate with each gate specification (3-2) an *exponentially-smoothed path* $\alpha(t)$, defined by the $(p-1)$ equations

$$\alpha_{\exp}(t)\equiv A_ke^{-\gamma(t-t_{k-1})}+B_ke^{\gamma(t-t_{k-1})} \tag{4-10}$$

in the intervals

$$t_{k-1}\leq t\leq t_k, \quad k=1,\cdots,p-1,$$

where γ is defined by (2-10). The $2p$ constants A_k, B_k are to be determined by the conditions

$$\alpha_{\exp}(t_k)=\alpha^{(k)}, \quad k=1,\cdots,p. \tag{4-11}$$

There is a one-to-one correspondence between the gate specifications and the smoothed paths. One finds

$$A_k=(\rho_k^{-1}-\rho_k)^{-1}(\alpha^{(k-1)}\rho_k^{-1}-\alpha^{(k)}),$$
$$B_k=(\rho_k^{-1}-\rho_k)^{-1}(\alpha^{(k)}-\alpha^{(k-1)}\rho_k), \tag{4-12}$$

where

$$\rho_k=e^{-\gamma\tau_k}, \quad \tau_k=t_k-t_{k-1}. \tag{4-13}$$

The exponentially-smoothed path enables one to take advantage of the remarkable "filtering" property of the OM function expressed by (4-9). We obtain

$$\int_{t_1}^{t_p}\mathcal{O}\{\alpha_{\exp}(t)\}dt=-\sum_{k=2}^{p}\Delta S_B(\tau_k), \tag{4-14}$$

where $S_B(\tau_k)$ is the entropy decrease associated with the reversed component of $\alpha(t)$ in the time interval τ_k.

The expression (4-14) therefore, is a kind of measure for the deviation of the specification (3-2) from phenomenological behavior. This, together with the remark following (4-5), motivates us to formulate the following postulate, in analogy to (4-1).

Postulate V: The conditional p.d.f. P_p of (3-11), (3-12) can be written as a function of the left-hand side of (4-14):

$$P_p(\alpha^{(p)},\tau_p;\cdots;\alpha^{(2)},\tau_2|\alpha^{(1)})=g\left(\int_{t_1}^{t_p}\mathcal{O}\{\alpha_{\exp}(t)dt\}\right). \tag{4-15}$$

Added to the four standard postulates introduced earlier, this postulate is sufficient for the derivation of the entire fluctuation theory. This circumstance provides its justification.

We proceed now to determine the function g. According to (3-12),

$$P_3(\alpha^{(3)},\tau_3;\alpha^{(2)},\tau_2|\alpha^{(1)})$$
$$=P_2(\alpha^{(3)},\tau_3|\alpha^{(2)})P_2(\alpha^{(2)},\tau_2|\alpha^{(1)}). \tag{4-16}$$

From (4-15) and (4-16)

$$g\left(\int_{t_1}^{t_2}\mathcal{O}(\alpha_{\exp})dt+\int_{t_2}^{t_3}\mathcal{O}(\alpha_{\exp})dt\right)$$
$$=g\left(\int_{t_1}^{t_2}\mathcal{O}(\alpha_{\exp})dt\right)g\left(\int_{t_2}^{t_3}\mathcal{O}(\alpha_{\exp})dt\right). \tag{4-17}$$

The only nonsingular solution of this functional equation is

$$P_2(\alpha^{(2)},\tau_2|\alpha^{(1)})\propto\exp\left(a\int_{t_1}^{t_2}\mathcal{O}(\alpha_{\exp})dt\right), \tag{4-18}$$

where a is a constant which remains to be determined.

We write (4-18) in a more explicit form by using (4-10), (4-12), and (4-9).

$$P_2(\alpha^{(2)},\tau_2|\alpha^{(1)})\propto\exp[\tfrac{1}{2}as(\alpha^{(2)}-\alpha^{(1)}\rho_2)^2(1-\rho_2^2)^{-1}]. \tag{4-19}$$

The ratio of time-reversed conditional probabilities is

$$P_2(\alpha^{(2)},\tau_1|\alpha^{(1)})/P_2(\alpha^{(1)},\tau_1|\alpha^{(2)})$$
$$=\exp\{\tfrac{1}{2}as[(\alpha^{(2)})^2-(\alpha^{(1)})^2]\}. \tag{4-20}$$

The same ratio is obtained also from (3-10) and (4-3) and the comparison yields $a=-1/k$. Thus we have, finally,

$$P_p(\alpha^{(p)},\tau_p;\cdots;\alpha^{(2)},\tau_2|\alpha^{(1)})$$
$$\propto\exp\left[-\frac{1}{k}\int_{t_1}^{t_p}\mathcal{O}(\alpha_{\exp})dt\right], \tag{4-21}$$

and

$$W_p\begin{pmatrix}\alpha^{(1)} & \cdots & \alpha^{(p)}\\ t_1 & \cdots & t_p\end{pmatrix}$$
$$\propto\exp\left[-\frac{1}{2k}s(\alpha^{(1)})^2-\frac{1}{k}\int_{1}^{t_p}\mathcal{O}(\alpha_{\exp})dt\right]. \tag{4-22}$$

Although phrased somewhat differently, Eqs. (4-21) and (4-22) describe essentially the distributions of Onsager and Machlup.[2] The latter expressed their result in a variational language. It is evident, however, from their discussion that the exponentially smoothed paths are the extremal functions of their problem. The concept of smoothed path is an artifice which enables one to formulate the distribution functions in closed form. Further discussion of this point is postponed to Sec. 8.

5. AVERAGES IN THE TEMPORAL REPRESENTATION

We will use the notation that $\bar{y}$ is the time average of the parameter y for a single system taken over an infinite time interval, and $\langle y\rangle$ is the average taken over an ensemble of systems at a particular instant of time. Since we are dealing with equilibrium ensembles, the ergodic property yields

$$\bar{y}=\langle y\rangle. \tag{5-1}$$

The distribution functions obtained in the last section can be used to compute averages of various types. For this purpose the explicit form (4-19) is a convenient starting point.

We write the conditional p.d.f. in the normalized form[18]

$$P_2(\alpha^{(2)},\tau|\alpha^{(1)})=(s/2\pi k)^{\frac{1}{2}}(1-\rho^2)^{-\frac{1}{2}} \times\exp[-(s/2k)(1-\rho^2)^{-1}(\alpha^{(2)}-\alpha^{(1)}\rho)^2]. \tag{5-2}$$

From P_2 and W_1 the W_p can be built up stepwise. We have

$$W_2\begin{pmatrix}\alpha^{(1)} & \alpha^{(2)}\\ t_1 & t_2\end{pmatrix}=(s/2\pi k)(1-\rho_{12}^2)^{-\frac{1}{2}} \times\exp\left[-\frac{s}{2k}\frac{(\alpha^{(1)})^2-2\alpha^{(1)}\alpha^{(2)}\rho_{12}+(\alpha^{(2)})^2}{1-\rho_{12}^2}\right], \tag{5-3}$$

where

$$\rho_{ik}=e^{-\gamma|t_i-t_k|}, \tag{5-4}$$

and

$$W_3\begin{pmatrix}\alpha^{(1)} & \alpha^{(2)} & \alpha^{(3)}\\ t_1 & t_2 & t_3\end{pmatrix}=\left(\frac{s}{2\pi k}\right)^{\frac{3}{2}}(1-\rho_{12}^2)^{-\frac{1}{2}}(1-\rho_{23}^2)^{-\frac{1}{2}} \times\exp\left\{-\frac{s}{2k}\left[\frac{(\alpha^{(1)})^2}{1-\rho_{12}^2}+\frac{(\alpha^{(2)})^2(1-\rho_{13}^2)}{(1-\rho_{12}^2)(1-\rho_{23}^2)}+\frac{(\alpha^{(3)})^2}{1-\rho_{23}^2}-\frac{2\alpha^{(1)}\alpha^{(2)}\rho_{12}}{1-\rho_{12}^2}-\frac{2\alpha^{(2)}\alpha^{(3)}\rho_{23}}{1-\rho_{23}^2}\right]\right\}. \tag{5-5}$$

From (5-5) the structure of the distribution function is apparent and it is easy to write down the general expression for W_p. Instead of doing this we prefer to characterize W_p by means of its characteristic function.

The characteristic function of the variables $z_1, \cdots z_p$ with respect to the distribution W_p is defined as the Fourier transform of the joint p.d.f.

$$g(z_1\cdots z_p)=\langle\exp(i\textstyle\sum_k\alpha^{(k)}z_k)\rangle =\int_{-\infty}^{+\infty}\cdots\int_{-\infty}^{+\infty}\exp(i\textstyle\sum_k\alpha^{(k)}z_k)W_pd\alpha^{(1)}\cdots d\alpha^{(p)}. \tag{5-6}$$

It is easy to verify that

$$g(z_1\cdots z_p)=\exp\left[-\frac{1}{2}\left(\frac{k}{s}\right)\left(\sum_1^p z_k^2+2\sum_{i>k}z_iz_k\rho_{ik}\right)\right]. \tag{5-7}$$

The proof consists essentially in showing that the matrix of the quadratic form in the parenthesis in (5-7) is the inverse of the form in the bracket of Eq. (5-5).

One makes use of the obvious relation

$$\rho_{ik}\rho_{kl}=\rho_{il},\quad t_i<t_k<t_l. \tag{5-8}$$

The use of the generating function has the advantage of allowing the straightforward calculation of averages. We have

$$\langle\alpha^{(i)}\alpha^{(k)}\rangle=-\left(\frac{\partial^2 g}{\partial z_i\partial z_k}\right)_{z_1=\cdots=z_p=0}. \tag{5-9}$$

Hence

$$\langle\alpha^{(i)}\alpha^{(k)}\rangle=ks^{-1}\rho_{ik}; \tag{5-10}$$

or, using (5-4),

$$\langle\alpha(t)\alpha(t+\tau)\rangle=\langle\alpha(t)\alpha(t-\tau)\rangle=\langle\alpha^2\rangle e^{-\gamma\tau}. \tag{5-11}$$

The equality of the first two terms of Eq. (5-11) is the usual expression of the principle of microscopic reversibility.

The conditional p.d.f. (5-2) immediately yields a description of the regression of fluctuations within an equilibrium ensemble. Focusing attention on those members of the ensemble for which

$$\alpha(t_1)=\alpha^{(1)}, \tag{5-11a}$$

we have

$$\langle\alpha(t_2)|\alpha^{(1)}(t_1)\rangle=\alpha^{(1)}\rho_{12}, \tag{5-12}$$

where the vertical bar indicates that the averaging is to be performed only over those ensemble members satisfying (5-11a).

Introduce the quantity

$$\Delta\alpha(t_2)=\alpha(t_2)-\alpha^{(1)}\rho_{12} \tag{5-13}$$

to denote the deviation of an actual path from the average regression curve (5-12) at the instant t_2. We find

$$\langle\Delta\alpha(t)^2\rangle=ks^{-1}(1-\rho_{12}^2). \tag{5-14}$$

We turn now to the discussion of systems undergoing an irreversible process. We consider an ensemble of

[18] This expression was first obtained by G. E. Uhlenbeck and L. S. Ornstein, Phys. Rev. **36**, 823 (1930). It served as a point of departure of Onsager and Machlup[2] for the derivation of (4-21).

systems in constrained equilibrium, each of whose members has $\alpha=\alpha^{(1)}$. At a given instant of time t_1 a "shutter" is opened in each system, and the entire simultaneously starts its approach to unconstrained equilibrium.

We see that the phenomenological regression given by (2-9) is identical to the average regression (5-12) of fluctuation theory. This statement, frequently advanced as a postulate,[2,7] is implied by our postulate V.

6. FLUCTUATIONS IN THE SPECTRAL REPRESENTATION‡

We turn now to the spectral description of fluctuation paths. Consider a very long time interval θ, and expand the fluctuation path of each member of the ensemble in a Fourier series[19]

$$\alpha(t)=\mathrm{Re}\sum_{n=1}^{\infty}\alpha_n e^{i\omega_n t}. \tag{6-1}$$

Here "Re" means "real part," and

$$\omega_n=(2\pi/\theta)n, \tag{6-2}$$

$$\alpha_n=a_n+ib_n. \tag{6-3}$$

In order to describe the statistical properties of the paths $\alpha(t)$ of (6-1), introduce the notation that

$$f_N(a_1,\cdots,a_N;b_1,\cdots,b_N)da_1,\cdots da_N db_1,\cdots db_N \tag{6-4}$$

be the probability that an observed $\alpha(t)$ satisfies the specifications that the real and imaginary parts of the first N Fourier coefficients of (6-1) lie within the intervals of width da_n, db_n centered around the values appearing in the argument of f_N. (We will call such a specification a *finite Fourier specification.*)

Analogous to the procedure of Sec. 4, we find it convenient to associate with every such path specification a *Fourier-smoothed path* $\tilde{\alpha}(t)$, defined by the equation

$$\tilde{\alpha}(t)=\mathrm{Re}\sum_{n=1}^{N}\alpha_n e^{i\omega_n t}. \tag{6-5}$$

The first N Fourier coefficients are determined by the specification associated with (6-4) and all other α_n's are set equal to zero.

On account of the differences in the associated specifications, it is not obvious that f_N should be very similar to W_p of the temporal description. However, it will be shown in the next section that for the stochastic process of Secs. 4 and 5

$$f_N=C_N\exp\left[-\frac{1}{k}\int_0^\theta \Theta\{\tilde{\alpha}(t)\}dt\right]. \tag{6-6}$$

In the above, C_N is a coefficient of normalization.

In this section we derive a few of the implications of the above formalism. On using (2-11), (6-1) and (6-4) we write explicitly for this probability density

$$f_N=C_N\exp\left[-\frac{\theta}{8k}\sum_{n=1}^{N}\frac{(a_n^2+b_n^2)\omega_n^2}{\sigma(\omega_n)}\right], \tag{6-7}$$

where

$$C_N=\left(\frac{\theta}{8\pi k}\right)^N\prod_{n=1}^{N}\omega_n^2/\sigma(\omega_n), \tag{6-7a}$$

and

$$\sigma(\omega)=R/(R^2+s^2/\omega^2) \tag{6-8}$$

is the real part of the admittance in the language of the electric analogy.

We concern ourselves at first with the autocorrelation coefficient. According to (5-1) we need not distinguish between time and ensemble averages. We have

$$\langle\alpha(t)\alpha(t+\tau)\rangle=\sum_{nm}\langle(\mathrm{Re}\,\alpha_n e^{i\omega_n t})(\mathrm{Re}\,\alpha_m e^{i\omega_m(t+\tau)})\rangle. \tag{6-9}$$

Computing expectation values with the probability densities (6-7), we find

$$\begin{aligned}&\langle a_n^2\rangle=\langle b_n^2\rangle=4k\sigma(\omega_n)/\theta\omega_n^2,\\ &\langle a_n a_m\rangle=\langle b_n b_m\rangle=0,\quad n\neq m,\\ &\langle a_n b_m\rangle=0.\end{aligned} \tag{6-10}$$

The above, therefore, becomes in the limit $\theta\to\infty$,

$$\langle\alpha(t)\alpha(t+\tau)\rangle=\frac{2k}{\pi}\int_0^\infty\frac{\sigma(\omega)}{\omega^2}\cos\omega\tau d\omega. \tag{6-11}$$

The integration yields

$$\langle\alpha(t)\alpha(t+\tau)\rangle=ks^{-1}e^{-\gamma\tau} \tag{6-12}$$

in agreement with (5-11). On setting $\tau=0$ we obtain a result identical with (4-4), which was derived by averaging in terms of W_1.

The spectral density for α^2 can be obtained by inspection of (6-11) with $\tau=0$ or, alternatively, by a direct calculation, using the p.d.f. (6-7), of the expectation value of the component of α^2 lying within a narrow frequency band. We have

$$G_\alpha(\omega)=(2k/\pi)\sigma(\omega)/\omega^2. \tag{6-13}$$

This is the fluctuation dissipation theorem of Callen and Greene.[8]

Let us introduce the quantity $\epsilon(t)$ defined by

$$R\dot{\alpha}+s\alpha=\epsilon. \tag{6-14}$$

On substituting (6-1) into the above and noting (6-7),

‡ *Note added in proof.*—The connection of the Onsager-Machlup formalism with the method of Rice (spectral representation) has been recently developed also by Hashitsume, Progr. Theoret. Phys. (Japan) **15**, 369 (1956).

[19] There should be a term $\alpha_0=(1/\theta)\int_0^\theta\alpha(t)dt$ included in (6-1) which is omitted for simplicity. The end results will remain unaffected since we will eventually allow $\theta\to\infty$.

we find that the coefficients in the Fourier expansion of $\epsilon(t)$ have Gaussian distributions. Furthermore, on making use of (6-13), we obtain the spectral density of ϵ as

$$G_\epsilon = 2kR/\pi. \tag{6-15}$$

Equation (6-14) is the Langevin equation, and the above properties are formal expressions of the Langevin postulates. The fact that these appear as theorems in our formulation is essentially a justification of our postulate V.

In the electric analogy (6-15) is the Nyquist relation[6] of thermal noise for an RC circuit. Changing from the entropy to the energy representation, we have

$$\epsilon = V/T, \quad R = \mathcal{R}/T, \quad \omega = 2\pi\nu, \tag{6-16}$$

where V is the voltage and $\mathcal{R}$ the conventional resistance. Hence

$$G_V(\nu) = 4kT\mathcal{R}. \tag{6-17}$$

7. EQUIVALENCE OF THE TEMPORAL AND SPECTRAL DISTRIBUTION FUNCTIONS

The averages obtained with the probability densities W_p from (3-1), (4-22) and f_N defined by (6-4), (6-6) were found to be consistent with each other. However, it still remains to show that the probability densities W_p actually imply the p.d.f.'s f_N. We first show that the inverse statement is true.

Using our f_N, we shall compute the characteristic function

$$h(z_1,\cdots z_p) = \left\langle \exp\left[i\sum_{k=1}^{p} z_k\alpha^{(k)} \right]\right\rangle, \tag{7-1}$$

and shall prove that the result is identical to the characteristic function $g(z_1,\cdots,z_p)$(5-7) computed by means of W_p. It is well known, however, that the characteristic function uniquely determines the distribution.[20] Therefore the process governed by the distributions f_N implies the process described by the W_p's.

For the sake of simplicity we carry out the calculation for the case of $p=2$. The generalization is entirely straightforward.

Let us set $t_1 = t$ and $t_2 = t+\tau$. We obtain from (6-1)

$$\begin{aligned} \alpha^{(1)} &= \alpha(t) = \operatorname{Re}\sum_{n=1}^{\infty} \alpha_n e^{i\omega_n t}, \\ \alpha^{(2)} &= \alpha(t+\tau) = \operatorname{Re}\sum_{n=1}^{\infty} \alpha_n e^{i\omega_n (t+\tau)}. \end{aligned} \tag{7-2}$$

We first compute

$$h_N(z_1 z_2) = \langle \exp[iz_1\alpha^{(1)} + iz_2\alpha^{(2)}]\rangle_N, \tag{7-3}$$

where the subscript N indicates that the averaging takes place only over the first N Fourier coefficients. On using for f_N the explicit form (6-7), we obtain

$$h_N = C_N \prod_{n=1}^{N} A_n B_n \exp\left[iz_1 \operatorname{Re}\sum_{N+1}^{\infty} \alpha_n e^{i\omega_n t} + iz_2 \operatorname{Re}\sum_{N+1}^{\infty} \alpha_n e^{i\omega_n (t+\tau)}\right], \tag{7-4}$$

where

$$\begin{aligned} A_n &= \int_{-\infty}^{+\infty} \exp\left\{ -\frac{\theta a_n^2 \omega_n^2}{8k\sigma(\omega_n)} + ia_n[z_1 \sin\omega_n t + z_2 \sin\omega_n(t+\tau)] \right\} da_n, \\ B_n &= \int_{-\infty}^{+\infty} \exp\left\{ -\frac{\theta b_n^2 \omega_n^2}{8k\sigma(\omega_n)} + ib_n[z_1 \cos\omega_n t + z_2 \cos\omega_n(t+\tau)] \right\} db_n, \end{aligned} \tag{7-5}$$

and C_N is given by (6-7a). We have

$$C_N \prod_{1}^{N} A_n B_n = \exp\left\{ \frac{-2k}{\theta} \sum_{n=1}^{N} [z_1^2 + z_2^2 + 2z_1 z_2 \cos\omega_n t] \frac{\sigma(\omega_n)}{\omega_n^2} \right\}. \tag{7-6}$$

We go now to the limit $\theta \to \infty$ and $N/\theta \approx \omega_N \to \infty$. Under this limiting process the residual factor in (7-4) can be replaced by unity,[21] and the negative of the exponent in (7-6) becomes

$$\begin{aligned} &\lim \frac{2k}{R\theta} \sum_{n=1}^{N} (z_1^2 + z_2^2 + 2z_1 z_2 \cos\omega_n t)(\omega_n^2 + \gamma^2)^{-1} \\ &= \frac{k}{\pi R} \int_0^\infty (z_1^2 + z_2^2 + 2z_1 z_2 \cos\omega t)(\omega^2+\gamma^2)^{-1} d\omega \\ &= \tfrac{1}{2} k s^{-1} (z_1^2 + z_2^2 + 2z_1 z_2 e^{-\gamma\tau}). \end{aligned} \tag{7-7}$$

Hence

$$h = \lim_{N\to\infty} h_N = \exp[-\tfrac{1}{2} k s^{-1} (z_1^2 + z_2^2 + 2z_1 z_2 e^{-\gamma\tau})]. \tag{7-8}$$

This is indeed identical to (5-7) for $p=2$. Since the extension to arbitrary p offers no difficulty, we have shown that the f_N imply the W_p.

That the inverse statement is also true follows from

[20] See, e.g., B. V. Gnedenko and A. N. Kolmogorov, *Limit Distributions for Sums of Independent Random Variables*, translation by K. L. Chung (Addison-Wesley Publishing Company, Inc., Cambridge, 1954), p. 50.

[21] Choose any $\epsilon > 0$ and consider the collection of all paths which make this factor differ from unity by an amount greater than ϵ. It can be shown that the probability measure of this collection tends to zero as $N \to \infty$. This follows directly from theorem 4.1 of J. L. Doob [*Stochastic Processes* (John Wiley and Sons, New York, 1953), p. 155] (make the substitution $y_n = \alpha_k e^{i\omega_k t}$), and from a property of convergence in the mean [Doob, p. 9, Eq. (4.6)].

a theorem due to Kolmogoroff.[22] It is a direct consequence of this theorem that any complete set of finite-dimensional p.d.f.'s (such as our f_N's) *uniquely* determines any other set of finite-dimensional p.d.f.'s for the same stochastic process (in our case, the W_p's).

8. DISCUSSION

a. Variational Principles and Smoothed Paths

The distributions in function space discussed in this paper can be considered as generalizations of the situation encountered in variational problems. Given an appropriate set of initial and boundary conditions a variational principle determines a single extremal function, i.e., a δ-function distribution in function space.

While variational problems are equivalent to their deterministic Euler-Lagrange equations, the distribution problem is equivalent to a stochastic differential equation.

It is interesting to consider variational principles which can be formulated in terms of the OM function. A variational principle which implies the phenomenological equation has been mentioned in Sec. 2b. We shall consider now another class of variational problems defined as follows: let us select the path $\alpha(t)$ which minimizes the time integral of the OM function

$$\int_{t_1}^{t_p} \mathcal{O}\{\alpha(t)\}dt = \min. \tag{8-1}$$

Depending on the constraints imposed on the family of trial functions, (8-1) generates a whole family of variational principles.

For example, impose the constraints that every $\alpha(t)$ in the family of trial functions satisfies the conditions

$$\alpha(t_k) = \alpha^{(k)}, \quad k = 1, \cdots p, \tag{8-2}$$

where $t_1 < \cdot < t_p$ are p instants of time and the $\alpha^{(k)}$ are given numbers. It is easily shown that the solution of this variational problem is the exponentially-smoothed path $\alpha_{\mathrm{exp}}(t)$ of (4-10).[23]

On the other hand, one can choose as trial functions all functions of the form (6-1) subject to the constraints that the first N Fourier coefficients have specified values. In this case one finds that the solution of the variational problem is the Fourier-smoothed path $\tilde{\alpha}(t)$ of (6-5).

We therefore have the following property common to both the temporal and Fourier descriptions. To every (gate, Fourier) specification there corresponds a set of constraints for (8-1). The solution of this variational problem is the (exponentially, Fourier) smoothed path. The probability density associated with the (gate, Fourier) specification is a function of the quantity obtained by inserting the solution of the variational problem back into the functional which is varied:

$$\left[\begin{array}{l} P_p = P_p\left\{\int_{t_1}^{t_p} \mathcal{O}(\alpha_{\mathrm{exp}}(t))dt\right\} \\[2ex] f_N = f_N\left\{\int_0^{\theta} \mathcal{O}(\tilde{\alpha}(t))dt\right\} \end{array}\right]. \tag{8-3}$$

In the foregoing discussion the exponentially smoothed fluctuation path has only a formal meaning; it serves to formulate the probability of a particular gate specification.

In addition, however, this concept can be given a more physical interpretation: the true fluctuation paths are clustered around the exponentially smoothed path. This statement can be cast into a precise form, as was pointed out to us by Siegel,[24] by computing the following problem:

Let $t_1 < t_2 < t_3$ be three instances of time and consider the specification $\alpha(t_1) = \alpha^{(1)}$, $\alpha(t_3) = \alpha^{(3)}$.

One can show by straightforward calculation that $\alpha(t_2)$ obeys a Gaussian distribution centered around $\alpha_{\mathrm{exp}}(t_2)$.

b. Conclusion

The basic problem of the time dependent theory of fluctuations is to combine the phenomenological kinetic theory and the stochastic theory of random processes into a consistent scheme. The main contribution of this paper is to achieve this junction in a novel fashion.

The conventional approach is characterized by certain assumptions which we call the Langevin postulates. The procedure is to augment the kinetic equation by a term corresponding to a random force specified in terms of statistical assumptions: the force is assumed to be a Gaussian random function with a white spectrum, the density of which is fixed by Boltzmann's principle [see Eq. (6-14)]. The existence of fluctuations makes it evident that the phenomenological kinetic equation is not exact, and the Langevin method "saves" the validity of the equation by adding to it a random perturbation. This is the approach taken, for example, by Onsager and Machlup in their derivation of the density functions W_p.

The point of view taken in this paper calls for a shift in emphasis from the differential equation to the set of functions which describe the fluctuation paths. Concrete physical situations are accounted for in terms of distribution functions over this set of paths, or in other words through p.d.f. in function space. These distributions are closely related to the kinetic differential equation; the solutions of the latter constitute the most probable region in function space. The connection

[22] A. N. Kolmogoroff, *Foundations of the Theory of Probability* (Chelsea Publishing Company, New York, 1950), Chap. III, Sec. 4. This theorem is also described in footnote 7 of reference 2.

[23] This result and its connection with p.d.f. W_p of (4-22) was discovered by Onsager and Machlup.[2]

[24] Armand Siegel (private communication).

between the kinetic equation and the p.d.f. in function space is stipulated by postulate V [Sec. 4].

Heuristically, this postulate was suggested by the Onsager-Machlup distribution function derived synthetically from the Langevin postulates. In the present logical structure it is sufficient to join postulate V to the standard assumption of the kinetic and the stochastic theories to develop the kinetic theory of fluctuations. In particular, we obtain both the OM distribution and the Langevin postulates as theorems. So far as results are concerned this presentation is found to be equivalent to the one conventionally followed. However, we believe that the present method has some appealing conceptual and formal features.

In the first place our formalism gives justice both to the reversible and to the irreversible aspects of time. The manifold of fluctuation paths is invariant with respect to time reversal[25] while the bias in favor of the forward direction is introduced by means of OM function, which exhibits a peculiar selectivity with respect to the sense of time.

Although the Langevin theory is consistent with the microscopic reversibility of fluctuation paths, it fails to provide any formal counterpart to this principle. Thus, the regression of fluctuations is described by the phenomenological equation and the reverse process, the building up of fluctuations is attributed to the random force.

A second point we may bring up in favor of the present approach is its economy. This is apparent in the postulational basis, where postulate V replaces the Langevin postulates and a number of others usually advanced without proofs. The economy appears also in the formal structure built on the postulational basis. We refer, for example, to the not inconsiderable simplification in the derivation of the Callen-Greene fluctuation dissipation theorem.[8]

Throughout this paper we have emphasized the parallel aspects of the time-dependent distributions with the statistical thermodynamics of equilibrium. It is in order to point out, therefore, that in its present stage of development the kinetic fluctuation theory is subject to considerable limitations. We shall indicate the crucial point in connection with the characteristic function defined by Eq. (5-6).

This relation is seen to be the analogue of the definition of the canonical ensemble in statistical mechanics.[26]

It can indeed be shown that this analogy has its merits although it is limited by the fact that our distributions are Gaussian. This approximation leads to the correct second fluctuation moments, but does not have the formally exact character of the canonical distribution.[27] In order to achieve such a generality it would be necessary to go beyond the Gaussian approximation.

While the solution of the last-mentioned problem would require essentially new ideas, there are a number of other generalizations which can be carried out with minor extensions of the present framework.

It is possible to consider systems with many variables, exhibiting inertial effects, spatial expansion and those which are subject to external forces. We hope to return to these questions later.

The authors are indebted to Dr. D. F. Falkoff, Dr. Bayard Rankin, and Dr. Armand Siegel for commenting on the manuscript.

[25] This means that time reversal transforms fluctuation paths into admissible fluctuation paths. The paths which are individually invariant under time reversal form a special subset.

[26] A. I. Khinchin, *Statistical Mechanics* (Dover Publications, New York, 1949), p. 76.

[27] R. F. Green and H. R. Callen, Phys. Rev. 83, 1231 (1951).

PART C

ON METHOD

LASZLO TISZA

THE LOGICAL STRUCTURE OF PHYSICS *

I. THE LOGIC OF CONCEPT FORMATION

Crystallizations of genuinely new concepts that are not combinations of concepts already established often mark turning points in the evolution of physics. In retrospect it is possible to view such processes of concept formation as interplays of two types of procedures. The first is what might be called *conceptual differentiation.* It arises as concepts distilled from past experience are brought to bear on experimental situations that are either novel, or at least more detailed and discriminating than those of the past. In such cases the original concept is usually modified and subdivided leading to a bifurcation or differentiation. An example is the splitting of the undifferentiated concept of *calor* (heat) into *temperature* and *caloric*, and the further subdivision of the latter into *energy*, *entropy* and *heat quantity*.

Counteracting the multiplication of concepts by differentiation is the occasional discovery of new fundamental concepts that enable us to represent concepts, that had been believed to be primitives, as constructions. An example is the electromagnetic field that made the unification of the theories of electricity, magnetism and optics possible. We note that whenever we speak of *concepts* we should include the *principles* formalizing their use. Thus the electromagnetic field plays its unifying role only if defined in terms of Maxwell's equations, including the displacement current.

The proper functioning of conceptual differentiation and integration is beset by characteristic difficulties. Differentiation may occur in a large number of small steps and the realization that a radical conceptual change has taken place may be long delayed. The difficulty is usually aggravated

* This paper is based on material presented on December 14, 1961 in the Boston Colloquium for the Philosophy of Science. A more detailed presentation is found in "The Conceptual Structure of Physics," January, 1963, *Reviews of Modern Physics.* We refer to this paper as *loc. cit.*

Synthese **14** (1962) 110–131; © *D. Reidel Pub. Co., Dordrecht, The Netherlands* by semantic confusion as concepts which are actually used in different senses in different contexts are still designated by the same term. It was recently pointed out by Karl Menger [1] that this difficulty arises even in contemporary mathematics. According to Menger "... mathematico-scientific methodology is in need – in fact, in urgent need – of a *separator* or a *prism* resolving conceptual mixtures into the spectra of their meanings". It would seem that the need for such a prism in physics is still more pressing.

The difficulties associated with conceptual integration are even greater. A limited field of experience can usually be represented in terms of several formal theories. The comparative merit of these formalisms often becomes apparent only by confrontation with phenomena belonging to quite remote specialties. Such a confrontation is always difficult, but it is becoming more so with the widening gap between specialties.

The main purpose of this paper is to outline a method of logical differentiation and integration that should enable us to proceed with the formation of concepts in a more controlled and systematic fashion. In particular, this method ought to supplement the subjectivity of historical evolution with a more objective appraisal of logical connections. In this method of analysis deductive systems play an essential role, a statement presumably surprising in view of the curious ambiguity surrounding the appreciation of deductive systems in physics.

The deductive method is quite prominent in the early development of mechanics by Galileo and Newton. Also the great systems of 19th century physics can be considered as somewhat loosely organized deductive systems. However more recently the use of deductive systems has declined. The impression seems to prevail that deductive systems are rigid, narrow and lead to the ossification of theories by lending authority to obsolescence. There is, of course, no denying that this description of the deductive method is often entirely accurate. However, the potentialities of this method are in striking contrast with its actual uses in the past. If properly handled, deductive systems can become instruments of radical innovation even while safeguarding the genuine achievements of the past. In essence two ideas are responsible for this change in outlook. In order to avoid rigidity, the postulational basis, or briefly, the basis of deductive systems has to be considered as hypothetical rather than an expression of self-evident truth. Therefore the basis is subject to change if this leads to an improvement of the system in accounting for experimental facts. This feedback from system to basis leads to the dynamic, evolutionary adjustment of the system to a widening range of experience. Instead of assuming that deductive systems have to be perfect in order not to collapse, the present method of analysis deals with imperfect systems. In fact, one of the tasks of the method is to locate and to eliminate imperfections. In view of this situation, the method can be aptly called *dynamic logical analysis*, or the *dynamics of deductive systems*.

While the idea of the hypothetical nature of deductive systems is, in principle, well known, it has not yet led to practical consequences for structuring the theories of physics. To achieve such applications we have to overcome the difficulties stemming from the narrowness of deductive systems. This narrowness is an irreducible fact of logic, and the scope of systems is usually further narrowed as the rigor and completeness of the formulations is improved. The cure suggested for this difficulty is the simultaneous use of many deductive systems. Whereas traditionally a deductive system was usually the entire universe of discourse, according to the present point of view deductive systems are themselves conceptual elements and their interrelations, particularly the conditions of mutual consistency become the main objects of study. In many important cases these relations exhibit typical patterns some of which we shall consider in the next section. An application of these ideas to the conceptual problems of quantum mechanics will be considered in Section III.

II. LOGICAL DIFFERENTIATION AND INTEGRATION

It is an important insight of mathematical logic that each deductive system contains primitive concepts that cannot be defined by conventional means, but are implicitly defined by the deductive system as a whole. This "definition by postulation" of primitives, so important in modern mathematics, turns out to be valuable also in physics. While recognizing the basic identity of the method in both disciplines one should also take note of a difference stemming from the fact that the deductive systems of physics are always supplemented by rules of correspondence which establish a link with experience. Mainly as a consequence of this connection, most concepts of physics evoke in the mind images or models, for which the postulates are presumably true statements. Although this intuitive aspect of the abstract concepts of physics is important for many reasons the relations between experiment, intuitive models and deductive systems are intricate, not unique, and hard to state in precise terms. The method of definition by postulation is an effective tool for clarifying the semantic ambiguities that arise out of this situation.

The fact that primitive concepts assume a precise meaning only relative to the deductive system into which they are incorporated, can be aptly referred to as the *relativity of concepts*.

Another interesting aspect of the present analysis is the substantial increase in the number of deductive systems. It appears that most of the basic disciplines, say classical mechanics, thermodynamics, quantum mechanics, are too ramified and diverse to be compressed within the confines of a rigorously built deductive system. This point has, of course, been advanced as an argument against the use of deductive systems in physics. Instead, we propose to overcome this difficulty by constructing a cluster of deductive systems, even for representing one basic discipline.

Since, as we have pointed out, each deductive system implicitly defines its own set of primitive concepts, the increase in the number of systems is tantamount to an increase in the number of fundamental concepts that are defined relative to a particular context specified in terms of a *deductive system*. For this reason, we designate the procedure leading to a collection of deductive systems as *logical differentiation*.

The conceptual wealth brought about by the method of logical differentiation provides us with all the freedom we may need to adjust our deductive systems to a variety of experimental situations.

However, if the analysis should stop at this point, one might feel that the freedom in the formation of deductive systems has been carried too far, since it seems to lead to a proliferation of basic concepts. Actually, however, this process is held in check by the second stage of our procedure, the *logical integration* of deductive systems. The conditions of mutual consistency of empirically interpreted deductive systems have empirical and logical aspects. A study of these problems should be classified as part of *logical empiricism*, although the present discussion goes beyond the traditional scope of this discipline. The possibility of a concise discussion of the general problem of mutual consistency hinges on the circumstance that the formal logical aspects of these compatibility conditions can be recognized within a model of transparent simplicity.

The same formal conditions can be given different concrete interpretations and used to handle most intricate situations.

The model in question is the practical (physical) geometry of a spherical surface, in other words geodesy. The physicist-geometer can avail himself of two abstract geometrical systems, to wit spherical, and plane geometry. From the point of view of the abstract mathematician the two systems are different and a discourse has to be completed in one *or* the other system. The consistent use of the two systems for the description of the real, physical sphere depends on a number of circumstances.

In the first place, we establish a hierarchy between the two systems: we stipulate that in case of conflict spherical geometry is the *dominant* system, and plane geometry is called either *derivative*, *subordinate*, or *degenerate*.

We shall refer to the relation of these systems as supplementary *relation*, or, briefly, *supplementarity*.

The second point is a clarification of what constitutes a "conflict" between the systems. In the mathematical sense a conflict always exists; however, in physical geometry we dismiss it as empirically irrelevant if it falls below the finite accuracy of the measurements which the theory is supposed to describe. The accuracy of measurement cannot be improved over the diffuseness set by thermal noise.

Somewhat more generally we proceed to formulate a principle to which the rules of coordination have to conform. Experiment provides us with the values of continuous variables within a certain accuracy. However, the analysis of mathematical physics would be extremely cumbersome and lose much of its precision if the finite width of continuous parameters corresponding to empirical quantities were to be observed in every step. In actual practice, the segments of finite widths are sharpened into definite points of the continuum for the purposes of the analysis. However, the results obtained have no physical meaning unless they are sufficiently insensitive to the actual unsharpness of the continuous parameters. Mathematical solutions satisfying this requirement will be called *regular*. Solutions that change their essential features when the fictitious sharpening is given up will be referred to as *singular*. They are devoid of physical meaning except as possible stepping stones to more realistic results. The requirement that mathematical solutions be regular in order to have a physical meaning will prove to be of great importance and deserves to be called a principle, the *principle of regularity*.

The second point concerns the question of *scale*. Spherical geometry is characterized by an absolute unit of length R, the radius of the sphere. If $R \to \infty$, the sphere degenerates into the plane, the dominant into the supplementary system. Let the linear dimension of a geometrical figure be a, then the parameter a/R is appropriate to express the deviation of spherical from plane geometry. It becomes meaningful to say that for small values of the parameter ($a/R \approx \epsilon \ll 1$,) where ϵ is the unsharpness of the parameter because of the inaccuracy of the measurement or because of noise, the subordinate system is asymptotically valid. Otherwise we have to fall back on the dominant system.

It is interesting to note that within plane geometry there is no absolute unit and it is meaningless to speak about small or large figures. All theorems are invariant under a transformation consisting of a change of the unit of length. We say also that plane geometry is *scale invariant*, or, it contains the theorem of similarity.

It is well known that deductive systems provide the implicit definitions of their primitive concepts. It is easily seen that the limiting process leading from the dominant to the derivative system profoundly modifies the conceptual structure of the deductive systems concerned. While the conceptual structure of spherical geometries, which differ only in the numerical value of the fundamental radius R, is identical, there is a discontinuous change in the conceptual structure as R becomes infinite. We are confronted here with the logical analog of nonuniform convergence. Paradoxically, the deceptive simplicity of the model may provide the greatest hurdle to an appreciation of the relation between the two systems. It seems to be almost as hard to recapture a feeling of puzzlement in connection with a paradox completely resolved, as to get rid of a feeling of discomfort when faced with essentially the same paradox in a novel context. Yet for those who believed that the earth was flat, the relativization of "up" and "down" and the existence of antipodes on the sphere was no less puzzling than the relativization of time for someone thinking in terms of classical kinematics. The logical and even the mathematical connection between the two situations are very close. Moreover, if one believes that the sum of the angles of triangles has to be π, then one will conclude that the concept of triangle "breaks down" on the sphere.

One of the conceptual problems to be solved in connection with the individual dominant systems is that of invariance. The transition from the subordinate to the dominant system marks the breakdown of scale invariance. At the same time, however, the dominant systems is characterized by an invariance of a novel type. In spherical geometry we have invariance of curvature.

In the course of historical evolution the dominant system has to be established by generalizing the subordinate system that reached what we shall call a *marginal breakdown*. In general, this is a task of great difficulty. While the converse procedure of arriving through a limiting process from the dominant to the subordinate system is a great deal more straightforward, even this transition is not without pitfalls.

Just as the sphere does not contain the infinite plane, so spherical geometry does not contain plane geometry. While the limiting process $R \rightarrow \infty$ transforms every theorem of spherical geometry into one of plane geometry, the converse is not true. Thus the important laws of similarity of triangles and other figures have no counterpart in spherical geometry. The derivation of such theorems is a new discovery, possible only in plane geometry. This situation is quite similar for other pairs of dominant-supplementary theories.

We shall say that the inconsistency between the subordinate and the dominant system is *under control*, or that the *inconsistency* is controlled, if the range of validity of the subordinate system is known, and within these limits the compatibility with the dominant system is ensured.

Fig. 1 contains a simplified, symbolic representation of the integrated structure of systems that results from the application of the principles discussed in this section. Although this structure, briefly, the *block diagram*, is still incomplete and tentative, it is adequate to provide the point of departure for our discussion. We have to remember that our method is set up to eliminate the shortcomings of the initially posited structure rather than to perpetuate them.

Maybe, the most significant aspect of the block diagram is the pluralistic character of the structure exhibited. Moreover, CMP [2] appears as derivative to several systems and dominant to none. This is in complete contrast with the logical structure hypothesized in classical physics in which CMP should have been dominant to all the other deductive systems of physics.

For even a sketchy discussion of the block diagram we have to refer to *loc. cit.* However in the next section we shall summarize some of the conclusions concerning the conceptual problems of quantum mechanics.

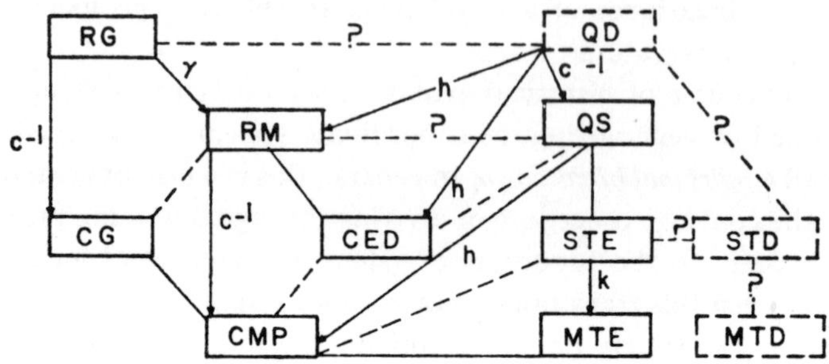

—— COMPATIBILITY - - - - CONTROLLED INCONSISTENCY,
→ SUPPLEMENTARY RELATION - - - ? - - - INCOMPLETELY UNDERSTOOD RELATION

TENTATIVE STRUCTURE OF PHYSICS

Fig. 1

c^{-1} inverse speed of light; h Planck's constant; k Boltzmann's constant; γ gravitational constant.

Systems in broken rectangles are essentially incomplete.

Symbols of systems:

CMP	Classical mechanics of particles	TD	Thermodynamics (generic term)
CMP(μ)	Classical mechanics of particles (microscopic interpretation of particles)	MTE	Macroscopic thermodynamics of equilibrium
		STE	Statistical thermodynamics of equilibrium
PhCM	Phenomenological classical mechanics	MTD	Macroscopic thermodynamics
CG	Classical gravitational theory	STD	Statistical Thermodynamics
CED	Classical Electrodynamics	QM	Quantum mechanics
RM	Relativistic Mechanics	QS	Quasi-static quantum mechanics
RG	Relativistic Gravitational-Theory	QD	Quantum dynamics

III. THE TWO PRINCIPLES OF CAUSALITY

The empirical basis of the physical sciences is provided by two major disciplines, astronomy and chemistry, the formalization of which gives rise to CMP and TD. The former adequately describes the motion of well separated macroscopic objects, whereas TD deals with intrinsic structural properties. Quantitative connections are much easier observable in astronomy than in chemistry. It was presumably unavoidable that CMP should be the first discipline to develop into a powerful and versatile formalism, while TD was late to start, slow to develop and is even now in an "underdeveloped" state measured in terms of its potentialities. Since the traditional unitary attitude, fostered by the unification of celestial and terrestial mechanics, could admit only one fundamental deductive system, it was taken for granted that, say TD would eventually be reduced to CMP. In contrast, according to the pluralistic view TD can be admitted as a semi-autonomous system with its *sui generis* primitive concepts, even if some of these are evidently not reducible in a rigorous sense to CMP.[3, 4] Thus TD and CMP implicitly define two conceptions of matter which are strictly speaking in conflict with each other. This inconsistency indicated by a broken line in the block diagram is brought under control by QM.[5] The application of the present method allow us to conclude that the conceptual structure of QM is much closer to TD than to CMP. It is for historical reasons that this fact did not become manifest thus far. QM has been insufficiently differentiated from CMP and insufficiently integrated with TD. This was made practically unavoidable by the incomplete development of TD.[6]

We turn now to a short summary of some of the features that make up the conceptual contrast, or rather complementarity of CMP and TD.

Instead of building up systems out of mass points, as in CMP, in TD one builds them out of "cells" that cover a certain region in space. We refer to the cells as *simple systems*, or *subsystems*, and to a collection of cells as a *composite system*. The independent variables of the theory are additive conserved quantities, briefly, additive invariants X_1, X_2, ... Composite systems are specified by sets of the X_i in the subsystems of the composite system.

Explicitly, the X_i are the volume V, the internal energy U, and the mole numbers of the independent chemical components N_1, N_2,... The characteristic coupling of thermodynamic systems comes about through the exchange of energy and of other additive invariants.

We turn now to the discussion of processes. In CMP all processes are alike, inasmuch as all of them are trajectories in phase space. In TD we have processes of different types. An important one is the transfer of a quantity X from one subsystem to another during an arbitrary, fixed time interval. The transfer quantity corresponding to the energy, with all of the other X's fixed, is called *heat*.

Walls are the boundaries separating two systems which completely prevent the passage of one or more additive invariants, but permit free passage to all others. The restrictions on a system imposed by its walls are referred to as *constraints*.

The concept of thermodynamic process is easily extended to chemical reactions. In this case the molecular potential barriers stabilizing chemical species play the role of walls.

We call *thermodynamic operations* any imposition or relaxation of a constraint through the uniting or subdividing of systems, or the alteration of the type of any of its walls.

The distinction of thermodynamic operations and processes is a characteristic feature of the present approach. The admission of thermodynamic operations into the framework of the theory allows one to account for experimental procedures that have no place within CMP because they correspond to transitions from one phase space to another.[7] Thus thermodynamics allows us to account formally for manipulations on thermodynamic systems. In particular, such manipulations may be performed on systems in the course of a measurement. Yet the insistence on operations as contrasted with processes does not necessarily require the intervention of an experimentalist, and does not introduce more subjective elements into the theory than are warranted by the nature of the situation considered. Thus in the case of chemical reactions and phase transitions, supplying catalysts or inhibitors that speed up or slow down the rates of reactions, is the corresponding thermodynamic operation. In an automatic chemical plant, and even more in biochemical processes, the "operations" become effective without the interference of an experimentalist and it may be more adequate to speak of parametric processes which *control* or *govern* other processes. This is quite in line with the use of these terms in cybernetics.[8]

In the present context the approach to equilibrium is described as follows. In a composite system in equilibrium the relaxation of an internal constraint (a thermodynamic operation) triggers a process leading to a new equilibrium, in which the partition of the additive invariants is uniquely determined by the nature of the system and by the existing constraints. Equilibrium is uninfluenced by past history, specified e.g. in terms of constraints that had been relaxed previously. This important result has been designated [3] as the *principle of thermostatic determinism*, suggesting a parallel with the *principle of mechanical determinism*. The latter has, of course, a temporal character and the connection between the two principles is traditionally established by reducing thermostatics to mechanical determinism. However it was pointed out *loc. cit.* that the sequences of thermodynamic operations and processes can be iterated to form composite processes. Due to the fact that the principle of thermostatic determinism assures us of the uniqueness of the final state of each step, the entire composite process assumes a deterministic character, provided that the operations are preset by some automatic device. Alternatively, if the operations are governed by the terminal states through feedback, the composite process becomes goal-directed.

The temporal determinism of mechanical processes depends on the *exhaustive specification* of the initial state of the system involved. If, as usual, the system is coupled to its environment, then the initial state of the environment has to be also completely specified; in practice this is an impossible task. Therefore mechanics accounts for the deterministic behavior of systems only if these are effectively decoupled from their environment. Evidently the case of planetary systems plays a uniquely favorable role in this respect. Being sheltered from the environment by vast reaches of space, the specification of its initial state can be practically accomplished with the accuracy required for macromechanical predictions extending over long intervals of time.

The impressive success of this procedure, together with the possibility of generalization to the differential equations of CED and QM, lead to the conviction that *all* causal chains in nature are conditioned by an exhaustive specification of the initial state. This conviction seems to be still quite widespread, although it is untenable from the point of view of experimental evidence. It is a fundamental fact in all experimental sciences that there is a selectivity of the specification of initial states that distinguishes between *essential* and *irrelevant* factors. It is this *selectivity* of specification that enables us to experiment at all, and to utilize the results of past experience under new conditions that are never exactly duplicated.

This fact becomes most impressive in living organisms: the seed of a plant develops into a well-determined organism, provided only that the environment to which it is coupled satisfies some very weak requirements concerning the range of temperature, pressure, and chemical potentials.

This principle of causality based on selective specification finds a theoretical foundation in thermodynamics which leads to the temporal determinism of composite processes just described. Thus we are led to *discern two principles of causality.* While the *first principle is based on exhaustive specification of the initial state*, the *second depends on selective specification.* The two principles are complementary: The first is *incompatible* with stochastic laws [9]; the second requires on such laws. Thus there is a built-in conflict between the theoretical and the experimental aspects of Newtonian natural philosophy, since they depend on the first and the second principles of causality. Critical philosophers such as George Berkeley and David Hume sensed these difficulties, but they were unable to resolve them. In contrast, physicists were too busy developing the rich implications of both methods to be slowed down by conceptual scruples that were not yet ripe for solution.

The foregoing consideration are considerably strengthened and sharpened when extended to quantum mechanics. I will have to confine myself to a few remarks and refer for further details to *loc. cit.* As pointed out above, in TD we consider transitions between equilibrium states. This ties in naturally with the transitions between quantum states in QM. Particularly interesting are the *pure states* which represent *potential* forms for the existence of objects. If such a state *actually* exists, we say the state is occupied, or realized. *The realizations of the same pure state in different regions of space time form a class of absolutely identical objects.*

The time-independent Schrödinger equation allows one to compute all pure states that can be constructed from a given assembly of particles. Thus, in principle, all of the pure states of structures built from a given set of particles can be computed on theoretical grounds without any recourse to experiment! Whenever this calculation has been actually carried out, the results have been verified by experiment to a high accuracy. There are undoubtedly huge classes of cases in which the same agreement could be achieved, provided that one overcomes the mathematical difficulties that increase enormously with the complexity of the systems.

Although the calculation of the properties of pure states by exclusively theoretical means is of the greatest philosophical interest, it is no less important that the taxonomical task of listing and specifying the set of pure states can be undertaken by judicious combination of theory and experiment. It can be shown that the "*retrodiction*" *of the discrete quantum numbers specifying pure states from the statistical evaluation of actual signals can be carried out in principle, and very often in practice, with absolute certainty.*

We may sum up the situation by saying that QS contains rigorous results of a taxonomical character concerning the existence of classes and their properties which are entirely foreign to classical physics. Remembering from Section II that dominant systems have their characteristic invariance property, we say that the absolute identity of any realization of the same pure state is the dominant invariance of QS, to be referred to as *morphic invariance*. It is the morphic invariance that brings it about that complex systems of well-defined identity can be built in nature through the mechanism of random processes.

A preliminary discussion (*loc. cit.*) shows that the combination of the concepts of morphic invariance with the two principles of causality provides a theoretical framework for the description of the functioning of organisms.

IV. PHILOSOPHICAL REFLECTIONS

The method of analysis of this paper is, I believe in agreement with the spirit and the underlying goals of the philosophical school that is alternately called analytic, positivist or empirical. I hope not to be too far off the mark in using these terms as synonyms. The concepts of logical differentiation and integration arose within the examination of the language of physics, which is a legitimate pursuit in this school of thought.

A point of agreement is the close attention paid to empirical requirements; more specifically, the approach to quantum physics as described *loc. cit.* is quite in line with the positivist program of developing the theory of the structure of matter in close parallel with experiment. The principle of regularity (Section II) is a methodological rule that serves as a safeguard against taking theoretical systems more seriously than their empirical support warrants.

At the same time, the present paper and its implications are at variance with some of the current practices of positivism. The clarification of these issues can be achieved as a by-product of the present analyses.

Let us consider first, the so-called *operational character* of the theoretial concepts. It goes without saying that the deductive systems of physics have to have a range of concordance with experience.

Yet, there is a growing awareness of the fact that there is something wrong, or at least lopsided about the great emphasis on "operationalism" which for several decades has dominated empirical philosophy.

We could do hardly better than quote from P. W. Bridgman's [10] penetrating reappraisal of the early ideas of operationalism. "If I were writing the Logic today I would change the emphasis so as to try to avoid what I regard as the one most serious misunderstanding. That is, I would emphasize more that the operations in terms of which a physical concept receives its meaning need not be, and as a matter of fact are not, exclusively the physical operations of the laboratory. The mistaken idea that the operations have to be physical or instrumental, combined with the dictum on page 5, 'The concept is synonymous with the corresponding set of operations', has in some cases led to disastrous misunderstanding. If I were writing again I would try to emphasize more the importance of the mental or paper-and-pencil operations. Among the very most important of the mental operations are the verbal operations. These play a much greater role than I realized at the time..."

The postulational definitions of the present approach are of course "paper-and-pencil" operations. Apart from being of equal importance with the instrumental operations, the two are related to each other in a well-structured manner. In the absence of precise postulational definitions the meaning of the instrumental operations cannot even be properly evaluated. The operational point of view considered as a sole criterion would always favor the low-level abstractions. However, the most spectacular advances in theory are connected with the discovery of abstractions of a very high level.

The constructive aspect of operationalism is appreciated particularly in the process of logical integration. If we have two theoretical systems, both of them confirmed by experiment, but inconsistent with each other, we usually find that the inconsistencies are produced by nonoperational assumptions that can be dropped without loosing the measure of experimental agreement already achieved. In fact, subsequent developments often result in an extraordinary expansion of the range of concordance.

This is what happened as CMP was generalized into the dominant RM. Einstein's operational analysis of the concept of simultaneity was a means to an end, it made the integration of CMP and CED possible. The concept of absolute simultaneity was abandoned because it blocked integration. The fact that it was not operational showed that this change in the foundation could be undertaken without damage.

However, "logical integration" has been, thus far, not in the vocabulary and the *purpose* of operationalism was not sufficiently appreciated. Erroneously it was taken for an end in itself, or, an ironical perversion of its real role, a means to forestall another "catastrophe" that would force us to re-examine our basic assumptions.

Although an occasional reappraisal of the foundations may seem a not entirely welcome interruption of the routine of research work, it is now recognized to be the unavoidable price to be paid for the continued use of new high-level abstractions. We again quote Bridgman: "To me now it seems incomprehensible that I should ever have thought it within my powers, or within the powers of the human race for that matter, to analyze so thoroughly the functioning of our thinking apparatus that I could confidently expect to exhaust the subject and eliminate the possibility of a bright new idea against which I would be defenseless."

We turn now to another difficulty of positivism, namely its predominantly restrictive character which bids us to dismiss most problems of traditional philosophy as "meaningless". Granted that the traditional methods are lacking in precision, we are faced with the unhappy choice of dealing with significant problems in an unsatisfactory manner or bringing to bear a precise method on problems the wider import of which is not immediately apparent.

There are at least two ways of breaking this deadlock. One could make a case for applying less rigid standards in appraising intuitive methods, or to extend the scope of analytic philosophy to more significant problems. It is only the second path that will be followed up at this point.

In order to understand the origin of the restrictiveness of positivism, we have to go back to the beginnings of mechanistic physics. With the brilliant success of Newtonian mechanics, it seemed tempting to brush away the complex and eroded conceptual system of scholastic philosophy. The proposition of temporarily restricting oneself to the conceptual framework of the new mechanics seems entirely sound, even in retrospect. However, the contention that this conceptual framework would be satisfactory at all times, was unwarranted, and turned out to be actually incorrect. During the mechanistic era it became customary to dismiss types of questions that did not fit into mechanistic systems as unscientific. As the crisis of classical physics revealed the limitation of the mechanistic conceptual scheme, the first inference was that the range of *legitimate scientific questions is even further limited*, since not even all the mechanistic questions are admissible.

The pluralistic character of the present approach brings two new elements into this picture. In the first place, each deductive system implies a characteristic set of precise questions. The number of interesting questions that become "meaningful" is particularly extensive in thermodynamics and quantum mechanics.

It is often stated that the concept of object breaks down in quantum mechanics. Actually, however, the opposite is true. As outlined in Section IV, of *loc. cit.* for the first time in QM we are in a position to give a formal representation of an object with many subtle ramifications and we can now resolve the Eleatic paradoxes stemming from the apparent inconsistency of identity and change. It seems that the new object concept is flexible enough to include living organisms that are entirely outside the mechanistic scheme.

The extension of meaningful conceptual problems in the present context proceeds in still another dimension. Not only do we have the concepts within each deductive system, but the deductive systems themselves are conceptual entities of distinct individual characteristics related to each other in quite specific fashion. These entities are of a logical type that is markedly different from that of the primitive concepts within the deductive systems.

DISCUSSION

Professors Philipp Frank, R. Mayberry, A. Siegel, A. Shimony, F. Steckerl, Rev. W. Wallace, O.P.

Professor Frank: I will only say something on the last part, on the question of operationalism here. As you know, Bridgman took his operationalism from Einstein, but there was from the beginning a certain discrepancy. Bridgman's original idea probably was that in any critical discourse every word should be defined by an operational meaning. And Einstein at the beginning probably agreed with this. But later, I remember the moment when Mr. Bridgman asked me to go for him to interview Einstein, and to ask him flatly, bluntly: "Did you change your opinion, or do you still believe that every concept in a scientific discourse should have an operational meaning?" And Einstein said: "You see, I wouldn't say this. I would not say that every word must have an operational meaning, but it should be required that *in the consequences of every word used there must be an operational meaning*." Also I may add on my part that sometimes the operational meaning isn't explicit, but implicit in the language. At any rate, whatever the intermediate non-operational steps may be, at the end there should be an operation meaning.

Professor Tisza: This is a very graphic statement of the problem and I agree, of course, with the revised formulation of operationalism. I believe that this agreement is general among physicists and, according to the above quotations, this includes also Bridgman in his later years. The problem at present is to catch up with the wide repercussions of the earlier exaggerated statements. In order to achieve this it may be helpful to point out that the early versions of operationalism stemmed from a very natural reaction to the stunning breakdown of the classical concepts. These were so firmly entrenched in tradition as to be attributed to the immutable forms of our thinking apparatus. The strict adherence to empiricism and distrust of a crippling conceptualization were the proper emergency measures in this situation. The development of the physical sciences during the last half century are indications of the success of the new departure, the boldness of which it is hard to appreciate at present. However, precisely the success of the last decades bids us to reconsider some of the attitudes borne out of the emergency. The situation seems more than ripe for a new conceptualization firmly rooted in the rich soil of the new experience, and freed from absolete patterns of thought.

Professor Mayberry: You outlined, in a sense, a hierarchy of determinisms or causalities, or better yet "unfoldings" of states: one, the mechanical predictions of a path; and then second, that which we have in a thermodynamics. I wonder if you could just go back over that again, especially in what sense you thought that the thermodynamic unfolding of states is a more precise or a more adequate causality than that which we usually associate with mechanical causality.

Professor Tisza: I would rather say that there are two complementary principles of causality. The first is the mechanical one, it depends on the exhaustive specification of the initial state and is adequate primarily in celestial mechanics. The second, depending on the selectivity of determining factors is indispensible for the understanding of chemical and biological phenomena, the main features of which exhibit quite stable causal features even though the systems are strongly coupled to their surroundings. This is incomprehensible in terms of the first principle of causality, according to which the slightest variation in the surroundings produces substantial changes in the course of events. Generally speaking the two principles of causality have to be applied jointly.

Dr Fritz Steckerl: In the long run do you think you will have a deductive system, one royal one – you know, a dominant one, spreading out over all possible statements; or may there be independent deductive systems, and which one you choose will just depend on the kind of question you ask?

Professor Tisza: Maybe, but I do not know. It is an important aspect of the present method that, in contrast to the traditions of speculative philosophy, we do not have to, in fact we should not commit ourselves regarding such deep problems. The important matter is the method that allows us to improve in a systematic fashion the logical structure of the existing theories. I believe that this method of analysis, in conjunction with new experiments, will lead to a reasonably unified logical structure. Strong beliefs concerning the specific features of this structure might interfere with the free functioning of the method.

Rev. William Wallace, O.P.: In your diagram the field theories show a type of duality and seem to converge to relativistic gravitational theory and to quantum dynamics. Have you thought about quantizing the gravitational field? Professor Wheeler at Princeton, has worked on this problem. He seems to provide an interesting synthesis of the two field theories.

Professor Tisza: I am not familiar with the attempts of quantizing the gravitational field. This is a fundamental problem that involves the integration of quantum theory with general relativity. It is natural that pioneering work should be done on it. However, I believe that a satisfactory solution may come only after the integration of quantum field theory with special relativity has been accomplished with conceptual clarity. In spite of impressive progress along this line, this aim has not yet been attained. However, your remark on the duality of the two field theories reminds me of a quite specific duality that exists between the geometrical formalism of RG and of STE. A short statement of this duality is given in footnote 4 of reference 3. It is suggestive that a geometrical duality should exists between the two field theories that deal with non-specific gravitational and with specific chemical interactions of matter.

Professor Siegel: I think I'm quoting Tisza when I refer to the kind of determinism in which two chemical reagents give rise to a third, and some kinds of processes involving biological determinism – you once called these morphic determinisms. Is this a kind of determinism whereby a given set of forms determines another set of forms?

Professor Tisza: The two concepts are not entirely the same. A few years ago I advanced the term "morphic determinism". Although I have not changed my mind on the ideas contained in this unpublished manuscript, I prefer now a different terminology. First we have "morphic invariance". This is implicit in QS, the part of quantum mechanics centered around the time-independent Schrödinger equation. A most succinct statement of this principle is as follows: The set of possible pure states of stable atomic systems is countable since it is specified by discrete quantum numbers. Moreover, different realizations of the same pure state are absolutely identical replicas of each other. As outlined *loc. cit.*, morphic invariance is largely responsible for the fact that there *is* an atomic physics, that precise inferences about atomic systems can be drawn from feasible experiments. Temporal concepts play only a marginal role in these considerations. A full quantum dynamics is not yet available for temporal problems. However, the two principles of causality account qualitatively for the observed temporal causal chains.

Professor Armand Siegel: I didn't mean to use your old terminology to commit you. However, I wish to emphasize the looseness of the temporal determinism you emphasize so much. If you seek support from cybernetics, this is a very strongly stochastic probablistic theory. And from the point of view which Einstein emphasized in his criticisms of quantum mechanics, he would have said that this was certainly a powerful theory, but it was an incomplete theory. That is, although quantum mechanics might be dominant to classical mechanics, if quantum mechanics by virtue of the stochastic element is incomplete, and if there is some more complete theory behind it, then it is not dominant and the question of dominance might have to be posed in terms of a deeper deterministic theory.

Professor Tisza: Unless you point out an inherent looseness in my preliminary argument (*loc. cit.*) expounding the second principle of causality, I must presume that your doubts are based on the widely accepted value judgment traceable e.g. to Boltzmann and reappearing in Einstein's critique of quantum mechanics, according to which stochastic theories are inherently second rate compared to strictly deterministic ones. This tradition, invested by great authority, brings it about that many significant contributions to statistical physics are supplemented by expressions of promise or hope, of doing better in the future, by reducing all statistical elements to rigorous deterministic dynamics. However, one of my main contentions is that this attitude originates in the lack of differentiation between the two principles of causality. Whereas the first is inconsistent with stochastic assumptions, the second is not. It seems to me that this circumstance removes a great deal of the obscurity surrounding statistical physics. As an example I will attempt to analyze the famous Einstein-Bohr controversy on the nature of quantum mechanics by breaking down the arguments into, what I believe to be their essential logical elements. According to Einstein:

E1. Nature exhibits both a space-time ordering and a causal ordering, and our theories should do likewise.

E2. The classical theories (CMP, CED, RM, RG) are the *only* ones to satisfy this requirement insofar as they operate in terms of differential equations, with the exclusion of stochastic elements.

E3. Accepting *E1* and *E2* we have to expect that quantum mechanics can be recast into a form similar to that of the classical theories.

The counterarguments of Bohr, shared by Copenhagen school and by a majority of physicists, can be broken up as follows:

B1. The proposition *E1* is dismissed as an arbitrary metaphysical requirement. There is no need to expect that theory should describe "nature" or "reality", whatever these terms may mean.

B2. Identical to *E2*.

B3. Our present knowledge of quantum mechanics is inconsistent with *E3*.

The essence of my own point of view is that by recognizing the bifurcation of the classical causality principle into two principles we are in a position to rejects *E2* = *B2*, since there are causal chains produced by stochastic laws. Granted that details have to be worked out, I believe that it will be possible to accept *E1* and *B3* and reject *B1* and *E3*. Of course, both the space-time order and the causal order have their limitations in theory and in experiment. Besides, the important morphic invariance would be impossible without these limitations.

Professor Abner Shimony: I think in addition to the question that Professor Siegel raised about the dominance of quantum mechanics over the classical mechanics, there are some other considerations which are very troublesome in a scheme like this. Intuitively one wants quantum mechanics (being a microscopic theory) to be dominant over classical mechanics. But in all the careful formulations of quantum mechanics it is not quite that, because in the foundations of quantum mechanics itself, classical concepts enter – particularly the notion of a classical measurement; which means that the place of quantum mechanics in the scheme as it stands now is a very peculiar one. I wonder whether you think that the carrying out of this scheme to its next stage will involve considerable transformation of quantum mechanics in the classic formulation.

Professor Tisza: This is a difficult question, but not all of the difficulties stem from the obscurities of quantum mechanics. I have argued *loc. cit.* that classical mechanics consists of logically heterogeneous parts. I believe that quantum mechanics is dominant over the Hamiltonian mechanics of point masses (CMP) and also over those parts of phenomenological classical mechanics (PhCM) that are needed for the traditional theory of quantum mechanical measurements. However, quantum mechanics still has to be extended to be dominant over some subtler aspects of PhCM.

Massachusetts Institute of Technology, Department of Physics, Cambridge, Massachusetts.

NOTES

1. Karl Menger, A counterpart of Occam's razor in pure and applied mathematics ontological uses, *Synthese* **12** (1960) 415.
2. Henceforth we shall designate the deductive systems by the abbreviations listed in the caption of Fig. 1.
3. L. Tisza, The Thermodynamics of phase equilibrium, *Ann. Phys.* **13** (1961) 1.

4. L. Tisza, and P. M. Quay, S. J., Statistical Thermodynamics of Equilibrium. To be published.
5. In the block diagram QM is differentiated into QS and QD. For a discussion of these finer points we refer to *loc. cit.*
6. It is interesting to compare this situation with another one closely analogous from the logical point of view. The inconsistency of CMP and CED was brought under control by RM. The space-time structure of this theory is consistent with that of CED, but distinct from CMP and dominant to it. However, in this case the common understanding of the situation is more satisfactory, presumably because early in this century CED was in a more advanced state of development than TD.
7. The term "mechanics" is generally used in an ambiguous fashion either to denote analytical dynamics or any of a variety of more phenomenological disciplines. We refer to Section IV B of *loc. cit.* where logical differentiation is used to resolve the conceptual spectrum. In the above discussion CMP is taken in a narrow sense as analytical dynamics.
8. N. Wiener, *Cybernetics*, 2nd edition, New York, J. Wiley and Sons, Inc., 1961.
9. These are probabilistic laws connecting the states of a system at different times with each other.
10. P. W. Bridgman, P. W. Bridgman's 'The Logic of Modern Physics' after Thirty Years, *Daedalus* **88** (1959) 518.

Printed in U. S. A.

REVIEWS OF MODERN PHYSICS VOLUME 35, NUMBER 1 JANUARY 1963

The Conceptual Structure of Physics*

LASZLO TISZA

Department of Physics and Research Laboratory of Electronics, Massachusetts Institute of Technology, Cambridge, Massachusetts

CONTENTS

INTRODUCTION

THE systematization of the theories of physics can be attempted along two different lines. A classification is either based on the logical conceptual structure, or else follows the pattern of classification of the objects that are of principal interest in the theory.

The most important example of the first procedure is provided by late 19th century physics which is divided into mechanics, electrodynamics, and thermodynamics. This logical structure, and particularly the program of ultimate unification by reduction to classical mechanics, was shattered by the crisis that marked the transition from classical to modern physics. At any rate, the practical needs of contemporary research shifted the emphasis to the second of the abovementioned methods of systematization, within which one speaks of the physics of elementary particles, nuclei, molecules, fluids, solids, plasma, and stars (to mention only some of the most important divisions). Further subdivisions have been considered which depend on the expanding range of knowledge. Thus, one has not only the physics of semiconductors and masers, but also the physics of elementary particles in various ranges of energy.

Both methods of organization satisfy specific and important needs. On the one hand, the classification by objects of interest is indispensable for making specialized knowledge possible. However, the totality of physics presented in this fashion has an encyclopedic character and is far beyond the grasp of an individual. On the other hand, the classical method of organization according to logical structure exhibits the simplifying and unifying power of high-level abstractions. This feature presumably accounts for the appeal of classical physics, which endures even in the face of breakdowns and limitations.

Of course, the unifying power of the classical logical structure is severely limited. The difficulties stem primarily from the fact that this structure centers around Newtonian mechanics as the basic discipline. The situation was characterized by Einstein[1] as follows:

"In spite of the fact that, today, we know positively that classical mechanics fails as a foundation dominating all physics, it still occupies the center of all of our thinking in physics. The reason for this lies in the fact that, regardless of important progress reached since the time of Newton, we have not yet arrived at a new foundation of physics concerning which we may be certain that the manifold of all investigated phenomena, and of successful partial theoretical systems, could be deduced logically from it."

During the quarter of a century that elapsed since Einstein wrote these lines, the discrepancy between the classical mechanical thinking and the formal development of physics has further increased. At the same time there is no decisive advance in the emergence of a new comprehensive foundation that can be accepted with assurance.

The purpose of this paper is to describe a new technique of logical analysis that is to bring about a more harmonious relation between conceptual thinking and formal developments. However, this is to be achieved within a program the scope of which is much more modest and manageable than the one hinted at by Einstein.

We propose to sort out and improve the logical structure of the existing, empirically verified theories. The establishment of a new foundation is the expected outcome rather than the prerequisite of this procedure.

* Supported in part by the U. S. Army Signal Corps, the Air Force Office of Scientific Research, and the Office of Naval Research; and in part by the U.S. Air Force (Office of Scientific Research, Air Research and Development Command) under Contract AF49(638)-95.

[1] A. Einstein, J. Franklin Inst. 221, 349 (1936). Reprinted in A. Einstein, *Ideas and Opinions* (Crown Publishers, Inc., New York, 1954), p. 300.

The problem is to take advantage of the use of high-level abstractions while minimizing the risks inherent in their use. Successful abstractions have a predictive value of unexpected scope that greatly exceeds the range of their original empirical foundation. This situation often produces an unjustified confidence in the absolute validity of the theory. Yet not even the most perfect agreement between theory and experiment, extending over a wide range, provides a guarantee against the appearance of an essential disagreement as the range of experience is extended even further. The resolution of such disagreements may require the reconceptualization even of those theories that have been confirmed by experiment.

As a protection against the ossification of successful theories, we propose to make the connection between basic assumptions and experimental predictions sufficiently manifest, in order to facilitate a continued readjustment of the basic assumptions to the expanding range of experience. The use of deductive systems is an important device for achieving this goal.

The basic disciplines of classical physics are indeed organized as deductive systems. However, during the 19th century when the three major classical disciplines assumed their definitive form, the nature of deductive systems was still very incompletely understood.

The modern conception of deductive systems to be used in this paper is in some of its main aspects diametrically opposed to the classical one. Traditionally, deductive systems are used to organize knowledge that is already well substantiated. It is deemed essential to start from a basis that is safe from any emergency that might call for its revision. Of course, the origin of this reliable basis was an ever unsolved mystery. For existing systems that have been accepted as entirely dependable, the problem seems to be merely a recondite philosophical quest for the justification of something about which there is no real doubt anyway. However, in the present state of physics the problem of foundations manifests itself in the painful absence of a dependable starting point for developing important theories of practical interest.

Accordingly, the present effort is directed mainly toward the clarification of the problem of foundation. We shall attempt neither to justify the traditional foundations, nor to advance any new foundations on the basis of *a priori* plausibility. The method to be used is the *explicit* recognition and the systematic improvement of the logical structure that is *implicit* in the existing experimentally supported theories.

We abandon the traditionally static, not to say dogmatic, relation between postulational basis and system. In the present approach the basis is tentative and subject to change if this leads to an improvement of the system in accounting for experimental facts. This feedback from system to basis leads to the dynamic, evolutionary adjustment of the system to a widening range of experience. Instead of assuming that deductive systems have to be perfect in order not to collapse, the present method of analysis deals with imperfect systems. In fact, one of the tasks of the method is to locate and eliminate imperfections. In view of this situation, we shall refer to this method as *dynamic logical analysis*, or the *dynamics of deductive systems*.

Section I of this paper contains a survey of the dynamic principles for the construction of deductive systems. These principles are not new, they have been developed in great detail in mathematical logic, and the empirical aspects of deductive systems have been investigated within logical positivism. The novelty of our presentation is therefore only a matter of selection and emphasis, and, last but not least, the contention, and, we hope, the demonstration that these principles are of practical use.

The actual construction of physically relevant deductive systems along these lines is, of course, a rather tedious enterprise. However, two instances have been actually produced thus far. First, the macroscopic thermodynamics of equilibrium,[2] briefly MTE, and the statistical thermodynamics of equilibrium,[3] briefly STE.

The axiomatizations of these well-known theories demonstrates the fact that the application of modern logical principles leads to a considerable restructuring—even of classical disciplines that seemed to have assumed a fairly rigid pattern.

It is an important insight of mathematical logic that each deductive system contains primitive concepts that cannot be defined by conventional means, but are implicitly defined by the deductive system as a whole. This "definition by postulation" of primitives, so important in modern mathematics, will be shown to be valuable also in physics. While recognizing the basic identity of the method in both disciplines, one should also take note of a difference stemming from the fact that the deductive systems of physics are always supplemented by rules of correspondence which establish a link with experience. Mainly as a consequence of this connection, most

[2] L. Tisza, Ann. Phys. (N. Y.) **13**, 1 (1961).
[3] L. Tisza and P. M. Quay (to be published).

concepts of physics evoke in the mind images or models, for which the postulates are presumably true statements. Although this intuitive aspect of the abstract concepts of physics is important for many reasons, the relations between experiment, intuitive models, and deductive systems are intricate, not unique, and hard to state in precise terms. The method of definition by postulation is an effective tool for clarifying the semantic ambiguities that arise out of this situation.

The fact that primitive concepts assume a precise meaning only relative to the deductive system into which they are incorporated, can be aptly referred to as the *relativity of concepts*.

An apparent difficulty for the deductive method is that most of the basic disciplines, say classical mechanics, thermodynamics, quantum mechanics, are too ramified and diverse to be compressed within the confines of a rigorously built deductive system. We propose to overcome this difficulty by constructing a cluster of deductive systems, even for representing one basic discipline.

Since, as we have pointed out, each deductive system implicitly defines its own set of primitive concepts, the increase in the number of systems is tantamount to an increase in the number of fundamental concepts that are defined relative to a particular context specified in terms of a *deductive system*. For this reason, we designate the procedure leading to a collection of deductive systems as *logical differentiation*.

The conceptual wealth brought about by the method of logical differentiation provides us with all the freedom we may need to adjust our deductive systems to a variety of experimental situations.

However, if the analysis should stop at this point, one might feel that the freedom in the formation of deductive systems has been carried too far, since it seems to lead to a proliferation of basic concepts. Actually, however, this process is held in check by the second stage of our procedure, the *logical integration* of deductive systems. This is a procedure which, to our knowledge, is advanced here for the first time.

The problem of logical integration arises from the methodological postulate that the various deductive systems are to express various aspects of the same reality. The reference to "reality" is apt to provoke queries of a philosophical nature. However, for the purposes of this paper it is satisfactory to interpret the statement above simply as follows. The separation of phenomena into "mechanic," "electric," "thermodynamic," etc., each described in terms of several deductive systems, is an abstraction. Real phenomena exhibit all of these aspects at once, and the various deductive systems should be applicable in conjunction with each other, and hence should be compatible.

The program of logical integration, pursued by ensuring compatibility or mutual consistency of deductive systems, is the central issue of this paper.

In Sec. II we examine the logical requirements under which two systems, strictly speaking inconsistent with each other, can nevertheless be made compatible for empirical purposes. In this procedure we assign a central role to a simple relation that is the prototype for such relations that exist, say, between relativistic mechanics (RM) and classical mechanics of particles (CMP), or quantum mechanics (QM) and CMP. Here and in what follows, the capitalized abbreviations refer to deductive systems, and are explained in the block diagram (Fig. 1). This diagram represents the integrated system of the major deductive systems of physics, obtained by the repeated application of our fundamental relation. We note that this relation is asymmetric: In an appropriate limiting case one of the systems, to be called *dominant* or *fundamental*, reduces to the *derivative system*. The trend towards the integration of the existing deductive systems can be conceived of as the driving force leading to such highly abstract and fundamental systems as RM and QM. In this integration process, the arbitrariness in the selection of single deductive systems is drastically reduced. It appears that taking cognizance of the *relative character* of *concepts* is the prerequisite for establishing increasingly *objective* and unique *integrated* logical *structures*.

We shall refer to the doctrine that calls for the establishment of such structures as are represented in Fig. 1 as *integrated* or *consistent pluralism*.

The problem of logical integration does not arise in mathematics. Even though some of the deductive systems are related to each other, in general, each deductive system of mathematics constitutes a universe of discourse of its own. Therefore, it is not surprising that the intersystem relation of Sec. II has not been discussed in mathematical logic.

We consider now the relation of our method of logical integration to that which is implicit in classical physics. The classical method of integration has a *monistic* character: The consistency of physics is achieved by singling out one of the existing deductive systems as "fundamental," while all other systems are considered "phenomenological," to be "justified" by reduction to the fundamental system. The concepts of the latter are the only ones directly related to "physical reality." Problems, which seem to play a key role in the reduction of phenomenological systems to the fundamental one, are accorded the status of *fundamental problems.*

From the time of Newton until the end of the 19th century the fundamental deductive system was assumed to be classical mechanics. The ideal monistic organization of physics was mechanistic. As the shortcomings of classical mechanics became evident, unified field theory and quantum mechanics were supposed to play the same fundamental role.

Although the monistic philosophy stimulated many brilliant achievements and scored interesting partial successes, the idea was never more than a program, its ultimate feasibility a matter of metaphysical conviction.

The monistic mechanistic metaphysics was indeed criticized from the positivist empiricist point of view. It was pointed out in essence that: (a) There is no reason why an expanding range of experience should be reducible to a system distilled from past experience, and (b) the narrowness of the alleged fundamental system is in striking contrast with the conceptual wealth of the systems to be reduced.[4]

The suggested alternative is the *positivist,* or *operational method,* according to which, abstract concepts should be introduced only if directly anchored in experimental operations. This procedure, at least in its more extreme versions, would discourage the introduction of high-level abstractions for the consistent organization of a wide range of experience. The positivist approach played an important liberating role in stimulating the development of modern physics along lines which were inconsistent with the mechanistic logical structure.

Nevertheless, operationalism has its unsatisfactory aspects. It leads to a highly fragmented state of physics that could be aptly described as *eclectic, nonintegrated pluralism.* Moreover, a close adherence to the operational principle is no more feasible than a strict adherence to the mechanistic organization. Actually, the practice of physics, at present, is an uneasy compromise between monistic unity and the diversity of eclectic pluralism. The exact nature of this compromise eludes precise formulation.

Even a most cursory inspection reveals that our integrated pluralism differs in essential respects from all versions of nonintegrated pluralism, and also from the classical monistic structure, although the difference is more subtle. It is possible, and the author finds it most likely, that the procedure of logical differentiation and integration will ultimately lead to a single system that dominates all other systems. Even if this conjecture should turn out to be correct, the present situation is very different from the one postulated in the monistic view according to which we are *already in possession of the most fundamental deductive system.* The present method has an open-ended character, greatly in contrast with the traditional idea of the deductive method, according to which everything is implicit in the already available basis. In a certain sense we are dealing with an *inverse deductive* method. Starting from the experimentally tested low-level abstractions, we search for the basis from which the former can be derived.

The first two sections of the paper are written in a manner which, we hope, is detailed enough to convey a reasonably definite idea of the logical equipment which we propose to use. The remainder of the paper is devoted to applications of this equipment. Since the potential scope of these applications seem to be quite extensive, our account of it is necessarily sketchy, it merely prepares the ground for further elaborations.

The applications can be, by and large, grouped into two categories; The first is only a better systematization of existing knowledge, whereas the second leads to genuinely new results. Although traditional logical analysis was brought to bear only on the first type of problem, in the present approach the two are closely related to each other.

As a first step we have a crude systematization of the major systems of physics represented in the block diagram of Fig. 1. The blocks symbolizing theories are connected by various symbols representing the logical relations defined in the first two sections. Strictly speaking, these symbols have a precise meaning only if the theories are represented in terms of reasonably rigorous deductive systems. Generally speaking, this condition is not satisfied, but some of the theories, say, those related to classical fields, are so well organized, that a more rigorous axiomatization is unlikely to affect the situation substantially. For cases of this sort, the block diagram can be considered as a table of contents for a novel type of encyclopedia of theoretical physics. Also, the first three sections should provide sufficient instructions for the specialist who might wish to axiomatize his own field of interest. The requirements of mutual consistency provides certain "boundary conditions" for this endeavor, but conversely he might generalize, widen, and certainly enrich the block diagram itself. Clearly, the latter does not constitute a rigid "grand

[4] The second argument is, at present, entirely obvious if the fundamental system is classical mechanics or unified field theory. The case of quantum mechanics is more subtle and we shall return to it in Sec. IV.

scheme" that one would have to accept in its totality at the outset.

A particularly interesting case falling into the second category is thermodynamics. Recently two axiomatizations of this discipline were developed, the first, MTE, leads to a restructuring of the macroscopic theory,[2] the second, STE, to an extension into the statistical domain.[3] It is our contention that implicit in these theories is a conception of the structure of matter that is in a way complementary to the mechanistic one implicit in CMP. At least potentially, this conception seems to be more microscopic than the mechanistic one and provides a more satisfactory transition to quantum mechanics than the historical path emphasizing the connection with CMP. The substantiation of these claims is, of course, a very extensive project the outline of which is contained in Sec. IV. The details are being developed in a series of papers, of which references 2 and 3 are the first two. At this point, we advance two remarks that might serve to explain our statement to some extent. CMP can be conceived as the conceptualization of astronomy, a macroscopic observational discipline, while thermodynamics is the conceptualization of chemistry, an experimental discipline dealing with structural transformations. The mathematical equipment of CMP is essentially the theory of differential equations, that of thermodynamics is mathematical statistics. To insert statistical elements into differential equations has a disruptive effect on the formalism. This seems to have been the reason for Einstein's objections to the statistical aspects of quantum mechanics. In contrast, differential equations may appear as limits of stochastic difference equations and can also be inserted in other ways into mathematical statistics. Hence this change in point of view opens up the possibility of a constructive solution of the famous Einstein–Bohr controversy.[5] In particular, we shall formulate a second principle of causality that is characteristic of the thermodynamic theories and that is in contrast to the traditional mechanistic principle.

The switch from differential equations to mathematical statistics profoundly affects the type of questions that have to be admitted as legitimate in physics. Whereas it is traditionally emphasized that quantum mechanics restricts the range of measurements that were thought to be possible within CMP, it has not been brought out that quantum mechanics renders many real measurements possible that were not even conceivable within the classical theories. It is apparent from these remarks that the present point of view is not without its philosophical implications. A few of these we shall discuss in Sec. V.

[5] *Albert Einstein: Philosopher-Scientist,* edited by P. A. Schilpp, (The Library of Living Philosophers, Evanston, Illinois, 1949).

I. DYNAMICS OF A SINGLE DEDUCTIVE SYSTEM

Deductive systems play a central part in the forthcoming discussion and we shall first summarize their properties, to the extent that these are relevant for our purposes.

A deductive system is a structure of concepts and propositions ordered according to the following principles. Complex concepts are explained in terms of simpler ones and, in the last analysis, the whole conceptual structure is derived from a few *primitive concepts,* or briefly, *primitives.* Definitions of this sort will be denoted as *synthetic definitions.* Similarly, complex propositions or theorems are inferred from simpler ones and are finally anchored in *postulates.*

The primitive concepts and postulates of the system will be briefly referred to as its *basis* or *foundation.* The *basis* has to be supplemented by rules governing the derivation of theorems and the formulation of synthetic definitions. In some respects a deductive system can be considered as a language; the aforementioned rules constitute its grammar.

The experimental relevance of the deductive systems of physics is conveyed by *rules of correspondence or coordination* that establish a relation between the conceptual elements on the one hand, and the objects of our common experience on the other. In general, the rules of coordination relate substantial parts of several deductive systems to extensive classes of measurements. It is seldom, if ever, possible to provide a convincing and direct experimental proof of the postulational basis.

In principle, we have to specify also the disciplines preceding the one under consideration, which are to be used freely. For the systems of physics, parts of logic and mathematics are such prior disciplines. However, these need not be explicitly listed for the current discussion.

We shall call the sequence of steps connecting the basis with a theorem a logical chain, or briefly, a chain. Symbolically,

$$T[C;P;D]$$

denotes a theorem T derived from the postulates P and involving the primitive concepts C, through the intermediary of a specified sequence of steps constituting a derivation D.

The logical content of the formula $T[C;P;D]$ can be expressed as follows: The theorem T is at least as true as the postulates entering its argument. In other words, the acceptance of P implies the acceptance of T, but the converse is not true. It is often possible to prove the same theorem within different systems.

In discussing systems as a whole we can take one of two possible attitudes symbolically represented in the following formulas:

$$S = S\{C;P\}\ , \tag{1.1}$$

and

$$S = S\{C;P;D\}\ , \tag{1.2}$$

respectively. We shall refer to (1.1) as the *logical composition* of the system. It contains the entire basis that generates the system, whereas in a chain we insert only those concepts and postulates that are necessary for the completion of the proof. In principle, the basis determines the complete logical potentialities of the system, and formula (1.1) expresses the important fact that the basis is, in a way, the nucleus that succinctly contains the entire system, which is often too extensive to be apprehended otherwise.

Evidently, deductive systems contribute most effectively to the economy of thought. Yet one should not lose sight of the fact that statements about the potentialities of a system have often a speculative character. It is not always obvious whether persistent difficulties of solving a problem are only of a technical–mathematical nature, or whether they stem from the inadequacy of the basis.

Therefore, in a more conservative vein, it is sometimes useful to refer to a system in terms of Eq. (1.2). In other words to consider the system as the collection of theorems that has actually been derived at a particular stage of the historical development.

The construction of a deductive system essentially consists of the selection of the basis and the derivation of chains leading from the basis to theorems which, in turn, are related to observations. Thus, three distinct fields are open for logical analysis. The first pertains to the origin of the chains and consists of the evaluation and justification of the basis; the second is concerned with the rules for the derivation of the chains from a given basis, while the third deals with the rules of coordination that relate the end points of the chains to experiments. Of the three possible lines of inquiry, we shall be concerned here with only the first.

The second line of study deals with the routine operations of deriving theorems from a fixed basis. This corresponds to the conventional idea of the "deductive method." We need not enter into the discussion of this aspect of the problem, partly because it has been thoroughly investigated within traditional logic, and partly because the standards of formal rigor emerging from these discussions are actually too high for the purposes of physics. For these purposes, the use of a reasonably precise mathematical formalism seems amply sufficient.

Although the third type of study, pertaining to the rules of coordination, is of considerable interest, we shall not enter into its discussion now. The rules of coordination are of a subtle nature; they contain abstract concepts on the one hand, and terms describing apparatus and objects of our common experience on the other. The disentangling of these various ingredients involves difficult epistemological problems. The more one ponders about these matters, the more surprising it seems how well these problems are handled in actual practice. However interesting it might be to clarify the logical basis of this highly intuitive activity, it seems unlikely that such a logical analysis could be of any real help to the experimental physicist at the present time.

The case for logical analysis is much more favorable when we turn to the first of the above mentioned problems and inquire into the meaning of the bases of deductive systems and the justification of their selection in preference to other possibilities.

In the discussion of the basis, the first problem to be considered is the clarification of the meaning of primitive concepts. Forming the point of departure for the synthetic definition of complex concepts, barring circularities, primitives cannot be defined in terms of even simpler concepts. Thus arises the paradoxical situation that the most important concepts of physics, or of any other science for that matter, cannot be defined along traditional lines.

The current attitude with respect to the paradox of undefinable concepts varies widely. Perhaps, the most frequent procedure is to cover up the difficulties by pseudo-definitions that fail to convey any precise meaning. At the other extreme is acceptance of the difficulty as inherent in the situation.[6]

There have also been attempts to explain the meaning of primitives in terms of "operational definitions," or rules of coordination. This program was never actually carried out in detail. In the author's opinion, the difficulties encountered are of a fundamental nature. The primitives form the beginning of the chains, while the connection with experi-

[6] "My own pet notion is that in the world of human thought generally, and in physical science particularly, the most important and most fruitful concepts are those to which it is impossible to attach a well-defined meaning." H. A. Kramers, Motto of *Collected Scientific Papers* (North-Holland Publishing Company, Amsterdam, 1956).

ment occurs at the end points. We clearly have two problems: the meaning of the primitives, and the experimental verification of the theorems. The power of experiment in deciding among competing assumptions will be greatly increased if the meaning of these assumptions is clear.

We propose to solve the problem of primitives by using the method of "definition by postulation" that is widely practiced in mathematics. According to this procedure, primitive concepts remain undefined in the synthetic sense, but their use is specified in terms of the postulates of the deductive system of which they are the ingredients. In fact, these postulates provide the instructions for the use of the primitives in forming the chains, and hence the observable prediction of the system.

Thus the precise meaning of the "undefined" primitive concepts is contained implicitly in the entire deductive system symbolically represented by its "formula" $S = S\{C;P\}$. We shall refer to this formula as the *analytic definition* of the concepts C.

Of course, we cannot expect a system to define a satisfactory set of concepts unless the formalism of the system is consistent.

The method of analytical definition is undoubtedly very abstract and is usually supplemented by the introduction of ideal objects, or intuitive "models" used as the interpretations of the primitive concepts. In this interpretation the postulates have to be true statements. Such intuitive notions are very useful and are instrumental in suggesting the experimental methods for the verification of the theory. Nevertheless, this use of models has many pitfalls. Their meaning is often not as clear as one might wish, and models that are intuitively accepted, or at least favored, are often oversimplified as compared with the theory to be interpreted.

In the presence of such difficulties, falling back on the method of analytic definitions may be the only recourse. In this way, we can establish the precise meaning of such concepts as "space," "time," "particle," etc. The problem of defining these by conventional means is complicated by the fact that these terms are currently used in the context of different deductive systems, such as classical, relativistic, and quantum mechanics. The method of analytic definitions makes it evident that the meaning of words denoting primitive concepts undergoes substantial changes as the context changes from one deductive system to another, or even as a deductive system is modified under the impact of expanding experience. Analytic definitions are most appropriate for handling these semantic difficulties which are quite baffling on the level of verbal arguments. It appears that primitive concepts have a definite meaning only *relative* to a particular *context* characterized with precision by a deductive system and its formula.

We now turn to the problem of the justification of the postulational basis. First of all, we recognize that the basis is hypothetical, selected on intuitive grounds without any claim of *a priori* justification. Being arbitrary, the basis is not unique and there are in general several competing possibilities to account for the same class of phenomena. The relative merits of the various bases are evaluated *a posteriori* on the strength of the system generated by them. While the theorems within a system are justified by reduction to the basis, this situation is now reversed and the basis is justified by the system as a whole. If need arises, the basis is improved by feedback coupling from the developed system. Thus the circularity of reasoning, carefully eliminated from the intrinsic structure of the system, reappears "in the large" in the dependence of the basis on the system as a whole. This circularity is not "vicious," and can be handled to great advantage for the improvement of the system.

Concordance with experiment is of overriding importance for the evaluation of deductive systems. We shall not even consider systems unless concordance with experiment is ensured for a significant range.[7] Nevertheless, this empirical requirement is not sufficient by itself. The main reason is that the same experimental facts can be conceptualized in radically different ways, and we need criteria for selecting among the various possibilities.[8] The problem is to identify those highly abstract concepts that are most effective in creating order within the chaos of raw experimental facts, described in terms of low-level abstractions. In order to facilitate such judgments, it is desirable to have criteria for deciding in an objective fashion whether, and to what an extent a theory is *fundamental, phenomenological, ad hoc,* or merely an instance of curve fitting. It is also desirable that these criteria be more specific than a requirement of "simplicity."

Although we cannot expect to establish hard and fast rules for deciding such questions, we shall arrive

---

[7] Concordance is not an absolute concept, but is *relative* to a certain range. A further discussion of this point is found in the next section.

[8] This can be made plausible in terms of the following simple example. A finite portion of a plane can, within finite accuracy, be approximated by an infinite variety of surfaces of small curvature. Hence, plane geometry can be replaced for empirical purposes by an infinity of Riemannian geometries. This argument is easily adjusted to other deductive systems such as classical mechanics.

at workable criteria from the examination of the logical structure of the theories. This structure is evaluated in terms of such basic properties as the *consistency*, *independence*, and *comprehensiveness* of the axioms. We note that in virtue of formula (1.1) we can make such statements equally well about the basis and about the system. Moreover, these properties arise both in intrasystem and intersystem relations. The latter are of particular importance for the purposes of this paper.

If two chains have been found within a deductive system, one of which is the assertion, the other the negation of the same theorem, we are faced with an *inconsistency*. A system is consistent if it contains no inconsistency. The negative character of this definition renders the branch of mathematical logic dealing with the scope and limitations of proofs of absolute consistency extremely difficult.

Therefore, we hasten to point out that we shall accept the consistency of dependable deductive systems without known inconsistencies on pragmatic grounds. Our concern will be a much more manageable problem, the *mutual consistency* of two systems to be considered in the next section.

We turn now to the discussion of the *independence* of the postulates.

Let us suppose that a logical chain is discovered connecting a statement P_0, formerly believed to be a postulate, with the rest of the postulates: $P_0[C,P]$. In such a case we say that P_0 is dependent on the rest of the basis. If it is impossible to find such a chain, P_0 is independent.

It is seen that dependence is as tangible a relation as inconsistency, whereas independence shares the elusiveness of consistency. Economy requires that various postulates of a deductive system be independent of each other. However, under static conditions the dependence of an axiom on the rest of the basis is an esthetic flaw rather than a serious defect.

The situation is different under dynamic conditions. In the case of imperfect systems it is important to know that certain postulates are independent of each other and can be independently rejected or retained as the system is revised.

The problem of independence is particularly interesting as an intersystem relation. The question to be decided is whether or not one deductive system is reducible to another.

Another important property of a postulational basis is its *comprehensiveness*. By this we mean the explicit formulation of all of the *essential* assumption actually involved in the development of the system.

The application of the criterion of comprehensiveness clearly involves an important element of judgment. Comprehensiveness can be overdone and the proliferation of trivial postulates could degenerate into hair splitting. A positive sign that the addition of a heretofore hidden assumption to the basis is worthwhile, is obtained if the denial or modification of the postulate in question generates a new and interesting system. Significant examples of this sort were demonstrated by Hilbert[9] who found that Euclid's axiomatization of geometry is not comprehensive. He proved the independence and significance of the additional axioms by developing consistent geometries in which these postulates are negated.[10]

We shall see in Sec. IV that the situation is somewhat similar in thermodynamics. The new development of the subject is brought about by modification of postulates that were not explicitly formulated in the classical theory.

II. INTERSYSTEM DYNAMICS

The dynamic interaction between a deductive system and its foundation leads to occasional revision of the latter. Often the revised system is in every respect superior to the earlier one, and completely replaces it in the living structure of physics.

However, there arise situations in which the old system cannot be dismissed as a historic relic, but has to be used along with the improved system. The joint use of different deductive systems for describing different aspects of reality raises novel problems of logical integration.

In preparation for this discussion we consider in this section, two principles that ensure the compatibility of certain types of deductive systems. It is an interesting fact that the formal logical aspects of these compatibility conditions can be recognized within a model of transparent simplicity. The same formal conditions can be given different concrete interpretations and used to handle most intricate situations.

The model in question is the practical (physical) geometry of a spherical surface, in other words geodesy. The physicist–geometer can avail himself of two abstract geometrical systems, to wit spherical, and plane geometry. From the point of view of the abstract mathematician the two systems are different and a discourse has to be completed in one *or* the

[9] D. Hilbert, *Die Grundlagen der Geometrie* (B. G. Teubner, Liepzig, Germany, 1956), 8th revised edition. English translation of the first edition (The Open Court Publishing Company, LaSalle, Illinois, 1899, reprinted 1950).

[10] Consistency is used here in the relative sense. The new geometries can be mapped into Euclidian geometry. Any inconsistency in the former would imply an inconsistency in the latter.

other system. The consistent use of the two systems for the description of the real, physical sphere depends on a number of circumstances.

In the first place, we establish a hierarchy between the two systems: We stipulate that in case of conflict spherical geometry should have priority. We express this by saying that spherical geometry is the *dominant* system, and plane geometry is called either *derivative*, *subordinate*, or *degenerate*.

We shall refer to the relation of these systems as supplementary *relation*, or briefly, *supplementarity*.

The second point is a clarification of what constitutes a "conflict" between the systems. In the mathematical sense a conflict always exists; however, in physical geometry we dismiss it as empirically irrelevant if it falls below the finite accuracy of the measurements which the theory is supposed to describe. The accuracy of measurement cannot be improved over the diffuseness set by thermal noise.

Somewhat more generally, we proceed to formulate a principle to which the rules of coordination have to conform. Experiment provides us with the values of continuous variables within a certain accuracy. However, the analysis of mathematical physics would be extremely cumbersome and lose much of its precision if the finite width of continuous parameters corresponding to empirical quantities were to be observed at every step. In actual practice, the segments of finite widths are sharpened into definite points of the continuum for the purposes of the analysis. However, the results obtained have no physical meaning unless they are sufficiently insensitive to the actual unsharpness of the continuous parameters.[11]

Mathematical solutions satisfying this requirement will be called *regular*. Solutions that change their essential features when the fictitious sharpening is given up will be referred to as *singular*. They are devoid of physical meaning except as possible stepping stones to more realistic results. The requirement that mathematical solutions be regular in order to have a physical meaning will prove to be of great importance and deserves to be called a principle, the *principle of regularity*.

The second point concerns the question of *scale*. Spherical geometry is characterized by an absolute unit of length R, the radius of the sphere. If $R \to \infty$, the sphere degenerates into the plane, the dominant into the supplementary system. Let the linear dimension of a geometrical figure be a, then the parameter a/R appropriately expresses the deviation of spherical from plane geometry. It becomes meaningful to say that for small values of the parameter ($a/R \approx \epsilon \ll 1$, where ϵ is the unsharpness of the parameter because of the inaccuracy of the measurement or because of noise) the subordinate system is asymptotically valid. Otherwise we have to fall back on the dominant system.

It is interesting to note that within plane geometry there is no absolute unit and it is meaningless to speak about small or large figures. All theorems are invariant under a transformation consisting of a change of the unit of length. We say also that plane geometry is *scale invariant*, or, it contains the theorem of similarity.[12]

The transition from the subordinate to the dominant system marks the breakdown of scale invariance. At the same time, however, the dominant system is characterized by an invariance of a novel type. In spherical geometry we have invariance of curvature. The more subtle invariances of the dominant systems of physics will be considered in Secs. III and IV.

In the course of historical evolution the dominant system has to be established by generalizing the subordinate system that reached what we shall call a *marginal breakdown*. In general, this is a task of great difficulty. While the converse procedure of arriving through a limiting process from the dominant to the subordinate system is a great deal more straightforward, even this transition is not without pitfalls.

Just as the sphere does not contain the infinite plane, so spherical geometry does not contain plane geometry. While the limiting process $R \to \infty$ transforms every theorem of spherical geometry into one of plane geometry, the converse is not true. Thus the important laws of similarity of triangles and other figures have no counterpart in spherical geometry. The possibility of deriving such theorems is a new discovery, possible only in plane geometry. We shall see that the situation is quite similar for other pairs of dominant–supplementary theories.

We shall say that the inconsistency between the subordinate and the dominant system is *under control*, or that the *inconsistency* is controlled, if the range of validity of the subordinate system is known, and within these limits the compatibility with the dominant system is ensured.

[11] See R. Courant and D. Hilbert, *Methoden der Mathematischen Physik II* (Verlag Julius Springer, Berlin, 1937). p. 176. Léon Brillouin, *Science and Information Theory* (Academic Press Inc., New York, 1962), 2nd ed. Chapter 21.

[12] E. P. Wigner, Proc. Am. Phil. Soc. **93**, 521 (1949).

III. LOGICAL INTEGRATION

Table I contains a simplified, symbolic representation of the integrated structure of systems that results from the application of the principles discussed in the last two sections. Although this structure, briefly, the *block diagram*, is still incomplete and tentative, it is adequate to provide the point of departure for our discussion. We have to remember that our method is set up to eliminate the shortcomings of the initially posited structure rather than to perpetuate them.

The block diagram can be discussed in different degrees of detail. On the most superficial level the theories associated with the various "blocks" are interpreted along traditional lines as suggested by their designation listed in Fig. 1. It will be recognized that the theories connected by arrows are indeed in a supplementary relation with each other, and the limiting processes leading to the derivative systems

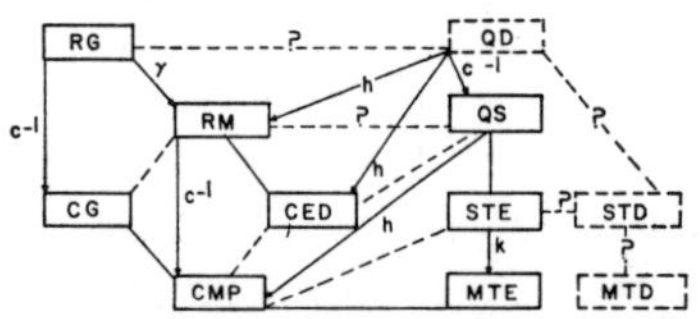

——COMPATIBILITY - - - - CONTROLLED INCONSISTENCY, ——→ SUPPLEMENTARY RELATION —?— INCOMPLETELY UNDERSTOOD RELATION

FIG. 1. Block diagram representing tentative structure of physics. Explanation of symbols designating deductive systems:

CMP	Classical mechanics of particles,
CMP (μ)	Classical mechanics of particles (microscopic interpretation of particles),
PhCM	Phenomenological classical mechanics,
CG	Classical gravitational theory,
CED	Classical electrodynamics,
RM	Relativistic mechanics,
RG	Relativistic gravitational theory,
TD	Thermodynamics,
MTE	Macroscopic thermodynamics of equilibrium,
STE	Statistical thermodynamics of equilibrium,
MTD	Macroscopic thermodynamics,
STD	Statistical thermodynamics,
QM	Quantum mechanics,
QS	Quasistatic quantum mechanics,
QD	Quantum dynamics,

are formally associated with the vanishing of the absolute constants indicated in the diagram.

Maybe, the most significant aspect of the block diagram is that it exhibits CMP as derivative to several systems and dominant to none. This is in complete contrast with the logical structure hypothesized in classical physics in which CMP should have been dominant to all the other deductive systems of physics. We shall use the discussion of the block diagram to sketch some of the refocusing that is made necessary by the changed role of mechanics.

In the current section we shall deal primarily with the relations involving CM, CED, and RM. These relations are fairly well understood and our brief discussion serves mainly to clarify our terminology in handling this situation; moreover it provides the prototype for the discussion of the more problematic relation among CMP, STE, and QM in Sec. IV.

If the discussion of the block diagram is to go beyond the broadly qualitative state, the theories have to be represented as deductive systems. This form of organization is not current in physics at the present time, and the traditional axiomatizations that are available, say for thermodynamics, do not satisfy the requirements of the present program.

Attempts to remedy this situation by up-to-date axiomatization bring to focus another complication. Most of the basic theories of Table I are too ramified and diverse to be compressed within the limitations of a rigorously built deductive system. However, these difficulties can be overcome by using a multiplicity of deductive systems, even within one basic theory. The establishment of this more intricate logical structure is in each case a major enterprise. In Sec. IV we summarize the specific problems that were encountered in MTE and STE.

In this section we first consider the case of CED, which is much simpler because the elaboration involves no departure from traditional lines of thought. A satisfactory axiomatization could be centered around Maxwell's equations, or an appropriate variational principle. The properties of material media are to be accounted for only phenomenologically; a microscopic description is outside the scope of CED.

If we let the light velocity $c \to \infty$, CED reduces to electrostatics (ES) which can be considered as a derivative system of the dominant CED. Even maintaining c finite, we obtain a series of theories classified in terms of λ/a and ω, where λ and ω are the wavelength and frequency of the electromagnetic disturbance, and a is a length specifying the size of the apparatus.

Additional possibilities arise if polarization and coherence properties are taken into account. However, we do not enter into these details at this point, and confine ourselves to the specification of a simple structure in Fig. 2.

It is apparent even without a detailed study that these theories are in supplementary relation to each other and CED is dominant to all. Thus the intrinsic logical structure of CED is somewhat of the type represented in Fig. 1 for physics as a whole, with the exception that the latter is not (yet?) derived from a single dominant system.

The block diagram of Fig. 1 represents, in a way, a structure of low "logical resolution," which, if the

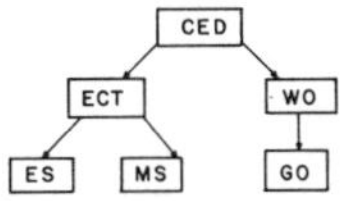

		ω	λ/a
ECT	QUASISTATIONARY CURRENTS, CIRCUIT THEORY	FINITE	>>1
MS	STATIONARY CURRENTS, MAGNETOSTATICS	0	∞
WO	WAVE OPTICS	FINITE	1
GO	GEOMETRICAL OPTICS	∞	0
ES	ELECTROSTATICS c = ∞	0	∞

FIG. 2. Structure of classical electrodynamics.

need arises, may be supplemented by a "logical fine structure."

Although the intrinsic logical structure of CED would deserve a detailed study, this is not important for the purposes of the present paper.

A characteristic feature of the block diagram is the occurrence of a "triangle" formed by three deductive systems, the relations of which are represented in one of two alternative forms:

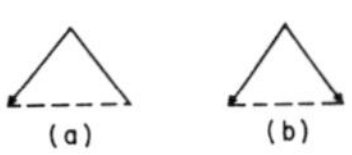

FIG. 3. Typical intersystem configurations for the resolution of inconsistencies.

In either case, the base of the triangle is formed by two systems that are mutually inconsistent with each other. This incompatibility is resolved by a third system which is either dominant to both of the original systems, or dominant to one and consistent with the other.

It is interesting to look at this logical constellation from the point of view of historical evolution. It is a historical fact that the incompatible systems are the first ones to appear on the scene. Their incompatibility, represented in the diagram by a broken line, constitutes a "line of stress," which is resolved by the discovery of the dominant system. The inconsistency is thus brought under control in the sense explained at the end of Sec. II. For the sake of concreteness, we single out the systems CM, CED, and RM for a more detailed discussion. These systems are related to each other according to the pattern of Fig. 3(a).

The fact that RM and CM are in a dominant-supplementary relation to each other is obvious. There is, in fact, a close mathematical analogy between the relativistic and Newtonian kinematics on the one hand, and spherical and plane geometry (see Sec. II) on the other. The absolute constant is the velocity of light c, the dominant invariance principle is *relativistic invariance*. The scale invariance of the supplementary system CM manifests itself, e.g., in the linear law of the vector addition of velocities.

On the other hand, RM is consistent with CED rather than dominant to it, since the velocity of light c maintains its finite value. The limiting process $c \to \infty$ leads to electrostatics, a highly degenerate part of CED.

The touchstone for the mutual consistency of CM and CED is the solution of two basic problems. The first is the fast motion of charged particles in the electromagnetic field. The second is the effect of the motion of the frame of reference of the observer on electromagnetic phenomens, including optics.

Historically, the study of both problems contributed essentially to the discovery of RM. In particular, Lorentz developed the theory of the electron in the electromagnetic field. However, this line of thought was complicated by uncertainties pertaining to the structure of the electron.

Einstein's great discovery was that the second of the aforementioned problems can be solved in phenomenological terms without involving assumptions concerning the structure of matter. The inconsistency of CM and CED was brought under control, in the sense explained in Sec. II, by giving up the space-time structure of CM in favor of that implicit in CED. *Thus the intersystem relation between CM and CED was inverted and it was the latter that assumed the dominant role.* More precisely, CED could be supplemented consistently by a mechanics (RM) that is dominant to CM. The expectation that all of physics, and in particular CED, is reducible to CM, was therewith disproved.

The establishment of RM led also to a consistent solution of Lorentz's problem of the motion of charged particles. However, it is not easy to delimit the phenomena in such a way as to stay clear of quantum effects and the subtler aspects of the consistency of the RM of the electron are still under discussion.[13]

The restructuring of physics connected with the recognition that the concepts of CMP do not have the expected fundamental character is a complex

[13] F. Rohrlich, Ann. Phys. (N. Y.) **13**, 93 (1961).

enterprise. For the purposes of discussion, it is convenient to divide the problem into three main divisions:

(i) Kinematics, or space-time structure,
(ii) Dynamics, the concept of force,
(iii) The structural properties of matter.

The abovementioned analysis leading to RM corresponds to item (i) of this classification.[14]

The Newtonian concept of *force* as an universal idea suffered its first setback as the action at a distance of electrostatics and magnetostatics was replaced by the Faraday–Maxwell field. This process can be considered as the resolution of the inconsistency of the partial theories of electricity and magnetism by the dominant CED.

This process was continued as the Newtonian theory of gravitation CG was replaced by the relativistic theory. We note that this replacement was necessary, since the inconsistency of CG and RM had to be resolved by the dominant RG.

In contrast, the present analysis does not require the development of a unified field theory in which RG and CED are consolidated into a single system. In the first place, there is no manifest inconsistency to be resolved in this case. And second, a unified theory of interactions cannot fail to consider chemical effects, which in turn are intimately connected with the structural problems of matter.[15] The discussion of these is our next objective.

Up to this point our analysis is not in disagreement with commonly held view. We conclude this section with an item that may be considered controversial.

It was emphasized above that Einstein's analysis leading to RM is phenomenological. However, the analysis has one aspect, the well-known inference about the nonexistence of the ether, which transcends this limitation. While it is correct to say that there is no place for the ether in RM and CED, this concept has other implications which require further study from the point of view of our present knowledge.

The rejection of the ether concept is based on the fact that the assumption of a local flow velocity is inconsistent with RM, inasmuch as it would allow us to assign velocities that moving bodies have with respect to an ether, stationary or otherwise. This, in turn, is contradicted by experiment.

It was pointed out by Dirac[16] that, though this inference seemed cogent in 1905, it is no longer convincing at present. The point is that within the classical theory a fluid is necessarily endowed with a local flow velocity, but this is not the case in QM. Thus an electron in an atomic s state does not possess a well-defined velocity.

It is instructive to discuss this matter in a more general context, since it lends itself to the illustration of our method of clarifying the meaning of primitive concepts. The problem of the "existence of the ether" is to be considered as a counterpart to the problem of the "existence of the atom." Texts of modern physics invariably argue that the answer to the first is in the negative, to the second, in the affirmative.

There is indeed convincing experimental evidence for the granular structure of matter. However, this is only a qualitative statement that is to be supplemented by the establishment of the laws governing the behavior of these "granules." Those denying the existence of atoms were right in the sense that the assumption of *c-atoms* obeying CMP are inconsistent with the facts of chemistry.[17] This statement is compatible with the existence of q-atoms obeying QM.[18]

The situation is similar with the ether. In an intuitive sense the ether was called upon to fill many roles. We may classify these along the threefold division invoked above in connection with the problems of mechanics:

(i) In kinematics the ether meant the existence of an absolute frame of reference. In view of RM the existence of such an ether is to be ruled out.

(ii) In dynamics, the most conspicuous role assigned to the ether was that of being the carrier of electromagnetic effects. Einstein pointed out that this notion is superfluous, although later, he came out in favor of the ether in connection with RG. We refer to the discussion of this point in a recent paper by Holton,[19] and in particular to the references given in his footnote 24.

(iii) The role of ether in the structural properties of matter is the most neglected aspect of this con-

[14] This statement is slightly oversimplified. The $E = mc^2$ relation involves in a broad sense the concept of matter.

[15] Chemical interactions occur on the molecular, atomic, nuclear, and elementary particle levels. We claim that these varied effects have some common features to justify the use of a generic terminology. We shall attempt to support this point of view in the next section. This point of view is by no means generally accepted. Molecular chemical forces are usually considered as mere instances of electromagnetic interaction. The point that we emphasize, however, is that these interactions assume quite novel features if injected into the quantum-mechanical theory of the structure of matter.

[16] P. A. M. Dirac, Nature **168**, 906 (1951).

[17] E. Mach, *Prinzipien der Wärmelehre* (Johann Ambrosius Barth, Leipzig, Germany, 1896), pp. 354–364.

[18] In the next section we shall subdivide QM into QS and QD, and accordingly we shall have to distinguish *qs* particles and *qd* particles.

[19] G. Holton, Am. J. Phys. **28**, 627 (1960).

cept, and the one that is of the greatest interest in the present context.

The point to be realized is that the problem of structure has two aspects: (i) Starting from elementary particles we may build up complex structures and in particular the pseudo-continuum of macroscopic bodies. (ii) The elementary particles in their turn may be built up as the excitons of a hypothetical continuum, the ether.

The attempt of accounting for atoms as vortex rings is a theory of this sort. The failure of this theory is inconclusive, since it is based on the classical *c* ether. It is an open question whether an ether concept could be defined within QD to describe the creation and destruction of elementary particles. To dismiss the possibility of a *qd* ether because of the inadequacy of the *c* ether would be as unwarranted as to declare ultimate *q* particles nonexistent because of the nonexistence of ultimate *c* particles.

IV. MECHANICS, THERMODYNAMICS, AND QUANTUM MECHANICS

A. Introduction

We now turn to the central theme of our discussion and consider the problems related to structure. The empirical discipline dealing with the structural forms of matter and their transformations is chemistry.[20] Thermodynamics, in the broadest sense of the word, is a conceptualization of this discipline in much the same way as classical mechanics is the conceptualization of astronomy.

Of course, the space-time detail provided by the description of structure in classical thermodynamics is severely limited. We claim that there are two complementary methods for overcoming this limitation. The first is the traditional method of mechanical models, the second proceeds by the gradual deepening and extension of the thermodynamic conceptual framework.

Our main purpose is to sketch the second method (Secs. IV.C and IV.D), but for contrast it is important to sum up the essence of the first (Sec. IV.B).

We denote the microscopic interpretation of classical point mechanics by CMP(μ), a symbol explained more precisely in Sec. IV.B. In this theory, matter is assumed to consist of invariant point particles that trace precisely defined trajectories. Hence the space-time detail built into the theory is the highest conceivable. This amount of detail is not only unnecessary, but it leads to implications that are contrary to experiment.

The present conceptual discussions of quantum mechanics are still decisively influenced by the accidents of its origin. Since the space-time detail is overstated in CMP(μ), quantum mechanics appears as a theory that imposes limitations on the possibilities of measurement.

We claim that the alternative approach that arrives at quantum mechanics by an increase of the space–time detail of thermodynamics conveys a more germane conception of the meaning of quantum theory as an enrichment and sharpening of common experience.[21]

Of course, even in order to consider this alternative approach we have to revise some of the generally accepted ideas concerning the nature of thermodynamics. We start with the noncontroversial statement that thermodynamics is the theory that is always in close contact with measurement. During the last century atomic phenomena were beyond the reach of the available experimental techniques, and hence it was concluded that thermodynamics should not involve any reference to microscopic entities. This attitude was correct under the existing conditions; however, the notion that thermodynamics is inherently macroscopic is an unwarranted generalization.

If thermodynamics is indeed a theory of measurement, it has, at present, to be substantially generalized in order to match the advances of experimentation into the microscopic domain.

This means that we have to give up the idea that thermodynamics is merely the mechanics of systems of very many degrees of freedom, and we have to identify the more genuinely distinctive features of the two disciplines. This we can achieve by the technique of logical differentiation, as applied particularly to thermodynamics.

As a first step in this direction, we have to recognize that the classical theory of equilibrium is a blend of two essentially different logical structures: We shall distinguish the theory of Clausius and Kelvin, on the one hand, from that of Gibbs, on the other.

In the first of these theories the thermodynamic system is considered as a "black box" and all relevant information is derived from the amount of energy absorbed or provided by idealized auxiliary devices, such as heat and work reservoirs, coupled to the

[20] We shall make a few remarks about biological problems in Sec. IV.E.

[21] Whether or not this alternative approach would have been historically possible is irrelevant for the present discussion. We structure the transition from thermodynamics to quantum mechanics by exploiting all of the advantages of hindsight.

system. The main achievement of this theory is the establishment of the concepts of internal energy and entropy from observable mechanical quantities.

This is the situation that serves as the point of departure for the Gibbs theory. Attention is now turned toward the system; the concepts of internal energy and entropy are taken for granted and are used to provide a more detailed description of the system in equilibrium, which includes its chemical and phase structure.

The logical mathematical structure of the Clausius–Kelvin theory was clarified in the axiomatic investigation of Carathéodory.[22] We shall speak of the Clausius–Kelvin–Carathéodory theory, briefly CKC.

Elsewhere[2] we have developed the axiomatics of the macroscopic thermodynamics of equilibrium, briefly MTE, that bears a somewhat similar relation to the Gibbs theory.[23] In that paper, the main emphasis was on the theory of phase equilibrium in which MTE represented improvements over the classical theory. In the present context these improvements appear only as incidental benefits of the improved logical structure. The main point is that basically the same experimental material and very nearly the same formalism are accounted for in a different conceptual language that is in some respects richer than either CKC or CMP(μ). Whereas MTE cannot be reduced to CMP(μ), it is, logically speaking, an "open system" that readily admits a consistent deepening and expansion in the statistical and quantum-mechanical direction and leads to the dominant systems STE and QS.

In statistical mechanics it is often assumed that the main aspects of the program of reducing thermodynamics to mechanics are unchanged as classical mechanics is replaced by quantum mechanics. This is, of course, a possible way of structuring the situation. We shall attempt to show, however, that an alternative approach is more illuminating.

In the first place, the conceptual basis of CMP(μ) is very simple, and the mechanistic reduction of MTE would require the elimination of much of its conceptual subtlety. In contrast, the transition from MTE to STE and QS involves an increase in conceptual wealth and sophistication.[24]

In Sec. IV.B we shall discuss the so-called fundamental problems of the mechanistic system. Our main thesis is that, in general, these problems are fundamental only in the context of CMP(μ). Moreover, we warn against the all too common tendency of identifying the mechanistic scheme with that of common experience.

Sections IV.C and IV.D deal with the conceptual structure of thermodynamics and quantum mechanics. The two sections differ inasmuch as Sec. IV.C is devoted to quasi-static problems, whereas the role of time is more explicit in Sec. IV.D.[25] The time-dependent problems are still incompletely understood and will be discussed here only in a heuristic vein.

The forthcoming discussion is only a program in which innumerable details have to be filled in. The present paper is confined to such arguments that arise from the conceptual reinterpretations of the existing formalism. However, the increased intuitive insight gained in this process is suggestive of new developments within the formalism itself. The process of logical integration leading to QD is clearly unfinished.

B. Classical Mechanics

According to arguments advanced in the general part of this paper, the term "mechanics" has a definite meaning only as the designation of a deductive system, or as a collection of deductive systems that arise as a result of logical differentiation.

The available axiomatizations of mechanics do not satisfy our requirements, since they have not been devised to differentiate between various disciplines that all go by the designation "mechanics," even though not all of them are consistent with each other. The following sketchy discussion provides only the raw material for a future axiomatics, but its precision is sufficient for the present exploratory study.

The starting point of our discussion is the analytical dynamics of mass points in the Hamiltonian, or some equivalent form to which alone we accord the status of a fundamental deductive system included in the block diagram (Fig. 1). We shall refer to this part of mechanics as the classical mechanics of particles, briefly CMP. This theory deals with isolated systems of invariant structureless particles that cannot be created or annihilated. These particles differ from geometrical points only in that they have masses and act as force centers. The kinematics of the

[22] C. Carathéodory, Math. Ann. **67**, 355 (1909).

[23] J. W. Gibbs, *Collected Works* (Yale University Press, New Haven, Connecticut, 1948), Vol. I.

[24] This state of affairs would seem to violate the traditional requirements of conceptual *simplicity* and *economy*. However, these requirements can easily be carried *ad absurdum*, if taken in an absolute sense rather than judged relatively to achievement.

[25] We use the term "quasistatic" in the sense that we consider primarily states of equilibrium and transitions between them. This meaning is different from that of the "quasistatic path" of CKC. We shall refer to the quasistatic and dynamic theories as QS and QD, respectively. Also, we use QM in a generic sense for quantum mechanics.

system is based on Newtonian space-time structure and obeys Galilean relativity. An instantaneous state is described as a point in phase space; a mechanical process is a trajectory in this space. The dynamics is specified in terms of the Hamiltonian, a function of the instantaneous phase-space coordinates of the particles; the forces are conservative, of the "action at a distance" type.

The concept of mass points is an idealization, but there are two kinds of rules of correspondence for connecting the theory with observation.

In the *macroscopic interpretation* the mass points arise through concentrating the masses of *well-separated* objects in their center of mass. CMP provides, then, the translational motion, while the intrinsic properties of the objects are either considered irrelevant, or are referred to other theories, such as chemistry and generalized thermodynamics.

In an essentially different *microscopic interpretation*, the formalism of CMP is combined with a rudimentary atomic theory. Using the terminology of Sec. III, we assume that matter consists of c particles. We shall refer to the theory that has the formalism of CMP, but in which the point masses are interpreted in this microscopic fashion as CMP(μ), and the symbol CMP will be used to designate the macroscopic interpretation only.

The reason that CMP is accorded a fundamental status is due to the validity of a series of important general theorems.

Using a plausible terminology, we say that the conceptual input of the theory is a set of invariant mass points with their instantaneous state in Γ space, while the output consists of additional invariants, the integrals of motion, and the principle of mechanical determinism provided explicitly as the Γ trajectory.

Among these theorems, the principle of mechanical determinism deserves special attention because its basic feature has been applied far beyond the scope of classical mechanics. A formulation that gives justice to this wider range of application states in essence that causal chains are conditioned by the *exhaustive specification* of the initial states of the systems involved.[26] In the cases of CMP and CMP(μ) the specification extends to the phase-space coordinates of all point masses of an entire isolated system. If, as usual, the system is coupled to its environment, then the initial state of the environment has to be also completely specified; in practice this is an impossible task. Therefore, in the strict sense, mechanical determinism is empirically meaningful only for systems that are effectively decoupled from their environment. Evidently the case of planetary systems plays a uniquely favorable role in this respect.

The wealth of general theorems of CMP is in curious contrast with the scarcity of soluble special problems. In fact, the case of two mass points interacting with central forces is the only rigorously soluble problem in the theory.

It is of obvious interest to extend the range of CMP by making it more flexible in handling concrete problems. Apart from such important technical devices as perturbation theory, this extension can be achieved very effectively by joining to the basis of CMP additional assumptions that are of a more or less phenomenological nature. We shall refer collectively to a number of deductive systems that are obtained by joining one or more of the following assumptions to the basis of CMP as phenomenological classical mechanics (PhCM):

(i) Potential energies specified phenomenologically rather than in terms of particle coordinates,

(ii) Continuous distribution of matter resulting in the mechanics of rigid bodies, fluids, and elastic media,

(iii) Dissipative forces,

(iv) Statistical assumptions,

(v) Assumptions of nonholonomic and rheonomic constraints.

While PhCM is of very great practical importance, this extension of the scope of the theory invalidates the general theorems of CMP. In fact, dissipative forces invalidate the constancy of the integrals of motion; and the manipulation of constraints substitutes one phase space for another, while a mechanical process is a trajectory in a fixed phase space.

The statistical assumptions that could conceivably be used in connection with CMP are of two sorts. First, one might consider an ensemble of identical systems with a probability distribution function of their representative points in Γ phase space at a given instant. The statistical assumptions of the second kind (stochastic assumptions) involve probabilities at a time t conditioned by the known states of the systems at an earlier time t_0 (transition probabilities). Although CMP is entirely neutral with respect to the distribution of initial states and any assumption of this sort can be consistently joined to it, stochastic assumptions are inconsistent with mechanical determinism.

[26] We are using the terms "causality" and "determinism" as synonyms. Such usage is encouraged by the fact that CMP has only a single mechanism for the precise interpretation of both concepts. In reality, the terms in question suggest different meanings, and indeed we shall provide distinct interpretations for them in the thermodynamic theory.

The feasibility of the classical reduction program hinges on the solution of a number of problems that are considered to be fundamental relative to the mechanistic program. The most important of these are: (a) stability; (b) hydrodynamics, including dissipative and stochastic effects; and (c) manipulation of constraints. These are, of course, closely connected with the assumptions of PhCM listed above.

The problem of stability was a stumbling block for the theory. Postulating appropriate phenomenological forces did yield stability for a limited number of structures without ever explaining a fraction of the facts of chemistry. After the discovery of the role of Coulomb forces among particles, even this limited result was found to be precluded by Earnshaw's theorem.[27]

The most successful contributions to the reduction program fall into class (b). The rigorous approach to these problems is based on the qualitative observation that systems of many degrees of freedom starting from an arbitrary initial state and evolving in a deterministic fashion seem to exhibit a randomization, as far as a macroscopic observer can ascertain. This randomization is only an *appearance* and the latent memory of the initial state manifests itself in the Poincaré cycles.

Although it is plausible that conservative systems do often behave in such a fashion, it is very difficult to state these ideas in quantitative terms. Moreover, the methods necessarily ignore the principle of regularity, and therefore often have an unrealistic character. In view of these difficulties, it became customary to develop theories of dissipation by adding stochastic assumptions to the theory. This procedure can be pragmatically justified by reference to the fact that real systems are always subject to random perturbations. Instead of reducing randomness and dissipation to dynamics, one studies the interplay of randomness and dynamics. Many of the existing theories that make use of randomness assumptions can be integrated into STD, although they do not contribute to the classical reduction problem.

We turn finally to the problem of constraints. At first sight this problem does not seem to arise in CMP(μ), since walls, other constraints, and the experimenter himself consists of atoms, and all could be included in the system and described in a super phase space. Hence the reducibility to mechanics can be upheld, but only at the price of giving up the separation of the object of experimentation from the experimenting subject.

[27] W. T. Scott, Ann. J. Phys. **27**, 418 (1959).

At this point we are reminded of the fact that CMP is the conceptualization of astronomy, an *observational* science.

We shall avoid all these paradoxes in thermodynamics and quantum mechanics which can be considered as conceptualizations of chemistry, an *experimental* discipline.

C. From Thermodynamics to Quantum Mechanics. Quasistatics

We shall develop the conceptual structure of a generalized thermodynamics that extends in successive approximations from a purely macroscopic theory (MTE) through statistics (STE) to quantum mechanics (QS). MTE and STE have been treated in more detail elsewhere.[2,3] At all three stages we confine ourselves to describing states of equilibrium, although we do allow for transitions among these states without describing them in detail. We shall refer to theories of this type as *quasi-statics*.

Equilibrium is a primitive concept analytically defined in terms of a deductive system, and thus will be redefined for each of the theories considered.

As we have mentioned in Sec. IV.A the conceptual structure of MTE is in practically every respect complementary to CMP(μ).

Instead of building up systems out of mass points, in MTE one builds them out of "cells" that cover a certain region in space. We refer to the cells as *simple systems*, or *subsystems*, and to a collection of cells as a *composite system*.

The independent variables of the theory are additive conserved quantities, briefly, additive invariants X_1, $X_2, \cdots$.[28] Composite systems are specified by sets of $X_i^{(a)}(i = 1, 2, \cdots a = 1, 2, \cdots)$, where the subscript i designates the types of variables, and the superscript a refers to the subsystems of the composite system.

Explicitly, the X_i are the volume V, the internal energy U, and the mole numbers of the independent chemical components $N_1, N_2, \cdots$. In spite of a superficial similarity, energy and mole numbers play a role in MTE that is very different from that in CMP(μ). In mechanics energy is not additive, and it is only the nonadditive contribution, the interaction potential that leads to the coupling of different systems. We shall see that in MTE the most characteristic coupling comes about through the exchange

[28] The number of the invariants X_i is significant; it involves the concept of thermodynamic degrees of freedom and leads to the phase rules. [2] We ignore these matters here for the sake of brevity.

of energy and of other additive invariants.[29]

The mole numbers N_i have an essentially different meaning in MTE and in CMP(μ). The most significant aspect of this difference is not in the shift from moles to molecules, since this is a simple scaling by Avogadro's number. In a context that does not involve numerical calculation, the same notation can be used interchangeably in both senses without risk of confusion. The important point is rather that the numbers of molecules of CMP(μ) are absolute invariants, the molecules themselves cannot be created or annihilated. In contrast, MTE deals with molecules that are subject to chemical change. Hence, the numbers of the molecules of chemical species, or briefly *species*, are in general not invariants.

The situation is different with the numbers specifying the amounts of the independent chemical components, or briefly *components*. These are constructed to be the invariants of the relevant chemical reactions and thus they qualify as the additive invariants of MTE.

Needless to say, "molecule" stands here for any kind of discrete lump of matter that has a definite identity and is subject to transformation into other "molecules." In addition to conventional molecules we have atoms, ions, radicals, nuclei, and elementary particles.

An important implication of the foregoing discussion is that the numbers N_i are, in general, only relative invariants and the entire formalism of MTE is only relative to the set of chemical reactions that have been used to define the independent components. The formalism is to be adjusted to any change brought about in the set of reactions.[30]

We turn now to the discussion of processes. In CMP(μ) all processes are alike, inasmuch as all of them are Γ trajectories. In MTE we have processes of different types and this variety is further increased in the deeper, dominant theories to be considered later.

An important type of process is the transfer of a quantity X from the subsystem a to b during an arbitrary, fixed time interval.

The transfer quantity corresponding to the energy, with all of the other X's fixed, is called *heat*.

Walls are the boundaries separating two systems which completely prevent the passage of one or more additive invariants, say X_λ, regardless of the nature of any system that may be adjacent, but permit free passage to all others, say X_k. The wall is *restrictive* of the X_λ and *nonrestrictive* of the X_k. Walls are said to exert *passive forces* on the distribution of the additive invariants. The restrictions on a system imposed by its walls are referred to as *constraints*.

It is most natural to think of a wall as a physical system with definite conductivity properties with respect to energy and matter (diffusion). Such systems are idealized as walls in MTE if their conductivities can be approximated as zero or infinite with respect to the time scale of the experiments considered. Alternatively, however, a wall may also be a mathematical surface. Thus a closed mathematical surface is restrictive of volume and allows us to define an open thermodynamic system that exchanges energy and mass with its surroundings. An important type of "wall" is provided by phase boundaries.[2]

Finally, the concept of thermodynamic process is easily extended to chemical reactions. In this case the molecular potential barriers stabilizing chemical species play the role of walls. As a last resort, all walls (except of course mathematical surfaces) depend on molecular potential barriers that may, or may not, be macroscopically organized.

We call *thermodynamic operations* any imposition or relaxation of a constraint through the uniting or subdividing of systems, or the altering of the type of any of its walls. In the presence of the existing constraints some of the variables X_λ^a are constant. We call these *fixed variables*. Those that are not entirely determined by the constraints are the *free variables* X_k^a. As a rule, we shall use Greek and Roman subscripts to denote fixed and free variables, respectively. The various admissible values of the free variables are called *virtual states*. The processes leading from one virtual state to another are *virtual processes*.

The distinction of thermodynamic operations and processes is a characteristic feature of the present approach. The admission of thermodynamic operations into the framework of the theory allows one to account for experimental procedures that have no place within CMP(μ) because they correspond to transitions from one Γ space to another. Thus

---

[29] A second type of coupling is conveyed by surface forces such as elastic and Maxwell stresses. All apparent action-at-a-distance forces should be put in this form. Thus, surface forces described by scalar pressure integrate easily into the formalism, and the displacement of a boundary between two systems can be considered as an exchange of volume. For stress tensors shape effects involving distant parts of the composite system may come into play which complicates matters considerably. We exclude these situations in the present discussion in which we aim only at the clarification of the most characteristic features of MTE.

[30] While there is no way of selecting absolute invariant species or components in molecular and nuclear chemistry, such absolute invariants are likely to govern the transformation of elementary particles.

thermodynamics allows us to formally account for manipulation on thermodynamic systems. In particular, such manipulations may be performed on systems in the course of a measurement.

Yet the insistence on operations as contrasted with processes does not necessarily require the intervention of an experimentalist, and does not introduce more subjective elements into the theory than are warranted by the nature of the situation considered.

Thus in the case of chemical reactions and phase transitions, supplying catalysts or inhibitors that speed up or slow down the rates of reactions is the corresponding thermodynamic operation. In an automatic chemical plant, and even more in biochemical processes, the "operations" become effective without the interference of an experimentalist and it may be more adequate to speak of parametric processes which *control* or *govern* other processes. This is quite in line with the use of these terms in cybernetics.[31]

According to the present approach, the relevance of the cybernetic concepts mentioned is not confined to applications in engineering and in biology. These concepts play a fundamental role, even in the microscopic approach to the dynamics of structure. This conclusion is confirmed and extended in our interpretation of quantum mechanics.

Another implication of the foregoing remarks is that we can use both the mole numbers of species, or those of components in our formalism, provided that we remember that the former are free and the latter are fixed variables. Of course, the distinction between these two classes is not absolute, but is relative to the specific conditions that govern reaction rates.

Although thus far we have been describing the conceptual raw material of MTE, we note that our discussion is equally valid with respect to STE, the statistical thermodynamics of equilibrium. However, the two theories diverge in their formalization of equilibrium.

We arrive at the concept of equilibrium by noting that the foregoing considerations leave the values of free variables undetermined. However, in isolated composite systems there are, in addition to the "passive forces" exerted by constraints, also "*active forces*" originating in the intrinsic dynamics of the system which bring about a *quiescent equilibrium* in which the free variables take on asymptotically constant values.

Although the trend toward equilibrium is the basic experimental fact underlying thermodynamics, the concept of equilibrium is nonetheless quite elusive, since there are no purely observational means for deciding whether an apparently quiescent system has actually reached equilibrium or is merely stranded in a nonequilibrium state while imperceptibly drifting toward equilibrium.

In the classical approach one meets this difficulty by advancing comparatively weak statements. Thus, in essence, the second law asserts the impossibility of processes in which systems drift away from equilibrium, but it does not claim that equilibrium is actually reached. The difficulty with this method is, however, that few if any significant results can be derived from these postulates without using a number of additional assumptions concerning the properties of material systems, such as the existence of homogeneous phases, validity of equations of state, and the like, which implicitly assume that equilibrium is reached. Since these assumptions are of limited validity, the rigor of the theory which the painstaking establishment of the universal principles was meant to ensure is considerably impaired.

In contrast to the classical procedure, we postulate in MTE that systems do reach their equilibrium and that the equilibrium values of the free variables are picked out from the set of virtual states by an extremum principle. The extremal function for this principle is the *entropy function;* the extremum is, by convention, a maximum.

For simple systems the entropy function is a first-order homogeneous function of the additive invariants

$$S = S(X_1, X_2, \cdots) \tag{4.1}$$

and satisfies certain regularity conditions. It is convenient to number the variables so that $X_1 = U$.

The entropy function of a composite system is the sum of the entropies of its subsystems. The entropy maximum principle is written formally as

$$S(X_\lambda) = \max_k\{S(X_k|X_\lambda\} = S(\overline{X}_k, X_\lambda), \tag{4.2}$$

where X_k and X_λ symbolize all of the free and fixed variables, respectively, and the $\overline{X}_k$ represent the values of the X_k for which the maximum in (4.2) is attained and which appear experimentally as the equilibrium values of the free variables.

The entropy is a primitive concept of MTE and is defined implicitly in terms of the maximum principle (4.2). At the same time, the validity of this principle is associated in the observational domain with the establishment of equilibrium. The aforementioned difficulties inherent in the equilibrium concept are now solved as follows. The implications of the

[31] N. Wiener, *Cybernetics* (J. Wiley & Sons, Inc., New York, 1961), 2nd ed.

formalism of MTE are not to be taken as the prediction of the *actual* behavior of a given system, but rather as a statement of the *normal equilibrium* behavior. The comparison of the observed behavior of systems with the theoretical normal equilibrium yields a set of practical criteria for deciding whether or not equilibrium prevails.

To be sure, within MTE these criteria are, strictly speaking, only of a *necessary* character. In order to clinch the argument by establishing *sufficient* criteria, we have to use the methods of STE and QMS. We shall return to this question below.

Relation (4.1) is usually called the fundamental equation (f.e.). It has the remarkable property of containing all of the thermostatic information about the system. It can be visualized as a surface in the space of the variables $S, X_1, X_2, \cdots$ to which we shall refer as the Gibbs surface in *Gibbs space*.

It is convenient to assume, first, that the f.e. is given, and to return later to the combined experimental-theoretical task of establishing it for concrete systems.[32] With the f.e. known, in principle, it is possible to develop the implications of the maximum principle (4.2). Let the entropy of a composite system in equilibrium be S_i. Consider any kind of a thermodynamic operation in which some of the internal constraints of the system are relaxed, and thus the manifold of virtual states is increased. With the increase of the set of comparison states, the maximum (4.2) either increases or remains unchanged. Denoting the entropy in the relaxed equilibrium state S_f, we have

$$\begin{aligned} &\text{(a)} \qquad S_f > S_i\,, \\ \text{or} \qquad & \qquad\qquad\qquad\qquad\qquad (4.3) \\ &\text{(b)} \qquad S_f = S_i\,. \end{aligned}$$

In case (a) the relaxation of constraints triggers a process leading to a redistribution of the additive invariants, that is, to a new equilibrium. This process involves the increase in entropy:

$$DS = S_f - S_i\,. \tag{4.4}$$

The symbol D indicates an actual change in contrast to virtual changes denoted by Δ.

In case (b) the relaxation of constraints leads to no process at all.

Simple as these statements are, they have broad implications that enable us to assess the relation of MTE with reality, that is, with experience understood in the broad sense of the word.

The fact that case (b) does indeed occur indicates that the value of the entropy and the entire formalism of MTE is insensitive to the distinction of whether or not a certain value of a variable is held constant by active or by passive forces. This feature of MTE is not rigorously concordant with experience, since the active forces that bring about the asymptotically constant values of the free variables also produce fluctuations, while the passive forces, or constraints, preclude fluctuations.

Remarkably enough, this limitation of validity hardly affects the usefulness of MTE. In the first place, fluctuations are often of negligible importance. Even more noteworthy is the fact that the marginal limitations of MTE are easily removed by refinement rather than disruption of its foundations.

Case (a) involving an increase of entropy leads to even more searching questions. How did the temporal concept "increase" suddenly emerge in the theory that does not even contain the concept of "time"?

Formally, the entropy maximum principle (4.2) is nothing but the well-known simple result of the calculus of variations according to which the maximum of a functional does not decrease as the set of trial functions is increased. The temporal aspect enters the picture through the interplay of operations and processes. If the relaxation of constraints leads to $S_f > S_i$, then this operation triggers a process in which the potentiality of a system for a new equilibrium is realized.[33]

The process itself has no counterpart in the formalism of MTE. The representation of this system reappears in this theory only after the new equilibrium is reached. This happens at a time that is *later* than the instant when the operation was performed. This is the point where the concept of temporal ordering enters the theory.

The description will continue to be valid until another operation triggers a new process. The representation of the system in disconnected time intervals is quite analogous to the spatial representation in terms of disjoint cells. All this may be summed up by saying that the system is described in a four-dimensional piecewise continuous space-time manifold. With each four-dimensional cell there is associated a point of Gibbs space, a set of numbers S, $X_1, X_2, \cdots$ specifying the equilibrium distribution of the relevant additive invariants.[34]

[32] In order to avoid repetition, we shall consider the empirical foundation only in the context of STE.

[33] At least this happens in what we call the normal equilibrium case. The alternative possibility is that the system remains stranded in the initial state that is now no longer an equilibrium, or in some intermediate frozen-in nonequilibrium state.

[34] This combination of the space-time manifold with Gibbs space constitutes the rudimentary elements of a nonlocal field theory.

The simple interplay between operation and process can be iterated to produce the description of complex situations that will be discussed somewhat further in Secs. IV.D and IV.E.

The above-mentioned classification of thermodynamic operations into classes (a) and (b) is important enough to be designated by special terms. We say that an *operation* is *irreversible* if it produces an entropy increase $DS = S_f - S_i$. This quantity is called the *measure of irreversibility*. An *operation is reversible* if it does not produce an entropy increase. Reversible operations occur in pairs; the relaxation and the reimposition of the same constraint that can be repeated indefinitely without producing any change in the system, except for the admission or suppression of fluctuations, a distinction not registered within MTE. We note that, in contrast to the usual procedure, we distinguish reversible and irreversible operations, rather than processes. The processes of the type considered thus far are all irreversible.

It is natural to ask two questions at this point. (i) How do we define irreversible thermodynamics, or, as we prefer to call it MTD, if irreversibility enters already so decisively in MTE? (ii) How do we arrive at the quasistatic processes of the CKC theory?

The answer to (i) is that the relaxation of an old to a new equilibrium is indeed considered both by MTE and by MTD, but the two theories differ in the detail of the description. In particular, MTD considers the relaxation time and to some extent the intermediate nonequilibrium states.[35]

The answer to (ii) is obtained by means of the following construction. A simple system can be coupled to a sequence of auxiliary systems, so that each operation triggers a process. A sequence of such operations and processes is called a *path* that is not uniquely specified by the initial and the final states. In particular, by increasing the number of well-chosen intermediate operations, the total measure of irreversibility can be decreased below an arbitrary threshold. In this limiting sense one can speak of reversible paths. These reversible paths, which are rather complex constructs in MTE, are the basic concepts in the CKC thermodynamics.

The entropy maximum principle states in essence that, given the nature of the systems involved in terms of their f.e., the fixed variables determine the equilibrium values of the free variables. Whether or not these equilibrium values are unique and under what conditions is not a matter of postulation but can be deduced from the theory. The clarification of this issue is indeed its main achievement. It appears that the abovementioned equilibrium values are almost unique with some qualifications that will be stated below. This result we have called[2] the principle of thermostatic determinism.

The entropy maximum principle is used to define in a well-known fashion the concepts of stable and metastable equilibrium.[36] The formalism, as it has been sketched thus far, is too scanty to carry us much further. It can be made considerably more flexible by transforming the f.e. to other sets of independent variables. Each of these transformations has an intuitive interpretation and extends the scope of the formalism to another class of experimental situations.

The first step in this transformation theory is to solve the f.e. $S = S(U, X_2, X_3, \cdots)$ for the energy

$$U = U(S, X_2, X_3, \cdots) \,. \qquad (4.5)$$

We shall refer to the formalism based on (4.1) and (4.5) as the *entropy* and *energy schemes*, respectively. Formally, the transition from (4.1) and (4.5) consists merely of the rotation of the Gibbs space, thereby resulting in the interchange of the roles of S and U as dependent and independent variables, and the two schemes are consistent with each other. In particular, they lead to equivalent stability criteria.

At the same time the transition to the energy scheme brings about a significant extension of the types of processes that are accounted for. The formalism is set up in terms of processes that conserve the independent variables summed over the composite system. Hence, in the entropy scheme we have energy conservation; the manipulation of the constraints do not involve any work and irreversible operations trigger irreversible processes. In the energy scheme, the total entropy is conserved, and hence operations[37] are reversible, but involve work performed on or by the environment.

We shall refer to the two kinds of processes as *entropic* and *energetic* processes, respectively.

An important extension of the formalism of MTE is the introduction of the intensities. These can be

[35] We note that MTD deals also with situations that do not connect an old with a new equilibrium, such as stationary currents and time-dependent environments.

[36] We note that the former is still not an *absolute* stability in the sense that every statement of MTE is relative to the choice of independent components, determined in turn by the set of relevant reactions. In principle, most if not all stable states can become metastable if the proper catalysts are added.

[37] In the energy scheme operations and processes are inseparably tied up. We may use the terms interchangeably.

defined in either scheme as the conjugate variables to the X_i:

$$P_1 = \frac{\partial U}{\partial X_1} = \frac{\partial U}{\partial S} = T\,, \quad P_i = \frac{\partial U}{\partial X_i}\,, \quad i \geqq 2\,, \tag{4.6}$$

$$\pi_1 = \frac{\partial S}{\partial X_1} = \frac{\partial S}{\partial U} = \frac{1}{T}\,, \quad \pi_i = \frac{\partial S}{\partial X_i}\,, \quad i \geqq 2\,. \tag{4.7}$$

Here, T is the thermodynamic temperature, the other P_i are negative pressure and the chemical potentials. Moreover,

$$\pi_i = -P_i/T, \qquad i \geqq 2\,. \tag{4.8}$$

In order to appreciate the significance of the intensive variables, it is convenient to single out one cell of a composite system for detailed consideration, and join all the others to provide the "environment" of this system. We require also that the over-all size of the environment be very large, in the limit infinite, compared with the system of interest. The environment (called also a *generalized reservoir*) is specified by the set of its intensities $P_1^0, P_2^0, \cdots$. This description is quite schematic; it ignores the detailed cell structure of the environment that may be quite complex. The specification is adequate, nevertheless, for deciding whether or not a system is in equilibrium when coupled to this environment. The condition for this is that the intensities of the system should equal those of the environment: $P_k = P_k^0$.

In general, for a given set of P_k the system will have a uniquely determined set of values for the conjugate X_k, and vice versa. The condition for this is

$$\frac{\partial(P_1,P_2,\cdots)}{\partial(X_1,X_2,\cdots)} \neq 0, \infty\,. \tag{4.9}$$

In the so-called regular points of Gibbs space where conditions (4.9) are satisfied the principle of thermostatic determinism takes on a very simple form: For a system in equilibrium with a reservoir $R\ (P_k^0)$ the free variables take on a unique set of values X_k^0 and vice versa. We shall say that the sets P_k and X_k are *matched* with each other.

This mutual determination of the invariants of a system and the intensities of a reservoir is confirmed by experiment. Nevertheless, the limitations imposed by conditions (4.9) are significant. The Jacobian in question vanishes at critical points and is infinite at absolute zero.

In the first case the initial intensities are inadequate for determining the values of the free variables (e.g., mole numbers per given volume) because of the huge critical fluctuations. Near absolute zero the responses X_k of the system become temperature independent and therefore are poor indicators of the temperature.

Thus the singularity of the formalism of MTE at critical points and at absolute zero has its counterpart in the experimental situation. Consequently, the limitation of MTE is rather different from the breakdown of, say CMP(μ). The information provided by the theory is incomplete rather than incorrect.

It is easy to point out the oversimplification in MTE that is responsible for its lack of detail, and to formulate a heuristic idea that guides us towards the deepening of its foundations. Whereas in MTE the virtual states have but a formal role, in the new theory they are assigned physical reality as fluctuations states. The free variables of a thermodynamic system are considered as random variables and the principle of thermostatic determinism is weakened to the extent that instead of the *values* of these variables, only their *probability distribution functions* (d.f.) are assumed to be fixed in equilibrium.

The injection of statistical ideas into the theory calls for an extension of the conceptual structure. First, we have to consider ensembles of systems $e\ (x_k|x_\lambda)$. An ensemble is defined as an unlimited set of systems describable in terms of the same selection of variables X; moreover the fixed quantities X_λ have the same value x_λ in each system, while the values of the free variables X_k take on random values x_k.

Generally speaking, the variation of the free variables will display a complex interplay of randomness with correlations stemming from molecular dynamics. One arrives at the simplest theory by considering independent random variables in which correlations are ignored.[38] In the theory of pure randomness the coupling between systems brings about a total loss of the "memory" of the state existing before the coupling was brought about. This point of departure leads to the theory of the statistical effects that take place in thermodynamic equilibrium. Accordingly, the theory obtained is called the statistical thermodynamics of equilibrium, briefly STE.

If an ensemble $e(x_k|x_\lambda)$ is in equilibrium with a reservoir $R(\pi_k)$ to which it is coupled by X_k exchange, the probabilities of the free variables are provided by the generalized canonical d.f.[39]

[38] To be more precise, the theory does not deal explicitly with correlations, but these are implicit in the set of fixed variables.

[39] This result is of course well known and can be safely made the point of departure for the following argument. However, we wish to give a few hints concerning a proof that has been given.[3] The derivation in question is based only on qualitative postulates concerning the existence of a d.f. conditioned by

$$dF(x_k,x_\lambda,\pi_k) = dG(x_k,x_\lambda) \exp[-\sum \pi_k x_k] Z^{-1}(\pi_k,x_\lambda) \quad (4.10)$$

The normalization of probabilities leads to

$$Z(\pi_k,x_\lambda) = \int_{x_k} \exp[-\sum \pi_k x_k] dG(x_k,x_\lambda) . \quad (4.11)$$

Here, $G(x_k,x_\lambda)$ is the *structure function* of the system and dG denotes the number of microstates within the range x_k and $x_k + dx_k$ of the free variables.[40] Its Laplace transform Z is the *partition function*, or *generating function*.

Although the formalism based on (4.10) and (4.11) is well known, we have some new angles to emphasize which arise within a systematic discussion of the three types of measurements that can be interpreted and evaluated in terms of this formalism:

(A) Coupling of a known system (x_λ) with a known environment $R(\pi_k)$. The predicted response of the system is the d.f. of X_k, in particular the canonical average $\bar{x}_k$,

(B) Measuring the response x_k of a known system (x_λ) and inferring the intensities π_k of the environment,

(C) Inferring the nature of an unknown system (x_λ) from its response x_k when coupled to a known environment $R(\pi_k)$.

Case (A) corresponds to the standard interpretation of the formalism. It leads at once to the f.e. of MTE if

$$\bar{x}_k = -\partial \ln Z/\partial \pi_k \quad (4.12)$$

is identified with the corresponding X_k, and the entropy is defined as

$$S = s(\bar{x}_k) , \quad (4.13)$$

where the random entropy function is

$$s = \ln Z + \sum \pi_k x_k . \quad (4.14)$$

Boltzmann's constant is chosen as unity. Thus relation (4.7) is now replaced by

$$\partial s(\bar{x}_k)/\partial \bar{x}_k = \pi_k . \quad (4.15)$$

The measurement of the $\bar{x}_k$ as a function of the π_k allows us to arrive, through integration, at the f.e.

This procedure leaves an integration constant in the entropy undetermined. This constant is far from irrelevant and its handling warrants a brief comment. The determination of the constant by the use of Nernst's law (lim $S = 0$ for $T \to 0$) is objectionable because one cannot be sure (i) whether the temperature is low enough for the entropy to vanish, and (ii) whether the system is actually in equilibrium. The correct procedure is to calculate the entropy constant in the ideal gas (low density) limit with the help of QS. If the results of this very reliable calculation are combined with the experimental values of the specific heat one obtains $S(T_0)$, where T_0 is lowest temperature actually reached. If this value is practically zero, we have a positive proof that equilibrium is established. Otherwise, one of the conditions (i) or (ii) is not satisfied and the case requires further study.[41]

We turn now to the experimental situation listed under (B). It was recently recognized by Mandelbrot[39] that the inversion of the conventional probabilistic problem (A) has an interesting physical interpretation. Suppose we make a finite number of measurements of the random variable X_k, say, the energy. We may use this information to arrive at a guess concerning the parameters π_k, in particular, the temperature of the reservoir with which the system has been in contact before the measurement of its energy. This, Mandelbrot points out, is precisely the problem of thermometry. Although one usually measures some convenient thermometric property, this is easily calibrated in terms of the

the relevant parameters and uninfluenced by previous history. It is an important point in the present context that the system that is coupled with the infinite reservoir is not assumed to have many degrees of freedom and may consist even of a single particle. The essential idea of the method which is due to L. Szilard, Z. Physik **32**, 735 (1925), is quite out of line with the usual techniques of statistical mechanics. However, as shown by B. Mandelbrot, IRE Trans. on Inform. Theory **IT-2**, 190, (1956); Ann. Math. Stat. **33**, 1021 (1962), it can be considered as an instance of standard procedures of mathematical statistics.

[40] This is a slight modification of terminology introduced by A. I. Khinchin, *Mathematical Foundations of Statistical Mechanics*, translated by G. Gamow (Dover Publications, New York, 1949).

[41] Although the procedure outlined above is well known, the meaning of Nernst's law as an essential ingredient for the operational definition of equilibrium does not seem to be generally appreciated. In fact, the third law has been the subject of an extensive discussion which may have been complicated by a semantic ambiguity surrounding the term "absolute entropy." On the one hand, the entropy is "absolute," in the sense that it involves the *entropy constant* provided by quantum statistics and not only the empirical *entropy differences*. On the other hand, the entropy is *relative* to the choice of independent components. Far from causing difficulties, this circumstance renders the use of the entropy concept particularly effective in providing subtle structural information about the level of equilibrium, whether it is extended to the distribution of spins, isotopes, molecular orientation, and the like. See F. Simon, Ergeb. exak. Naturw. **9**, 222 (1930); L. Pauling, J. Am. Chem. Soc. **57**, 2680 (1935); K. Clusius, L. Popp, and A. Frank, Physica **4**, 1105 (1937); J. O. Clayton and W. F. Giauque, J. Am. Chem. Soc. **54**, 2610 (1932); F. E. Simon, "40th Guthrie Lecture," in *Year Book of the Physical Society, 1956* (The Physical Society, London, 1957). Of course, an exclusively macroscopic thermodynamics can deal only with entropy differences. However, such a theory has to accept the awkward fact that its basic concept, equilibrium, cannot be put on an operational basis—an impressive case to show that a theory such as MTE calls for a deeper theory to solve some of its main problems.

energy. We note that this formal conception of thermometry covers also those striking instances of "measurements" in which, say, the composition of fossils is used to infer the temperature of the ocean in early geological times.[42]

From the formal point of view, the inference from an observed value of a random variable to the parameter of the d.f. is a problem of parameter estimation which is dealt with in mathematical statistics, sometimes called also *retrodiction*.[43] It is a standard technique in the evaluation of experiments and will prove to be an essential ingredient of a realistic theory of measurement. The suggestion that the methods of mathematical statistics be used within the context of basic theory might produce at first blush some apprehension. The inversion of probability calculus is by no means as unique as, say, the inversion of differentiation. Hence, the problem of retrodiction is beset by ambiguities and disputes, involving not only the choice of mathematical techniques, but also methodological difficulties centering around the nature of induction. We hasten to point out that the uses that we wish to make of estimation theory are not affected by these difficulties. In fact, our analysis of situation (C) will lead to a considerable clarification of the problem of induction.

Meanwhile we return to the problem of the estimation of the intensities, as it arises, for instance, in thermometry. The formal aspects of this problem are very simple. It is satisfactory for our purposes to consider a single method that is plausible enough on its own right without requiring for its justification the technicalities of mathematical statistics and that yields an intuitive interpretation of known formulas of statistical mechanics. At the same time, as Mandelbrot points out, this method is a standard one in mathematical statistics and is called the *maximization of likelihood*.[44]

Suppose that in a single measurement the free variables X_k of a system (x_λ) are observed to have the values x_k. (For instance, in the case of thermometry the energy U of the small system, the thermometer, is found to have the value u.) We define the estimate $\hat{\pi}_k$ of the unknown intensities π_k by the requirement that this choice should maximize the probability of the event that has been actually measured, namely, that the variables X_k of the system exhibit the values x_k. This is achieved by maximizing the logarithm of the probability (10) with respect to π_k. The expression ln dF, taken as a function of the π_k, is usually called the *likelihood function*. This term is meant to be reminiscent of "probability" without implying that (4.10) is, in the technical sense, a probability for the parameters π_k. For such an assertion there is no basis and dF need not even by normalizable in the π_k.

We see from (4.10) and (4.14) that the maximization of the likelihood function is equivalent to the minimization of the random entropy $s(\pi_k,x_k)$ with respect to π_k. Hence,

$$[\partial s(\pi_k,x_k)/\partial \pi_k]_{\pi_k = \hat{\pi}_k} = 0 \qquad (4.16)$$

or

$$-(\partial \ln Z/\partial \pi_k)_{\pi_k = \hat{\pi}_k} = x_k\,. \qquad (4.16a)$$

The left-hand side of this equation is the average $\bar{x}_k$ computed with the canonical d.f. belonging to the intensities $\hat{\pi}_k$. This result is eminently plausible. We select those values $\hat{\pi}_k$ which produce a canonical d.f., with the average $\bar{x}_k(\hat{\pi}_k)$ being equal to the single observed value x_k. This procedure leads to consistent results if the observation is the average of a finite number of measurements.

The extremal value of the likelihood is a maximum, the quadratic form, with the matrix elements

$$\frac{\partial^2 s}{\partial \pi_i \partial \pi_k} = -\frac{\partial \bar{x}_i}{\partial \pi_k} \qquad (4.17)$$

evaluated for the estimated intensity values $\hat{\pi}_k$, is positive definite. It can be shown that this condition is identical to the condition of thermostatic stability in MTE.

We define now the "estimated entropy"

$$\hat{s}(x_k) = \min\,(\ln Z + \textstyle\sum \pi_k x_k) \qquad (4.18)$$

and obtain from (4.14), (4.16), and (4.17)

$$\partial s/\partial x_k = \pi_k\,. \qquad (4.19)$$

It can be shown that (4.18) and (4.19) form the basis of a complete formalism of MTE which is an alternative to (4.14) and (4.15). Moreover, the functional form of $s(\bar{x}_k)$ is the same as that of $\hat{s}(x_k)$.

Relation (18) has been used to great advantage by Fowler[45] and Khinchin.[40] But the formal relation gains an intuitive interpretation as a maximization of likelihood.

We see now that the transition from MTE to the

[42] H. C. Urey, H. A. Lowenstam, S. Epstein, and C. R. McKinney, Bull. Geol. Soc. Am. **62**, 399 (1951); S. Epstein, R. Buschbaum, H. A. Lowenstam, and H. C. Urey, *ibid.* **64**, 1315 (1953).

[43] This term was used by S. Watanabe, Rev. Mod. Phys. **27**, 179 (1955), however, in a sense that is essentially different from the above interpretation: Present observations are used to retrodict past observations.

[44] R. A. Fisher, Phil. Trans. Roy. Soc. **222**, 309 (1921); R. A. Fisher, *Contributions to Mathematical Statistics*, (J. Wiley & Sons, Inc., New York, 1950).

[45] R. H. Fowler, *Statistical Mechanics* (Cambridge University Press, New York, 1936).

dominant STE is remarkably simple. The f.e. and the resulting formalism remain valid, but there is a change in the conceptual interpretation. More precisely, we have two conceptual interpretations corresponding to the basic experimental situations (A) and (B). In the first case we have an ensemble with definite intensities π_k and canonically distributed x_k. In the second case we have X_k fixed and estimated intensities $\hat{\pi}_k$. When Boltzmann's constant goes to zero (or Avogadro's number to infinity) the two versions of the f.e degenerate into the single f.e. of MTE. We have $\bar{x}_k = X_k$ and $\pi_k = \hat{\pi}_k$. Moreover the requirement for the maximization of likelihood provides an intuitive interpretation for the basic extremum postulate of MTE.

We turn now to the third experimental situation listed above under (C), in which we attempt to identify an object specified in terms of its fixed variables x_λ, assuming that the response x_k in the known environment $R(\pi_k)$ has been measured. In this wording the problem greatly exceeds the scope of the formula (4.10) and it is in this wider sense that we shall discuss it. We are confronted with the "problem of object" that is of great practical importance for every experimental science. Nevertheless, one might doubt the wisdom of injecting it into basic theoretical physics, and indeed, this has not been done up to the present time. The point is that there is no satisfactory method for estimating the nature of an object from its observed responses without prior knowledge of the possible set of objects available for selection. In practice, such prior taxonomical information on possible objects is available by induction from past experience. However, neither the justification of induction nor an insight into its limitation was provided by classical physics. Attempts at discussing this problem within the classical context were abstruse and devoid of interesting results. The whole question gradually lost respectability and it seemed to have escaped notice that the solution of the problem of the existence and observability of definite objects is implicit in the practical procedures of quantum physics. We shall attempt to identify and to formulate explicitly the conceptual aspects of the theory that are responsible for this success. We note that this view is the opposite of the accepted opinion according to which the concept of *object* "breaks down" in QM.

The scope of quantum physics is extremely wide. In keeping with our program of logical differentiation we ought to subdivide the field into precisely formulated deductive systems. However, this we are not yet prepared to do. Instead, we shall start by considering at first a restricted and well-understood part of the theory and proceed by introducing additional conceptual elements.

The simplest part of quantum physics that serves as our point of departure deals with the discrete stationary states of stable structures of particles only. We shall refer to this theory briefly as QS. This symbol can be read also as "quantum statics" and should remind one of "thermostatics" rather than of "mechanical statics." In fact, QS can be conceived as the deepening of STE. The concept of "time" plays a similar, and somewhat rudimentary role in both of these theories. We shall consider the generalizations of QS that involve the concept of time in a more elaborate manner in Sec. IV.D.

We shall list now the features of QS that are essential for our argument.

(i) QS is a theory of objects, or structures, constructed of distinct classes of identical particles. Within the domain of molecular and low-energy nuclear chemistry (with the exclusion of β decay), we have only the three classes of electrons, protons, and neutrons.

(ii) Both particles and their structures are capable of existing in *pure states* represented by state vectors in Hilbert space, or wave functions specified by a set of quantum numbers. The pure states represent *potential* forms for the existence of objects. If such a state *actually* exists, we say the state is occupied, or realized. *The realizations of the same pure state* in different regions of space time form a class of absolutely identical objects.

(iii) Changes in occupation are called *transitions*, in the course of which a formerly occupied state is *annihilated* while another one is *created*. Transitions give rise to observable effects, the calculation of which belongs into the time-dependent theory. However, we anticipate that the predictions of observable effects are of a statistical nature, and that the probabilities of the random variables which constitute the measured signal depend parametrically on the quantum numbers of the initial and the final states.

(iv) The time-independent Schrödinger equation allows one to compute all pure states that can be constructed from a given assembly of particles.

(v) Sets of pure states provide so-called mixtures that can be identified with the ensembles of STE and hence with the equilibrium states of the systems of MTE.

Item (v) is parallel to the similar situation in statistical mechanics, provided that the pure states are considered as the analogies of the points of Γ phase space. However, here the similarity ends, since items (i) through (iv) are entirely without classical analog. These are precisely the features that allow us to put the problem of object on a new firm foundation.

From the philosophical point of view item (iv) is of the greatest interest. It means that, in principle, all of the pure states of structures built from a given set of particles can be computed on theoretical grounds without any recourse to experiment! Whenever this calculation has been actually carried out, the results have been verified by experiment to a high accuracy. There are undoubtedly huge classes of cases in which the same agreement could be achieved, provided that one over comes the mathematical difficulties that increase enormously with the complexity of the systems. We may at this point leave open the question whether or not the scope of QS extends to systems of all degrees of complexity.

Although the calculation of the properties of pure states by exclusively theoretical means is of the greatest interest, it is no less important from the practical point of view that the taxonomical task of listing and specifying the set of pure states can be undertaken by judicious combination of theory and experiment. In fact, we state: *The retrodiction of the discrete quantum numbers specifying pure states from the statistical evaluation of actual signals can be carried out in principle, and very often in practice, with absolute certainty.*

The feasibility of this task hinges on items (i) through (iii). In the first place, the fact that different representatives of the same pure state are absolutely identical allows us to repeat the "same" experiment an unlimited number of times. [Items (i) and (ii).] Therefore, the statistics of the observed signal [item (iii)] can be made sufficiently good to resolve the predictions corresponding to different discrete values of the parameters to be estimated.

An example is the spectroscopy of a system emitting a line spectrum. Many emission processes take place between the "same" initial and final states and yield the "same" emission within the limits of the line width. The selection rules, particularly in a magnetic field (Zeeman effect) enable one to analyze the experimental data to yield the quantum numbers of the atomic states.

In spectroscopy the statistical element of the procedure plays a somewhat subordinate role. Therefore, we mention another case, although it is, strictly speaking, outside the scope of QS and should be taken up only in the next section; the problem of determining the spin, or some other discrete intrinsic property of a particle from scattering experiments. The tentative assumptions of different discrete spin values lead to different theoretical predictions of an observable signal, say, the angular distribution of scattering. Since the "same" experiment can be repeated an indefinite number of times on identical replicas of the same class, it is, in general, possible to resolve the different predictions and eliminate all but one of the competing possibilities.

It is interesting to compare these considerations with the well-known Bohr-Heisenberg theory of measurement[46] which insists that the space-time specification required by CMP(μ) cannot be achieved even under ideal conditions in the total absence of noise, and is limited by the uncertainty principle. In this context the term "measurement" is not supposed to refer to real measurements, but constitutes an ingenious conceptual device to clarify the limitations of mechanical models. These considerations are indispensable whenever such models are used, either for heuristic purposes or for the intuitive visualization of a situation.

In contrast, the present reasoning proposes to sort out the factors that brought about the extension of actual experimentation in the microscopic domain. The performance even of qualitative experiments is remarkable in view of the fact that this was held to be impossible as recently as the beginning of this century. This pessimistic forecast was by no means unreasonable from the point of view of classical continuum physics.[47] The factors that allow us to make error-free retrodiction from noisy measurements are precisely those discontinuous features of quantum physics that have no classical analog and that appeared disruptive within the fabric of continuum physics.

The essential point is the discovery of the concept of *absolute identity* of different replicas realizing the same *pure state*. This is in strong contrast with the concept of identity in common experience. Macroscopic objects are never exactly identical. However, we accept two objects as "identical" in a vague sort of way if we cannot discern any difference between them, or if their differences are within specifications. Hence the "identity" of common experience is a concept that is only meaningful relative to the precision of measurement or to conventional specification. Alleging the identity of two objects within

[46] Niels Bohr, *Atomic Theory and the Description of Nature*, (Cambridge University Press, New York, 1934).

[47] If the apparent continuity of the laws governing the behavior of macroscopic objects would extend to the microscopic domain, we could infer nothing about the properties of single atoms by coupling them to macroscopic measuring devices.

continuum physics would seem merely to admit one's inability to discern the existing differences.

We may sum up the situation by saying that QS contains rigorous results of a taxonomical character concerning the existence of classes and their properties which are entirely foreign to classical physics. Remembering from Sec. II that dominant systems have their characteristic invariance property, we say that the absolute identity of any realization of the same pure state is the dominant invariance of QS, to be referred to as *morphic invariance*. It is the morphic invariance that brings it about that complex systems of well-defined identity can be built in nature through the mechanism of random processes.

Up to this point we have referred to the quantum numbers only as identifying tags. We now give a short summary of their physical meaning. The first step in describing a finite stable structure is the establishment of an orthogonal set of energy eigenstates. If macroscopic translation is separated off, this spectrum is discrete, and falls within our definition of QS. The knowledge of the number of states g_i, having the energy eigenvalue E_i, provides us with the structure function and leads to the general formalism of STE.

The actual calculation of g_i is based on a specification of the additional quantum numbers. First we consider the quantum numbers associated with the total angular momentum and one of its components. The dynamic interpretation suggested by this designation establishes a connection with CMP. It is important to realize however, that these quantum numbers admit also an intuitive geometrical interpretation in terms of the shape of the wave function. The idea of this geometrical interpretation ties in with the "morphic" aspect of QS emphasized above. The geometrical representation of the shape of the wave function[48] is more realistic and is also a more intuitive picture of stationary atomic states than the historically motivated tracing of planetary orbits.

The morphic effects associated with angular momentum are geometrically represented in terms of the spherical harmonics. The same functions arise also in the classical vibration of macroscopic objects. It should be emphasized, therefore, that morphic invariance does not extend to macroscopic vibration. The difference becomes evident in terms of the radial functions. In the case of the hydrogen atom the scale of the radial function, say, the location of its first maximum, is expressed in terms of universal constants; it constitutes a significant scientific statement concerning the class of all hydrogen atoms. In contrast, the scale of the radial wave function of a macroscopic object specifies only one individual object, it is not a significant scientific statement.

The importance of the morphic interpretation of the angular momentum is further underlined by the existence of such nonclassical integrals as parity and statistics,[49] which have no classical dynamic analog but which admit a morphic interpretation in a fashion quite similar to angular momentum.

Finally, we turn to the discussion of the parameters that specify the configurations of molecules. The last term is meant here in a generic sense, it includes macromolecules and macroscopic objects also. These parameters specifying configuration are often accepted as classical ones, since they can be visualized in terms of localized potentials. However, these potentials have a phenomenological character (PhCM of Sec. IV.B) and cannot be classified among the dynamical integrals of CMP(μ). We claim that the parameters describing localization are expressions of the morphic aspects of QS on a molecular and macroscopic level. In fact, the quantum effects involved in these discussions can become quite subtle and are still incompletely understood in some marginal, but presumably not unimportant, cases.

The problem of molecular configurations was first solved in great generality by Born and Oppenheimer,[50] who initiated the method usually referred to as the adiabatic approximation. Consider an assembly of electrons and invariant nuclei. The Schrödinger equation of the system is derived, as usual, from the Hamiltonian; we shall call it more specifically the *particle Hamiltonian*, a function of the coordinates and momenta of the particles:

$$H = T_n + T_e + V(x,X), \tag{4.20}$$

where T_n and T_e are the nuclear and the electronic kinetic energy, and V the total potential energy; x and X represent all of the electronic and nuclear coordinates, respectively.

The method exploits the large mass ratio of nuclei

[48] H. E. White, *Introduction to Atomic Spectra* (McGraw-Hill Book Company, Inc., New York, 1934), pp. 143 and 146.

[49] The term "statistics" is a misnomer. It would be more logical to say that bosons and fermions have an even and odd *permutational parity*, respectively, thus emphasizing the obvious analogy with the conventional *inversion parity*, a symmetry property. To be sure, permutational parity affects the statistical properties of many-particle systems. However, this effect is largest at absolute zero where the system is in a single state and there is no statistical situation in the proper sense of the word. The distinction between even and odd permutational parity disappears in the high-temperature, low-density limit, where the important effect is statistics.

[50] M. Born and J. R. Oppenheimer, Ann. Physik **84**, 457 (1927).

and electrons. As a first step we solve the electronic problem at fixed nuclear positions. We can think of situation as being brought about by letting the nuclear masses tend to infinity implying that T_n tends to zero. The Schrödinger equation in this case is

$$(T_e + V)\phi(x,X) = W(X)\phi(x,X), \qquad (4.21)$$

where the eigenvalue W and the eigenfunction ϕ depend parametrically on the fixed nuclear configuration X. We confine our attention to the neighborhood of the configuration X^0 for which the lowest eigenvalue has a stable minimum. Further, we assume that in this neighborhood the lowest eigenvalue W^0 (X^0) is nondegenerate.[51] We return to the justification and limitation of this assumption below.

The solutions of the original problem of finite nuclear masses is sought in the form

$$H\phi(x,X)\psi_{ns}(X) = E_n\phi(x,X)\psi_{ns}(X), \qquad (4.22)$$

where the eigenfunctions ψ_{ns} of the nuclear motion satisfy to a good approximation the equation

$$\mathcal{H}\psi_{ns} = E_n\psi_{ns} \qquad (4.23)$$

with

$$\mathcal{H} = T_n + W^0(X) \,. \qquad (4.24)$$

Evidently (23) has the form of a Schrödinger equation for the nuclear motion alone. The electrons that convey the binding appear only implicitly in the "potential" W^0 (X). We shall refer to $\mathcal{H}$ as the quasi-Hamiltonian. In spite of its formal analogy with the particle Hamiltonian H, there are differences that are fundamental for the present argument. Although the transition from H to $\mathcal{H}$ involve a number of approximations, a matter to which we shall return, it would be nevertheless misleading to conceive of this transition merely as the substitution of a rigorous formalism by an approximate one. Instead, we propose to think of the procedure as[52] a transformation between formalisms of different types. While the particle Hamiltonian is firmly rooted in mechanics, the quasi-Hamiltonian has direct ties with the thermodynamic description of the system.[53] H is a mechanical concept inherited from CMP, it is expressed in terms of the coordinates of all of the particles. This is no longer true of $\mathcal{H}$. Not only have the electronic coordinates been eliminated, but the nuclear coordinates appear in such collective terms as Eulerian angles and normal coordinates of the vibration around an equilibrium. The existence of an equilibrium configuration brings it about that $\mathcal{H}$ has geometrical symmetries; a point group for molecules and a space group for crystals, whereas the particle Hamiltonian has only the permutational symmetry, if we ignore the translation and rotation of the system as a whole. A single-particle Hamiltonian may generate a huge number of quasi-Hamiltonians. Consider, say, an assembly of a large number of carbon and hydrogen atoms. There are thousands of ways of combining these into molecules, each of which is described by its own quasi-Hamiltonian. It is the Schrödinger equation based on $\mathcal{H}$ that leads to the relevant structure function and to the fundamental equation of STE and MTE. In particular, the configurational symmetry of the quasi-Hamiltonian is essential for the derivation of the phase rules.[2]

A final statement to round out the usefulness of the adiabatic method is that the quasi-Hamiltonian and its energy spectrum can be empirically determined from spectroscopic data.

We turn now to a critical discussion of our assumption that the lowest electronic state associated with equilibrium configuration X^0 is nondegenerate. A systematic reason for degeneracy would arise only for symmetric configurations. However, the Jahn–Teller theorem[54] assures us that degenerate states of nonlinear polyatomic molecules are unstable with respect to deformation of the symmetric configuration that removes the degeneracy. For this reason, the assumption of the absence of degeneracy in the adiabatic approximation would seem to be generally justified. There are nevertheless exceptional situations that require an essential deepening of the theory. It may happen that an unstable symmetric configuration splits into, say, two configurations of lower symmetry which are separated only by a potential barrier that is sufficiently low to permit a quantum-mechanical resonance between these states. Thus we are led to the entirely novel situation of having a resonance between two quasi-Hamiltonians, the so-called dynamic Jahn–Teller effect.[55] Even though the

[51] We exclude even an approximate degeneracy, i.e., the lowest excited level W' (X) should be much higher than the nuclear energies E_n of (4.22).

[52] The existence of spin casts doubt on the point of view that would consider the Schrödinger equation derived from (4.20) as the absolutely correct statement of the problem. The correct account of the spin requires a relativistic treatment, and it would seem to be impossible to keep out the complications of relativistic field theory.

[53] In Sec. IV.A we formulated the program of finding the qualitative differences between the mechanical and the thermodynamic viewpoints. A partial answer to this question is implicit in the relation of H and $\mathcal{H}$.

[54] H. A. Jahn and E. Teller, Proc. Roy. Soc. (London) **A164** 117 (1938).

[55] See W. Moffitt and A. D. Liehr, Phys. Rev. **106**, 1195 (1957); W. Moffitt and W. Thorson, *ibid.* **108**, 1251 (1957);

occurrence of this phenomenon is rather exceptional, it indicates that the existence of the quasi-Hamiltonian (4.24), and the corresponding Schrödinger equation for nuclear motion must not be taken for granted.

D. Dynamics of Structures

The "static" theories MTE, STE, and QS that deal with problems of structure have their "dynamic," or "kinetic" counterparts MTD, STD, and QD. In the block diagram (Fig. 1) the symbols of these systems are in broken rectangles to indicate that they are still essentially incomplete. In particular, QD is supposed to be a dominant theory in which the conceptual aspects of QS and of RM are brought in harmony with each other. The value of the present method will be ultimately judged by its success, or the lack of it, in bringing about such a dominant theory. Meanwhile we shall merely sketch a few ideas that can be discussed at the present status of the theory.

It is indicative of the difficulties of the dynamics of structures that even the traditional dichotomy between statics and dynamics is an oversimplification. The static theories have significant temporal aspects and the transition to dynamics requires a number of distinct steps, each of which leads to a deepening of the existing theories. The eventual situation will be presumably analogous to, even though more complicated than, the logical fine structure of CED represented in Fig. 2. Mechanistic analogies are of little help in disentagling this situation. In the first place, CMP has no significant static limit, since time-independent conditions can be maintained only in the trivial case of systems consisting of noninteracting particles.[56] Second, the concept of time in CMP plays the same standard role in all conceivable problems, a deceptive simplicity that is in strong contrast with the multiplicity of roles played by the concept of time in thermodynamics and quantum mechanics. We shall occasionally refer to these theories by the generic name *quantum thermodynamics*. This term is to denote a whole cluster of deductive systems, the most important of which are still incomplete. The diversity of the temporal processes in these theories leads to a greatly increased flexibility in the representation of reality, the ramifications of which are far from being fully explored.

We shall concentrate on two closely related issues and emphasize a duality that is characteristic of quantum thermodynamics and is in contrast with mechanistic monism. This means that we claim an independent status for pairs of concepts when the mechanistic philosophy calls for a reduction of one to the other. The first duality is that of two principles of causality, while the second contrasts energetic and entropic processes.

The great predictive success of CMP and of CED has led to the conviction that all causal chains in nature are accounted for by differential equations and rely on the *complete specification* of the initial state of the system. Actually in thermodynamics we encounter another situation in which causal chains are determined by a *selective* specification of the initial conditions. We shall speak of a *second principle of causality* based on *selective memory*, that is to be distinguished from the *first principle of causality* based on *exhaustive memory*.[57]

As a convenient point of departure we consider the composite systems of MTE. We have seen in Sec. IV.C that the concept of thermodynamic operations, *followed* by entropic processes triggered by these operations, implicitly contains an element of temporal sequence. Such sequences can be repeated to form what we shall call *composite* (*entropic*) *processes*. Due to the fact that the principle of thermostatic determinism assures us of the uniqueness of the final state of each step, the entire composite process assumes a deterministic character, provided that the operations are preset by some automatic device. Alternatively, if the operations are governed by the terminal states through feedback, the composite process becomes goal directed. Of course, while composite processes consist of temporally ordered sequences, they lack a definite time scale as long as one is confined to using MTE. An important step toward providing a time scale is achieved in MTD in which the approach to equilibrium appears as a relaxation process characterized by definite relaxation times. However, in a purely macroscopic theory the completion of a relaxation process requires an infinite amount of time. This

H. C. Longuet-Higgins and K. L. McEwen, J. Chem. Phys. **26**, 719 (1957); H. C. Longuet-Higgins, U. Öpik, M. H. L. Pryce, and R. A. Sack, Proc. Roy Soc. (London) **A244**, 1 (1958); A. D. Pierce, "Electron Lattice Interaction and Generalized Born–Oppenheimer Approximation," Ph.D. Thesis Massachusetts Institute of Technology, January 1962 (unpublished).

[56] This statement is not inconsistent with the existence of mechanical statics but points once more to the semantic confusion surrounding the term "mechanics." The statics in question belongs to PhCM rather than to CMP (see Sec. IV.B).

[57] This terminology would be ambiguous in the case of the systems with "real" memory, that is, for computers and organisms. These have been, of course, tacitly excluded thus far, but we shall briefly discuss them in the next section.

difficulty is finally removed in STD. In this theory equilibrium includes thermal noise and a process terminates in a finite stretch of time as the static equilibrium is approached within the average noise level.

The composite entropic processes provide examples for the second principle of causality and we shall describe in some detail the distinction between selective and exhaustive memory. The first aspect of selectivity is that a system may be coupled to its environment by the exchange of energy and matter, and yet it suffices to specify the environment in terms of a few intensity parameters. This specification is adequate to determine the effect of the environment on the state of the system, although it would have to be enormously increased to satisfy mechanical requirements according to CMP(μ) or even QM. However, this additional detail is irrelevant in the sense that it has no effect on the observed behavior of the system. The second aspect of selectivity is that the variables of the system are classified into free and fixed variables. Simple as this idea is, it is highly nonmechanical, since in CMP(μ) the energy of an isolated system in rest is the only invariant, apart from the particles the invariance of which has been postulated. The role of the fixed variables is to identify the system. In particular, the fixed variables along with the intensities of the environment determine the probability distribution function of the free variables. If some of the constraints imposed on the system are relaxed, the entropic process triggered by this operation brings about the loss of the memory of the formerly fixed variables. This loss of memory is a prerequisite for the uniqueness of the final equilibrium.

Whereas the first two aspects of selectivity emphasize the constructive role of "forgetting," the third aspect deals with the *persistency* of the values of the fixed variables. We emphasize that the actual preservation of the fixed variables under realistic "noisy" conditions, as a result of quantum-mechanical stability, is physically meaningful. This is in contrast with the exhaustive but fragile memory of the initial state in CMP(μ) that is so often erased by the slightest perturbation and is denied physical meaning by the principle of regularity. It takes only some reflection to realize that the existence of identifiable objects, which respond in a predictable fashion under "identical" conditions, depends on the second principle of causality; the object is permanent within its lifetime and the conditions are identical only because the relevant specification is selective.

We are now in a position to clarify the relation of causality and determinism. In common parlance these terms suggest different meanings, and indeed we can provide distinct interpretations for them in the thermodynamic theory. The choice of thermodynamic operations that govern the composite process may, or may not, depend on our decision, and accordingly we may speak either of causal or deterministic chains of events. The former would arise in experiments, the latter, say, in the astronomical setting, or in the evolution of an organism determined by the genetic code. Of course, the delimitation of the two situations is by no means fixed, but shifts with the advances of experimental techniques.

In order to complete the description of real situations, we have to supplement entropic composite processes with energetic ones. Since unbalanced forces produce accelerations, these processes can be treated in MTE or STE only in the limiting case for which the accelerations are negligible. It is interesting that entropic processes are not subject to such a restriction, their treatment is perfectly germane to the equilibrium theories.[58] As soon as accelerations are admitted, we have to use a dynamic theory. The problem is complicated by the fact that energetic processes are almost always tied up with entropic processes, and it is, in fact, the interplay of the two that is our main concern in the present context. This problem presents itself in the sharpest form within quantum mechanics and this is the only approach that we wish to consider here.[59]

In quantum mechanics energetic processes are described in terms of the time-dependent Schrödinger equation that implies the temporal evolution of the

[58] This is in contrast with the usual contention that thermostatics deals with reversible processes and irreversible thermodynamics with irreversible ones.

[59] A few parenthetical remarks concerning the role of energetic and entropic processes in MTE and STE are meant only to outline the meaning of some of our terms by linking them with traditional designations. The hydrodynamics of ideal fluids and the phenomenological theory of relaxation processes are limiting cases of MTD which deal with energetic and entropic processes, respectively. The theory of Markov chains is a limiting case of STD dealing with entropic processes. This is essentially an instance of mathematical probability theory in which the postulated probabilities for elementary events allow us to compute probabilities of complex events (that is, sets of chains of elementary events). There are many phenomenological theories, such as the hydrodynamics of viscous fluids and thermohydrodynamics, that deal with the joint effects of energetic and entropic processes. These theories are important from the practical point of view but provide little fundamental insight. A much more detailed description of the interplay of energetic and entropic processes for a particularly simple type of system is provided by the Boltzmann equation of the kinetic theory of gases. The achievement of this theory is somewhat obscured by the fact that, according to the tradition set by Boltzmann the aim is a rigorous derivation of this equation from Liouville's equation. In other words, in conformity with the mechanistic program one proposes to reduce entropic processes to energetic ones (Boltzmann's problem).

wave function in terms of a unitary transformation

$$\psi(t) = U(t - t_0)\psi(t_0) . \quad (4.25)$$

The conservation of entropy appears as the unitary invariance of the number of dimensions of the subspaces of Hilbert space spanned by the state function or, more generally, by the density matrix. Another aspect of the entropy conservation is the maintenance of strict phase relations, a characteristic property of wave propagation in continuous media. Accordingly, we shall briefly refer to energetic processes as *propagations*. Of course, in the case of stationary states the phase does not propagate, but varies synchronously over all space.

We introduce now the concept of *events* to denote: (i) transitions between stationary discrete quantum states, (ii) interaction processes described in terms of the S-matrix formalism, including the events of high-energy physics. In all of these cases we have the same type of situation as in a chemical reaction.[60] An initially existing state disappears, whereas another state appears. This is described formally in terms of annihilation and creation operators. The difference between quantum events and chemical reactions is that the former are the elementary indivisible constituents out of which chemical reactions and, more generally, entropic processes, are constructed. Quantum events exhibit in a particularly clean-cut fashion the characteristic features (selective memory) of entropic processes that we have summed up under the designation of the second principle of causality. First, there is the peculiar persistency of quantum states. As long as the system is in a particular pure state, the corresponding quantum numbers have the character of fixed variables. Also the negative aspect of the selectivity of memory is very significant. A system in a pure state does not "remember" where it came from, nor the instant when the transition into the state occurred; pure states do not "age," although they have a definite lifetime. All of these factors are essential prerequisites of the morphic invariance of the pure state which, in turn, ensures error-free retrodiction, as we have already seen.

Events bring about a measure of randomization of the phase relations. The random aspect of events stems from the unpredictability of the instant at which an individual event is initiated. Otherwise, of course, events are by no means so unstructured as, say, the steps of the "random-walk" problem. Since their description consists in the specification of the initial and final states, events have their individual characteristics just as the states involved have.

The instant of an event is in some respects analogous to the instant of an operation that triggers an entropic process, although without any subject or even a molecular device as a catalyst, to perform the operation. At the same time genuine operations involving manipulations of the boundary conditions also have their legitimate role in quantum mechanics. In particular, relaxation of an internal constraint in a composite system changes the basic set of eigenfunctions of the system. Therefore, even while the wave function is instantaneously unchanged, the result is the creation of a huge number of phase relations. Subsequently these are destroyed in a series of events that constantly interrupt the propagations. This is the mechanism by which the entropy increases.

It is obvious that both propagations and events are indispensable for a descriptive account of quantum phenomena, and both concepts are essential ingredients of the formalism. This is particularly evident in the perturbation expansion in which propagations and events appear intertwined. From the formal point of view, this technique can be considered as the generalization of the basic problem of mathematical probability theory. Again the probability of complex events is computed from that of elementary events. However, the theory operates in terms of probability amplitudes rather than in terms of probabilities. This conceptual shift is comparable to the one that is instrumental in effecting the transition from MTE to STE. Once more the theory is deepened, our probabilities are now capable of interference, in other words, stochastic events can be combined with propagations expressed in terms of differential equations. Whereas entropic processes cannot be joined to CMP without disrupting the formalism, energetic processes can be joined to this modified probability theory.

Potentially we have a most harmonious correspondence between theory and experiment. However, these potentialities are only incompletely realized in the existing presentations of quantum mechanics. There are two major reasons for this situation. First, there is the more or less accidental historical link between quantum mechanics and CMP(μ) which induces us to assign a privileged status to propagations over events. The rule that requires the reduction of events to propagations has been transferred from CMP(μ) to quantum mechanics, even though the reasons for this program that were imperative in the mechanistic scheme do not prevail in quantum thermodynamics. The fact that the rule is often

[60] A. Einstein, Verhandl. deut. Physik Ges. 16, 820 (1914).

broken or circumvented does not reestablish logical clarity. The second, and deeper reason is that the theory in which events and propagations are freely compounded for the construction of complex events is, in spite of important achievements, still in an incomplete state. This is particularly true of that part of quantum thermodynamics which we have called QD in the block diagram. The present method of analysis enables us to dispose of the first of these difficulties and, by sorting out the elements of the situation, we hope to prepare the ground for attacking the second.

It is well known that the propagations (4.25) are insufficient to account for all of the coupling effects of quantum mechanics, and the processes of measurement have to be joined as independent conceptual entities, a point expressed with particular precision by von Neumann.[61] We claim that the present suggestion of supplementing propagations by events as irreducible primitive concepts is the generalization of the traditional view. The crucial point is that many events are measurable. Atomic physics became possible because of the discovery of a long line of devices, from scintillation counters to bubble chambers, that enable us to retrodict events from observed signals. However, events occur whether or not we observe them. Therefore, the supplementation of propagations cannot be relegated to a special class of studies devoted to the theory of measurement, but such additional elements have to be standard parts of the formalism. This point of view enables us to avoid some of the paradoxes that arise from the overemphasis of the subjective elements in the quantum formalism. As an example of this sort we consider, with von Neumann, the transformation of a pure state into a mixture of increased entropy because of the coupling with a measuring device. A typical apparatus would be a screen with a number of slits and a counting device in the path of the beam of particles. In our language, we consider the coupling with the measuring device a thermodynamic operation in which certain constraints are enforced.[62] In this procedure the splitting of the beam is a propagation, the absorption or scattering in the counter at a location corresponding to a particular split beam is an event in which one state is actualized to the exclusion of other potentialities.[63]

In the course of this process the assembly of particles making up the beam is transformed from a pure state into a mixture with a corresponding increase of entropy. We have said nothing, thus far, of measurement. However, if the events are observed and recorded, we have gained information, this gain can be considered as partially balancing the increase of entropy.[64] Thus measurement is associated with an increase of information, whereas the parallel increase in entropy takes place independently of any subjective act of measurement. We add that the typical entropy increases of MTE arise in the course of relaxations rather than of enforcement of constraints.

The suggested new relation between propagations and events raises a number of problems that have to be solved before the theory can be reconstructed along the lines indicated. A problem of paramount importance concerns the validity of the Schrödinger equation. In the traditional theory it is often taken for granted that the Schrödinger equation is entirely correct and contains all legitimate quantum-mechanical information about systems of any complexity.[65] How is this statement modified by the fact that events are accorded an independent status? We propose the tentative rule that the results deduced from the Schrödinger equation are correct whenever they satisfy the principle of regularity (see Sec. II). This principle infringes sufficiently on the validity of (4.25) to allow us to join entropic processes to the formalism without in any way impairing its consistency. We shall apply the principle of regularity to both of the abovementioned criteria of entropy conservation.

The most obvious infringement of the definiteness of results of (4.25) arises in case of degeneracy. Consider, for instance, a time-independent Hamiltonian. The eigenfunctions of this operator should be constant for eigenfunctions associated with nondegener-

[61] J. von Neumann, *Mathematische Grundlagen der Quantenmechanik*, (Verlag Julius Springer, Berlin, 1932), Chap. V.

[62] Note that in systems in equilibrium considered in MTE the enforcement of a constraint produces no entropy increase. The present case is different, since we start from a pure state, a nonequilibrium situation.

[63] This collapse of the wave function is a standard occurrence and is the inevitable consequence of the fact that the wave function has a dual reference to classes and to individual systems. Events refer to an individual "choice" out of the set of possibilities determined by the evolution of the wave function of the class. This duality is foreshadowed in MTE, the formalism of which deals with the state of *normal* equilibrium to which an *actual* system may, or may not, conform. The widespread reluctance to make full use of the collapse of the wave function is sufficiently explained by the fact that this phenomenon seems paradoxical in terms of the classical framework of CMP and CED or if one attempts to interpret the Schrödinger equation as a classical wave equation.

[64] L. Szilard, Z. Physik **53**, 840 (1929); L. Brillouin, reference 10, Chap. 13.

[65] The attitude reflected in this assumption is obviously the same one that generated similarly sweeping claims for the formalism of CMP(μ). It would be tempting to subject this claim to critical analysis that could center around the uncertainties in the choice of the appropriate Hamiltonian discussed at the end of Sec. IV.C. However, the line of thought pursued above is more constructive at this juncture.

ate eigenvalues. However, in the case of degeneracy this result is upset even by the smallest perturbation. Hence, the principle of regularity requires that the implications of the strictly deterministic conclusions of (4.25) should not be considered physically meaningful in this case. Thus there arises a gap that is filled by the random events resulting in entropic processes.

We consider now the entropy conservation of energetic processes in terms of the criterion of coherent phase relations. If Eq. (4.25) is taken seriously as an entirely rigorous description of the system, such coherent phase relations would extend over infinite space-time. In reality, however, coherence can be expected to prevail only in a finite space-time domain. In many simple problems involving, say, a single collision, this artificial extension of the range of coherence is a harmless idealization. Not so in composite processes, for which the extent of coherence is often significant and should not be summarily decided by postulate without specific examination. In composite systems, the interruption of coherence occurs at the walls constituting the cell boundaries. However, just as phase boundaries play the role of "natural" walls, it seems that wave functions corresponding to pure states have their "natural" space-time extension of coherence. Such intrinsically coherent units should become basic entities of the suggested theory. They are to be joined to each other either incoherently or coherently.

We turn now to the discussion of the well-known controversy between Einstein and Bohr[5] concerning the desirable and attainable features of quantum mechanics. The essence of this argument is the compatibility of causality with a complete space-time description of systems. Bohr emphasized the complementarity of the two descriptions, whereas Einstein held up CMP(μ) and CED as ideal theories in which the two requirements are entirely compatible. Accordingly, the argument always turned around the point, whether or not quantum mechanics could be cast into a form that agrees more closely with the structure of the classical theory.

This argument has been deadlocked for a long time. We propose that the reason for this impasse is that both participants tacitly assumed that there is only one type of causality depending on the exhaustive specification of the initial conditions. In contrast, we have argued that most of the causal chains expressing the regularities in nature are of a second kind and are based on the selective specification of composite processes in composite systems. Although the quantum thermodynamics of these systems is still to be developed in a systematic manner, there seems to be no incompatibility between the piecewise continuous space-time description of these systems and the existence of causal chains depending on selective specification. An exacting test for such a view is the development of theoretical constructs that are to represent the behavior of living organisms.

E. Physics of Organisms

Living organisms exhibit complex morphological and functional properties that are very different from the properties of systems investigated in physics. Nevertheless, physics is supposed to have a fundamental character and it is a widely entertained hope that in some sense, still to be clarified, biology is reducible to physics. Let us assume that physics reaches a degree of development in which the principles governing the building of complex structures from elementary entities is well-enough understood. In this not unlikely situation we should be able to synthetize also the conceptual material that is necessary for the understanding of the properties of structures of increasing complexity. For nuclei, atoms, and molecules this has already been achieved to a not inconsiderable degree. According to the point of view that is optimistic with respect to the reducibility of biology, this gradual extension of our understanding need not stop at any sharp boundary separating inanimate and living objects. We may set aside the more speculative aspects of this question by appraising the scope and limitations of the reducibility program at three stages of development of the physics of structures: (i) mechanistic physics based on CMP(μ), (ii) traditional quantum mechanics, and (iii) the quantum thermodynamics of composite systems outlined in Sec. IV.E.

The discussion based on CMP(μ) is only of historical interest. The obvious failure of this discipline to account for anything resembling an organism produced a strong case for vitalism, a doctrine that postulates *ad hoc*, irreducible, vital forces to fill the gap left by mechanistic physics. This argument can now be dismissed as inconclusive, since not even phenomenological mechanics is "mechanistic" in this technical sense of the word (see Sec. IV.B). This state of affairs changed radically with the development of quantum mechanics and the subsequent clarification of chemical stability and reactivity. In the terminology of Sec. IV.C we may say that the morphic invariance of QS provides at least the prerequisites for the explanation of the morphological aspects of biology. If, in addition, we remember the recent spectacular advances of molecular biology it seems hard to maintain an attitude that would hold any specific aspect of biology as out of bounds for physical-chemical investigations. At the same time, we are still far from a satisfactory solution of the reduction problems, the scope and limitations of which we shall attempt to spell out as follows. Reduction seems to proceed most satisfactorily with respect to problems "in the small," whereas it has not even started for problems "in the large." This contrast between the "atomistic" and "holistic" attitudes of physics and biology is well known, it is the real stumbling block of the reduction program. It is our contention that this inability to cope with problems of organization is only a property of mechanistic physics, rather than of physics in general. In particular, this limitation, still present in traditional quantum mechanics, stems from the historical link with CMP(μ) (see Secs. IV.C and IV.D). While a huge amount of detail remains to be worked out, it seems that the quantum thermodynamics of composite systems has the necessary flexibility to account for the holistic problems of organisms.

We note that this theory incorporates a number of concepts that are evidently indispensable for any theory of organisms. We have the distinction between free and fixed variables, processes and operations; processes are controlled by fixed variables, whereas the fixed variables are changed by operations or parametric processes. The combination of processes and operations leads to the concept of composite processes and to the second principle of causality, which depend on selective specification of the initial conditions. In MTE one usually considers composite systems consisting of two subsystems. However, the complexity of composite systems can be increased in a variety of ways. The first step is to increase the number of subsystems, then, by appropriate connections, couplings may be introduced between any two of them, not necessarily adjacent in space. The result may be an entity with complex topological properties. The restrictivity of these connections may be regulated, just as chemical reactions are catalyzed, or poisoned. The system reaches a measure of autonomy if the entire control system is operated automatically by feedback from the environment and/or the state of the system. The second principle of causality ensures a mixed causal, stochastic response of the system. For systems of a certain complexity the code containing the instructions for its response may become rather complex and be in need of logical organization to ensure a reasonably unique operation. The logical structure of the block diagram (Fig. 1) exhibits one possibility of a consistent organization of a complex logical structure in which potential inconsistencies among different sets of instructions are kept under control. A further step leading toward complexity is that composite systems are united into a supersystem. The coupling may involve the exchange of additive invariant and/or the exchange of information. All this can be repeated to yield a complicated hierarchy of structures.

Beyond a certain complexity the code of instructions plays the role of a rudimentary "mind." Moreover, the logical structure of this mind and the physical structure of the system attuned to each other for smooth functioning are expressions of a psychophysical parallelism.

Summing up, we may say that the basic problem of the physics of organism is to establish the connection between the built-in patterns of constraint and freedom on the one hand, and the functioning of the organism on the other. This concept of constraint and freedom is entirely foreign to CMP(μ), but is present in all aspects of quantum thermodynamics. In fact, these concepts are present even in PhCM. This is the conceptual basis of the fact that machines such as automata and computers can simulate certain functions of organisms. By providing standardized interchangeable parts, the makers of the machines imitate the morphic invariance of nature. This imitation can never reach the perfection of the original, and jamming and overheating certainly set a limit to the complexity of devices made out of macromechanical elements.

V. PHILOSOPHICAL REFLECTIONS

The method of analysis of this paper is, I believe, in agreement with the spirit and the underlying goals of the philosophical school that is alternately called analytic, positivist, or empirical. I hope not to be too far off the mark in using these terms as synonyms.

The concepts of logical differentiation and integration arose within the examination of the language of physics, which is a legitimate pursuit in this school of thought.

A point of agreement is the close attention paid to empirical requirements; more specifically, the approach to quantum physics as described in Sec. IV is quite in line with the positivist program of developing the theory of the structure of matter in close parallel with experiment.

Moreover, the principle of regularity (Sec. II) is a methodological rule that serves as a safeguard against taking theoretical systems more seriously than their empirical support warrants.

At the same time, the present paper and its implications are at variance with some of the current practices of positivism. The clarification of these issues can be achieved as a by-product of the present analysis.

Let us consider first the so-called *operational character* of the theoretical concepts. It goes without saying that the deductive systems of physics have to have a range of concordance with experience.

Yet, there is a growing awareness of the fact that there is something wrong, or at least lopsided, about the great emphasis on "operationalism" which for several decades has dominated empirical philosophy.

We could do hardly better than quote from Bridgman's[66] penetrating reappraisal of the early ideas of operationalism. "If I were writing the 'Logic' today I would change the emphasis so as to try to avoid what I regard as the one most serious misunderstanding. That is, I would emphasize more that the operations in terms of which a physical concept receives its meaning need not be, and as a matter of fact are not, exclusively the physical operations of the laboratory. The mistaken idea that the operations have to be physical or instrumental, combined with the dictum . . . , 'The concept is synonymous with the corresponding set of operations,' has in some cases led to disastrous misunderstanding. If I were writing, again I would try to emphasize more the importance of the mental or paper-and-pencil operations. Among the very most important of the mental operations are the verbal operations. These play a much greater role than I realized at the time. . . ."

The analytic definitions of the present approach are of course "paper-and-pencil" operations. Apart from being of equal importance with the instrumental operations, the two are related to each other in a well-structured manner. In the absence of precise analytic definitions the meaning of the instrumental operations cannot even be properly evaluated.

The operational point of view considered as a sole criterion would always favor the low-level abstractions. However, the most spectacular advances in theory are connected with the discovery of abstractions of a very high level.

The constructive aspect of operationalism is appreciated particularly in the process of logical integration. If we have two theoretical systems, both of them confirmed by experiment but inconsistent with each other, we usually find that the inconsistencies are produced by nonoperational assumptions that can be dropped without loosing the measure of experimental agreement already achieved. In fact, subsequent developments often result in an extraordinary expansion of the range of concordance.

This is what happened as CMP was generalized into the dominant RM. Einstein's operational analysis of the concept of simultaneity was a means to an end, it made the integration of CMP and CED possible. The concept of absolute simultaneity was abandoned because it blocked integration. The fact that it was not operational showed that this change in the foundation could be undertaken without damage.

However, "logical integration" has been, thus far, not in the vocabulary and the *purpose* of operationalism was not sufficiently appreciated. Erroneously it was taken for an end in itself, or, an ironical perversion of its real role, a means to forestall another "catastrophe" that would force us to re-examine our basic assumptions.

Although an occasional reappraisal of the foundations may seem a not entirely welcome interruption of the routine of research work, it is now recognized to be the unavoidable price to be paid for the continued use of new high-level abstractions. We again quote Bridgman: "To me now it seems incomprehensible that I should ever have thought it within my powers, or within the powers of the human race for that matter, to analyze so thoroughly the functioning of our thinking apparatus that I could confidently expect to exhaust the subject and eliminate the possibility of a bright new idea against which I would be defenseless."

We turn now to another difficulty of positivism, namely, its predominantly restrictive character which bids us to dismiss most problems of traditional philosophy as "meaningless." Granted that the traditional methods are lacking in precision, we are faced with the unhappy choice of dealing with significant problems in an unsatisfactory manner or bringing to bear a precise method on problems the wider import of which is not immediately apparent.

There are at least two ways of breaking this deadlock. One could make a case for applying less rigid standards in appraising intuitive methods, or to extend the scope of analytic philosophy to more significant problems. It is only the second path that will be followed up at this point.

In order to understand the origin of the restrictiveness of positivism, we have to go back to the beginnings of mechanistic physics. With the brilliant suc-

[66] P. W. Bridgman, Daedalus **88**, 518 (1959).

cess of Newtonian mechanics, it seemed tempting to brush away the complex and eroded conceptual system of scholastic philosophy. The proposition of temporarily restricting oneself to the conceptual framework of the new mechanics seems entirely sound, even in retrospect. However, the contention that this conceptual framework would be satisfactory at all times was unwarranted, and turned out to be actually incorrect.

During the mechanistic era it became customary to dismiss types of questions that did not fit into mechanistic systems as unscientific.

As the crisis of classical physics revealed the limitation of the mechanistic conceptual scheme, the first inference was that the range of *legitimate scientific questions is even further limited*, since not even the mechanistic questions are admissible.

The pluralistic character of the present approach brings two new elements into this picture. In the first place, each deductive system implies a characteristic set of precise questions. The number of interesting questions that become "meaningful" is particularly extensive in thermodynamics and quantum mechanics.

It is often stated that the concept of object breaks down in quantum mechanics. Actually, however, the opposite is true. As we have seen in Sec. IV, for the first time in QM we are in a position to give a formal representation of an object with many subtle ramifications and we can now solve the related philosophical puzzles that have been unresolved since their discovery by the Eleatic philosophers. It seems that the new object concept is flexible enough to include living organisms that are entirely outside the mechanistic scheme.

The extension of meaningful conceptual problems in the present context proceeds in still another dimension. Not only do we have the concepts within each deductive system, but the deductive systems themselves are conceptual entities of distinct individual characteristics related to each other in quite specific fashion. These entities are of a logical type that is markedly different from that of the primitive concepts within the deductive systems.

INDEX

Index